D1237505

Military Aid and Counterrevolution in the Third World

Military Aid and Counterrevolution in the Third World

Miles D. Wolpin
University of New Mexico

Lexington Books
D.C. Heath and Company
Lexington, Massachusetts
Toronto London

Library of Congress Cataloging in Publication Data
Wolpin, Miles D
Military aid and counterrevolution in the Third World.
Bibliography
1. Military assistance, American. 2. Military
education–United States. I. Title.
UA12.W64 322'.5 72-7066
ISBN 0-669-84640-6

Published simultaneously in Canada.

Printed in the United States of America.

International Standard Book Number: 0-669-84640-6

Library of Congress Catalog Card Number: 72-7066

To Sylva and Arthur,
for their years of love

Contents

List of Tables and Figures

Preface

This study was undertaken because of a notable void in the literature on military aid. Considering the prevalence of military intervention in the Third World, it was quite disconcerting to discover the paucity of serious and systematic investigation of linkages between officers in these countries and their counterparts in hegemonious superpowers. Whether the former were defined as agencies of modernization or the structural status quo, it seemed that such transnational relationships might well have some behavioral relevance.

In congressional hearings, military writing, and scholarly monographs there was an abundance of source material. This is not to say that we have not been forced to leave many questions unanswered and to venture informed conjectures on others. To be sure, a grave though we think not necessarily mortal deficiency of this work is the absence of interview data from either trainees or those administering the program. Although the author was advised by a friendly Pentagon colonel that such interviews were best left unsought, the primary reason is that no established foundation or research institute was willing to even receive an application for investigative funding.

It is to be hoped that some independently minded scholar will follow up and supplement this research by testing some of our interpretations and propositions through survey research and perhaps more refined analysis than we have offered here. We think that the neocolonial dependency relationships delineated in the following chapters will convince others that further study is vitally needed.

For their generous support and interest in furthering our own investigation, special gratitude is owed to Harry Magdoff and the Louis M. Rabinowitz Foundation, David Riesman and the Ann Parsons Educational Trust, and John W. MacArthur of Marlboro College.

The author is heavily indebted to a number of congressmen and officers in the United States Army. They provided invaluable assistance in obtaining unpublished and generally unavailable official data—none of which, however, was of a classified nature.

Criticism of the manuscript by Lila Walz, Stan Newens, and Rosalind Boyd highlighted more than a few shortcomings in earlier drafts. They bear no responsibility for the remaining deficiencies.

Particular appreciation is reserved for Susan Jackson Wolpin for both her tolerance and loving encouragement. Both Ken Fleet of the Bertrand Russell Peace Foundation and Rosalind Boyd were also extremely helpful in nurturing this project toward completion.

Others who have aided the investigation or preparation of the manuscript have been Nick Hadden, Paul Ferrini, Sister Jean Harris of St. Francis Xavier University, Ann-Marie Harrison, Beverly MacIssac, Judi LeBlanc, and Margaret Brown.

Military Aid and Counterrevolution in the Third World

1 Introduction

It should be no surprise to anyone that our military interests permeate the far corners of the globe. We spend this money very freely, even if it is not wanted, to force our own purposes upon the recipient.[1]

Congressmen Silvio O. Conte, Jeffrey Cohelan, Edward P. Boland and Sidney R. Yates
December 1969

About fifteen years ago, a specialist on military aid suggested that a "chasm" existed between the publicly stated goals and the real purposes of the Military Assistance Program or MAP.[2] Not long afterward a presidentially appointed task force, which had been commissioned to evaluate the MAP, submitted a number of recommendations intended to strengthen the program. The Draper Committee's Report also distinguished between America's "public policy" and its "action policy" in various foreign countries.

Official statements commonly declare that military assistance is furnished upon request in order to strengthen allied or friendly countries in the face of Communist aggression or subversion. But like the Nixon administration's contention that it was only seeking $409,000,000 for military aid in fiscal year 1972, this represents only "the tip of the iceberg." Thus, in the course of Senator Proxmire's investigation of the MAP during January 1971:

Two Pentagon officials, pressed for a complete total at the hearings, got out their pencils in a moment of high comedy, disagreed between themselves and finally came up with a total of nearly $4.9 billion, or almost ten times as much. Fulbright introduced a table of his own showing more than $6.9 billion in military assistance and sales for the next fiscal year. He estimated that total military outflow from the U.S. since 1945 might add up to $175 billion![a]

---

[a]And these expenditures have financed American training of more than 320,000 foreign military men from seventy "independent" nations. *I.F. Stone's Bi-Weekly*, January 25, 1971, p. 1.

Both the dollar amounts and the training figures are only approximations. With regard to the latter, thousands of Defense Department foreign trainees are not funded under the MAP and therefore are excluded from its tally. As for appropriations, the same is true. Other programs such as economic aid have been used as a cover for military purposes. Thus, one finding of the Proxmire investigation was that "the most shocking concealment of military assistance funds in terms of rhetorical techniques concerns the Food for Peace Program. According to the budget document, 'This program, closely linked to the AID program, combats hunger and malnutrition, promotes economic growth in developing countries, and develops and expands export markets for U.S. commodities.' That is a direct

While the exact amount expended may be indeterminate, scrutiny of the actual goals promoted reveals the existence of several generally unacknowledged objectives. And there are even grounds for uncertainty with regard to that traditional marketplace idol—the threat of Soviet or Communist aggression. Well before either the Marshall Plan (1947) or the North Atlantic Pact (1949) were launched, the Russians had voluntarily *withdrawn* their troops from northern Iran, the large Danish island off Bornholm, Norwegian Spitzbergen, and Finland. This is not to deny that Moscow was determined to have friendly governments in Eastern Europe through which it had been invaded three times in a little over a century. But it does raise serious doubts about Soviet intentions of invading Western Europe. Worth recalling is the fact that the Swedes declined to enter NATO while the French demanded that the United States finance their war to regain Indochina as the price of committing twelve divisions to NATO.[3] Obviously their fear of a Soviet sweep to the Atlantic could not have been overwhelming. Finally, we now know that the Truman administration consciously exaggerated the Soviet military threat in order to secure domestic support for its postwar overseas commitments.[4] The Eisenhower administration was equally indifferent to the ethics of manipulating public fears.[b]

The "Open Door" and Third World Nationalism

An understanding of the less visible objectives of military aid must take account of a limited number of economic and historical factors. These explain why in many regions of the world the United States is viewed as imperialistic rather than as a disinterested benefactor.

Since the transformation of the American socioeconomic order to an industrial one during the last quarter of the nineteenth century, it has been afflicted by chronic unemployment and excess productive capacity. The corporate and political elites who have set national policies viewed this situation as a potential source of social instability, i.e., a threat to their positions and/or privileges. In order to stimulate domestic economic demand, they championed a vigorous policy of overseas economic expansion.[5] Thus, Secretary of State Dean

quotation from the act. Yet, we find that in the past six years nearly $700 million of Food for Peace funds have been channeled into military assistance programs. In 1970, $108 million worth of Food for Peace has gone for military aid to foreign governments. To continue using Food for Peace funds for military purposes is, to say the least, a corruption of the English language." Ibid.

[b]Senator Fulbright has recently pointed to the fact that U.S. military aid to many Third World countries was *not a response to Soviet intrusion.* On the contrary the U.S. penetrated these societies first and thus catalyzed the USSR to make its own offers. Similarly, a high Kennedy-Johnson administration official has testified that Chinese entry into the Korean War was primarily reactive and defensive. U.S., Congress, Joint Economic Committee, *Economic Issues in Military Assistance, Hearings*, before the Subcommittee on Economy in Government of the Joint Economic Committee, 92d Cong., 1st sess., January-February, 1971, pp. 48, 86, Cf., p. 179.

Acheson—a Wall Street lawyer who was one of the leading architects of America's cold war strategy—summarized the conservative basis of this view before the House Committee on Post-War Economic Policy:

> If you wish to control the entire trade and income of the United States, which means the life of the people, you could probably fix it so that everything produced here would be consumed here, but that would completely change our Constitution, our relations to property, human liberty, our very conception of law. And nobody contemplates that. Therefore, you find you must look to other markets and those markets are abroad.[6]

The search for overseas markets has been traditionally known as the Open Door policy. Although often rationalized in terms of reciprocity and mutual advantage, its effects are actually more akin to an elephant dancing with a flock of chickens. That is, it has constituted one of the more important obstacles to economic development in the Third World—Africa, Asia, and Latin America.

Trade expansion has generally been associated with a drive for tariff reduction and nondiscrimination against foreign investors. This creates enormous difficulties for "infant industries" in underdeveloped countries. Either they are driven out of business or those in potentially dynamic sectors may be taken over by American corporations. This is why the United States erected high tariffs against imported manufactures during the early nineteenth century. Once developed, however, America sought to deny a similar option to political elites in the Third World.

For those who wish not only to control their own economies but to also develop, there has been an even more serious impediment. The Open Door policy has been traditionally defined by Washington as encompassing not merely goods but also freedom to invest. While such investments have undoubtedly created jobs and been a source of foreign exchange earnings, their net effects in much of the Third World have been negative. First, when foreign corporations own the most dynamic sectors of the economy, the decisions of these transnational firms sharply limit the *de facto* sovereignty of the government. Thus, if the latter were to raise corporate income taxes, the firms might well curtail their plans to expand productive facilities. Such investment is vital to offset unemployment—a constant problem for most governments. Second, the foreign firms plan production and set prices to maximize profits or advantage for their worldwide operation, not for the country concerned. And finally, as the Christian Democratic Foreign Minister of Chile informed President Nixon in the Cabinet Room of the White House on June 12, 1969:

> It is generally believed that our continent receives real financial aid. The data show the opposite. We can affirm that Latin America is making a contribution to financing the development of the United States and of other industrialized countries. Private investment has meant and does mean for Latin America that the sums taken out of our continent are several times higher than those that are

invested. Our potential capital declines. The benefits of invested capital grow and multiply themselves enormously, though not in our countries but abroad. The so-called aid, with all its well known conditions, means markets and greater development for the developed countries, but has not in fact managed to compensate for the money that leaves Latin America in payment of the external debt and as a result of the profits generated by direct private investment. In one word, we know that Latin America gives more than it receives. On these realities it is not possible to base any solidarity or even any stable or positive cooperation.[7]

That this is not the raving of an emotional "Latino" is suggested by his political affiliation and the audience of ambassadors who witnessed him making these remarks.[c] Data confirming his contentions were presented by the president of Charles Pfizer & Co. a year earlier in the course of an address to the American Management Association:

Between 1950 and 1966 . . . corporations and private citizens brought into the country $59.0 billion in excess of all private dollar outflows. . . . direct investments have returned substantial income to their companies in the United States, far greater than the direct investment outflows. . . . From 1950 to 1966 these investments returned in dividends and royalties and fees alone $20 billion in excess of all outflows. . . . Recently Professor Behrman argued before the Joint Economic Committee that the payback period for outflows of U.S. dollars for manufacturing investment abroad is about 2½ years on the average. If this is right—and I must say, this estimate comes close to my own experience—this is a very short term indeed.[d]

[c]Secretary of State Rogers is aware of the fact that many underdeveloped countries in Latin America *do not want* U.S. corporate investments. And it is known to himself and members of the Foreign Affairs Committee that the firms are repatriating more than they are investing in these economies. U.S., Congress, House, Committee on Foreign Affairs, *Foreign Assistance Act of 1969, Hearings*, before the Committee on Foreign Affairs, House of Representatives, June 1969, 91st Cong., 1st sess., pp. 1064, 1068-69.

[d]Excerpted from a speech by John J. Powers, Jr., president of Charles Pfizer and Co., delivered at an American Management Association special briefing on "New Foreign Investment Controls," in New York City on April 10, 1968, reprinted in the *NACLA Newsletter* 2 (November 1968), pp. 8, 10.

The relative and growing importance of overseas investments by American corporations is suggested by the following data: "Between 1950 and 1964 sales abroad via exports increased 250 percent (from $10 to $25 billion). At the same time output abroad from U.S. investments increased 325 percent (from $44 to $143 billion). These figures gain added significance when they are compared to the performance of the U.S. domestic economy. In this period the sales of domestic manufactures increased only 225 percent. Significant as these figures are, they tell only part of the story. Between 1950 and 1965, the earnings on foreign investment as a percentage of profits of domestic nonfinancial corporations increased by two-thirds (from a very impressive 22 percent to a truly remarkable 36 percent). As these data suggest, overseas investment is becoming an ever more crucial source of investment opportunity. Foreign plant and equipment expenditures as a percentage of domestic expenditures increased from 8 percent in 1957 to 17 percent in 1965. The conclusion is inescapable: our economy is maintained by foreign economic activities; it could not survive without them." Robert I. Rhodes, "Review of *The Age of Imperialism: The Economics of U.S. Foreign Policy*, by Harry Magdoff, New York, Monthly Review Press, 1969," *Science & Society*, 33 (Fall-Winter 1969), 488-89.

These figures suggest why some Third World leaders who aspire for national self-determination and development in the underdeveloped areas frequently advocate expropriation and other measures intended to at least partially close the open door.[e]

What is not generally known is that the United States has traditionally opposed and endeavored to subvert social radicals and economic nationalists in other countries. The emergence of such a pre-cold war posture began with the U.S. decision to occupy Cuba and the Philippines in the 1890s rather than simply grant recognition and military aid to the revolutionary nationalist forces on those islands. The suppression of Aguinaldo's Philippine rebellion in 1900-02, imposition of the Platt Amendment and armed invasions of Cuba during the first two decades of this century, opposition to Article 27 of Mexico's 1917 Constitution and intervention during that country's social revolution, the suppression of Sandino in Nicaragua, and the American invasion of Russia (1918-20) marked the consolidation of this counterrevolutionary pattern. Military assistance was first extended in the 1920s to such American occupied countries as the Dominican Republic, Haiti, and Nicaragua. During the early 1930s internal order and the protection of investments but not civil liberties were the objects of U.S. military aid to Latin America.[8] Thus until its viability became problematic, Washington supported the terroristic Machado dictatorship in Cuba. After the regime's overthrow by radical nationalists in 1933, American naval units were sent to the Cuban coast and recognition was withheld–but not from a concern for political freedom. Roosevelt's Ambassador Extraordinary Sumner Welles reported "pessimistically" that it was "within the bounds of possibility that the social revolution which is under way cannot be checked."

> Grau San Martin met several times with Welles and the political leaders, and the Ambassador arranged for Adolph A. Berle, Jr. to explain the 'financial and economic picture' to Grau San Martin. Berle told the Cuban President that American companies–public utilities, importers, and sugar mills–could not and would not do business under existing conditions. . . . Welles continued to meet with Batista to try to persuade him to take a more vigorous stand against Grau San Martin.[f]

---

[e]The mobilization and allocation of capital are key variables in the development equation–which is not to imply that others such as efficiency, a propensity to save, and the existence of natural resources are unimportant. A developing Third World nation would be one which was measurably closing the per capita GNP gap between itself and the North Atlantic Community mean. An alternative and more social indicator might be mean per capita square feet of living space in dwellings with running water for those whose family income approached the median. Regardless of our definition, the gap is widening rather than closing for most Third World nations. See, for example, Pierre Jalee, *The Third World in World Economy* (New York: Monthly Review Press, 1969).

[f]And five years later the Roosevelt administration supported oil and silver sanctions against the Mexican government when it nationalized American oil companies which had refused to conform with Mexican laws governing labor relations. Robert F. Smith, *The United States and Cuba: Business and Diplomacy, 1917-1960* New Haven, Conn. College and University

Our purpose in recounting these events is simply to suggest that two key patterns antedated the cold war: (1) hostility to self-determination when the "wrong" people were leading the movement; and (2) indifference to dictatorial regimes. Naturally we are simplifying the events and problems such as those associated with developing economies. Our purpose has been to abstract only those facets which are most important to explaining America's cold war role.

We have, of course, largely neglected the domestic policy-making process. Both major party leaderships—financially and often professionally related to the corporate-owning upper class—have at the national level advocated or acquiesed in this orientation. Thus the pattern preceded, coincided with, and succeeded the New Deal. Policy initiation was vested with *civilians* in the executive branch. A recent study by historian Gabriel Kolko focused upon the backgrounds and careers of policy-making officials "in the State, Defense, or War, Treasury and Commerce Departments, plus certain relevant executive level agencies." Included were

> 234 individuals with all their positions in government during 1944-60, comprising the lesser posts if an individual attained the highest executive level. As a total, these key leaders held 678 posts and nearly all of them were high level and policy-making.
>
> In the aggregate, men who came from big business, investment, and law held 59.6 percent of the posts, with only forty-five of them filling 32.4 percent of all posts.[9]

According to the author, this pattern of elite recruitment has remained stable since the New Deal era. As for officials from more humble origins, Kolko argues that co-optation and exclusion have operated to maintain consensus.

Hence, it should come as no surprise that men with such backgrounds, identifications, career patterns, and economic interests would perceive the post-World War II emergence of the Soviet Union as a great power as a distinct threat. For years, indeed until the early fifties, the view was common that the Bolshevik regime would shatter and collapse of its own contradictions—for socialism was believed to be inherently unworkable. Such illusions were fed by a fierce antipathy for Communists who not only closed the open door, but did so on principle and with relatively high organizational effectiveness. Washington's policy-makers were well aware that authoritarianism under Communist governments was no more draconian than that exercised by such "free world" stalwarts as France, Chiang Kai-shek, Syngman Rhee, Rafael Trujillo, and Haile Selassi. Thus an examination of congressional hearings on military aid during the sixties and early seventies reveals almost universal indifference to political liberties in

---

Press, 1960, pp. 152-53. Gardner, *Economic Aspects*, pp. 116-18. Cf., Robert F. Smith, "The Formation and Development of the International Bankers' Committee on Mexico," *Journal of Economic History* 23 (December 1963), 574-86.

the Third World—so long as the regime is not nationalistic, radical and, of course, "Communist."[g] The fact that many "enemy" regimes of this stripe have done more to equalize social opportunities for the bulk of their citizenry is likewise never mentioned, let alone balanced against Washington's determination to force an Open Door policy upon these countries.[h]

There is evidence that even during the Second World War the United States sought to influence internal politics in the West European countries to the detriment of leftist resistance forces who had so valiantly led the fight against Nazi invaders.[10] In Germany itself, American military authorities discriminated against socialistic economic policies advocated by the only party that had voted against conferring dictatorial powers upon Hitler—the Social Democrats. While there was residual or projective fear that the Soviet Union might unleash pro-Moscow Communist parties in Western Europe, in general Stalin abided by his wartime agreements *until* the declaration of the Truman Doctrine in 1947. There was less to fear in the Anglo-French territories where the United States generally supported the restoration of these colonies. Pressure was, of course, exerted against both the British and the French to open the door on equal terms to American trade and investments.[11]

But these empires had been fatally weakened by two world wars. At the same time, radicals and nationalists began to struggle for independence. Although the United States has preferred that former imperial powers "carry the ball" whenever possible in their ex-dependencies, there has been a trend towards increasing American involvement.[12] Meanwhile, though not always with success, the prewar pattern of disrespect for national self-determination has been

[g]According to a former State Department official, even "the word 'communist' has been applied so liberally and so loosely to revolutionary or radical regimes that any government risks being so characterized if it adopts one or more of the following policies which the State Department finds distasteful: nationalization of private industry, particularly foreign-owned corporations, radical land reform, autarchic trade policies, acceptance of Soviet or Chinese aid, insistence upon following an anti-American or nonaligned foreign policy, among others. Thus the American ambassador to Cuba at the time of the brief Grau San Martin Government in 1933 found it to be 'communistic.' In 1937 Cordell Hull privately spoke of the Mexican government, which was nationalizing U.S.-owned oil properties, as 'these communists down there,' but made no public charge. Since the Second World War, however, the term 'communist' has been used to justify U.S. intervention against a variety of regimes with widely differing ideologies and relationships with the Soviet Union, including Arevalo's Guatemala, Mossadeq's Iran, Goulart's Brazil, Sukarno's Indonesia, Caamano's Dominican revolutionary junta, as well as insurgent movements in Latin America, Africa, and Southeast Asia." Richard J. Barnet. *Intervention and Revolution* (New York: World, 1968), p. 9. Cf., Michael Parenti, *The Anti-Communist Impulse* (New York: Random House, 1969).

[h]For an instructive case study of the measures being implemented by a freely elected government which the CIA and the US Ambassador were instrumental in overthrowing, see Eduardo Galeano's *Guatemala: Occupied Country* (New York: Monthly Review Press, 1967). Cf.: Leslie Dewart, *Christianity and Revolution: The Lesson of Cuba* (New York: Herder & Herder, 1963; Maurice Zeitlin and Robert Scheer, *Cuba: Tragedy in Our Hemisphere* (New York: Grove Press, 1963); Viator (pseud.), "Cuba Revisited after Ten Years of Castro," *Foreign Affairs*, 48 (January 1970), 312-21.

maintained *when the governments in question were not receptive to corporate investor interests and/or aspired to an independent or* nonaligned foreign policy.[i] Operating under a partially valid "domino theory," American political leaders were willing to devote large resources to the preservation of capitalism in the smallest of countries (e.g., Vietnam) where at the particular point in time there might be few actual American investments. Similarly, pressure was brought upon the British in 1963 to deny Guyana independence because the wrong party had won the elections.[13] More clearly manifesting this neocolonial orientation was direct or indirect U.S. involvement in the overthrow of left-wing nationalist governments in Iran (1953), Guatemala (1954), the Congo (1960), Laos (1962), the Dominican Republic (1963), Brazil (1964), Indonesia (1966) and probably Cambodia (1970).[14] That these interventions have been neither unplanned nor simple products of a conspiratorial and uncontrolled CIA is amply intimated by the following colloquy between two prominent liberals—a former vice-president of the United States and the then secretary of defense:

> Senator Sparkman. At a time when Indonesia was kicking up pretty badly—when we were getting a lot of criticism for continuing military aid—at that time we could not say what that military aid was for. Is it secret any more?
>
> Secretary McNamara. I think in retrospect, that the aid was well justified.
>
> Senator Sparkman. You think it paid dividends?
>
> Secretary McNamara. I do, sir.
>
> Senator Sparkman. I believe that is all, Mr. Chairman.[15]

Among the "dividends" was the military supervised—with U.S. advice—assassination of 300,000 supporters of the previous government, the deposition of that country's neutralist leader Sukarno, and the provision of generous subsidies to foreign investing corporations.[j]

Military Training and External Reference Groups

The object of our inquiry here is to assess in what degree military aid is an effective foreign policy instrumentality for influencing the internal politics of countries in the underdeveloped areas. In their studies of Third World political systems, many social scientists have tended to wholly ignore or depreciate the significance of external intervention by great powers.[16] This is especially true

[i]Hence, when Guinea demanded complete independence from French tutelage, "America was slow in coming to the aid of Guinea at the time that France cut off all economic assistance to that country. . . ." The goal of subverting that country's government was frustrated when it sought and obtained aid from the USSR. Quotation is from: U.S., Congress, House, *Report of a Special Study Mission to West and Central Africa*, March 29th to April 27th, 1970, by Hon. Charles C. Diggs, Chairman, Subcommittee on Africa of the Committee on Foreign Affairs, House of Representatives, 91st Cong., 2d sess., p. 12.

[j]These events are treated in Chapter seven.

with respect to military aid.[17] As the following colloquy and subsequent chapters make clear, this "aid" may, upon occasion, not even be desired by the recipient government and is extended primarily to support or force a change in the politics of particular foreign governments.

Mr. Passman. If I remember correctly, several years ago Syria was one of the few nations around the world that refused to accept our aid program in any way, shape, form, or fashion. We finally got in with a little program but immediately got out, and then we went for many years without any program in Syria.

Six or seven years ago I was in Lebanon and went to Syria where I had lunch with our consul general. The man was so elated I thought he was intoxicated. He said, "It looks as if we are going to finally get Syria to accept a little of our aid program. We are starting a Public Law 480 Program."

He was really an enthusiast. I see now we have finally prevailed upon them to let us start a military training program. Did it take much persuasion to get them to let us in?

Admiral Heinz. No, sir; we are not . . . [censored].

Mr. Passman. How long had you been there with a training program? I don't think you had a military aid program until the early sixties.

Admiral Heinz. The first training program was in fiscal year 1963.

Mr. Passman. What has been the total for this country?

Admiral Heinz. It has been about $100,000 since 1963.[k]

As implied by the preceding quotation, the focus here will be upon training rather than the provision of war materials. Of the approximately 320,000 MAP trainees between 1950 and 1970, almost 80 percent have come from Third World nations. This includes nearly 150,000 from East Asia and about 50,000 each from Latin America and the Near East/South Asia region. A much smaller though steadily increasing number of African military men have also been trained under the program.

Rather than dwell upon figures at this point, we shall briefly outline our thesis. The sources cited in the preceding section establish reasonably well a pattern of American hostility to both economic nationalism and social egalitarianism in the underdeveloped areas of the world since the 1890s. We

---

[k]U.S., Congress, House, Committee on Appropriations, *Foreign Assistance and Related Appropriations for 1968, Hearings*, before a subcommittee of the Committee on Appropriations, House of Representatives, 90th Cong., 1st sess., p. 691. With regard to "unwilling" recipients in the Middle East, see: Furniss, *Perspectives*, p. 31.

Perhaps the most poignant expression of this imperialistic attitude occurred in the course of the following exchange:

"Mr. Passman. In many of the nations where you have an economic aid program, you also have a military program, and some of those nations are determined to have military forces; are they not?

Secretary Rusk. We are determined that some of them have a military program." U.S., Congress, House, Committee on Appropriations, *Foreign Assistance and Related Agencies Appropriations for 1966, Hearings*, before a subcommittee of the Committee on Appropriations, House of Representatives, 89th Cong., 1st sess., p. 509.

hypothesize but will not test the proposition that antipathy has varied proportionately to the degree of principled and sustained commitment by nationalists to developmental policies which were incompatible with a "healthy investment climate" for U.S. economic interests.[l] The post-World War II alliance systems were structured in large measure to safeguard existing and potential resource exploitation in the Third World for Western transnational corporations by deterring or offsetting Sino-Soviet aid to revolutionary nationalist movements. The latter could not even begin to close the door without some semblance of nonalignment. Our second hypothesis which we shall also treat as an assumption is that an independent foreign policy and an ability to translate economic nationalism into effective policies are closely linked variables.[m]

Although a variety of other foreign policy instrumentalities have been and continue to be so utilized, we shall focus upon the role of nontechnical aspects of MAP training as a source of optimally American and secondarily Western reference group identifications by officers from Third World countries. It will be

---

[l]Even though we would only predict a general posture rather than the selection of particular means to effect a modification in the target country's policy process, it would be exceedingly difficult to validate our contention without access to classified Pentagon and CIA memoranda. Action on loan applications and content analysis of the *Department of State Bulletin* or congressional hearings might function as alternative indices.

[m]The analysis would have to be refined by specifying both directional movement, lead times and degrees of *de facto* cold war or pro-American alignment. Other heady complications would arise from the necessity to elaborate criteria which would classify the content of economically nationalistic policy orientations. It would be fanciful to assume that the latter preclude all foreign investments regardless of imposed conditions. Thus the Allende administration in Chile has exempted selected European, Japanese, and even U.S. investors from expropriation. Yet few would dispute the economically nationalistic character of the regime nor Washington's hostility towards its policies and very existence. Thus in late 1970 when American economic aid was in the process of being phased out, subcommittee chairman Zablocki cautioned: "here is an instance where the [sic] political end certainly should weigh heavy in the balance as to the continuation or discontinuation of the military training program." The subsequently articulated reasons for its continuation were censored from the hearing record available to the public. A couple of years later Secretary of State Rogers opined that he found it "interesting that in the case of Latin America we are still providing some military assistance to Chile, for the reasons we think it would be better not to have a complete break with them." In other hearings during the year preceding the Allende electoral victory, there is extensive censorship of the U.S. position and likely intervention at the time. U.S., Congress, House, Committee on Foreign Affairs, *Reports of the Special Study Mission to Latin America on Military Assistance Training and Developmental Television*, Submitted by Hon. Clement J. Zablocki, Hon. James G. Fulton and Hon. Paul Findley to the Subcommittee on National Security Policy and Scientific Developments of the Committee on Foreign Affairs, House of Representatives, 91st Cong., May 7, 1970, p. 31. U.S., Congress, House, Committee on Foreign Affairs, *Foreign Assistance Act of 1972, Hearings*, before the Committee on Foreign Affairs, House of Representatives, 92d Cong., 2d sess., March 1972, p. 39; U.S., Congress, Senate, Committee on Foreign Relations, *Rockefeller Report on Latin America, Hearings*, before the Subcommittee on Western Hemisphere Affairs of the Committee on Foreign Relations, United States Senate, 91st Cong., 1st sess., November 1969, pp. 8-11, 26, 35-6, 45, 51, 85-93, 117-19; U.S., Congress, House, Committee on Foreign Affairs, *Cuba and the Caribbean, Hearings*, before the Subcommittee on Inter-American Affairs of the Committee on Foreign Affairs, House of Representatives, 91st Cong., 2d sess., July-August 1970, pp. 91-2, 97-9.

shown that the political indoctrination and social interaction which have been integral concomitants of the training experience were intended by American policy-makers to: (1) develop a propensity to solicit and/or acquiesce in American policy suggestions; (2) structure a definition of national interest which precludes nonalignment; and (3) inculcate an ideology of development which stresses subsidies and hospitality to transnational corporations. We shall also demonstrate that U.S. officialdom is fully cognizant of the active roles which these officers play in Third World political systems and that this is a key justification for assigning training a higher priority than the provision of military equipment.

The degree of psychological dependency and ideological harmony which is essential for a smoothly functioning neocolonial system is seldom, if ever, attained in practice. Not only are there intra-Western rivalries for subspheres of influence, but the socializing process has been rendered incomplete by such factors as: (1) inexperience and a lack of systematization during the 1940s and 50s; (2) the formidable problem of concealing racist and culturally chauvinistic attitudes which many American military men manifest upon occasion; (3) particular traditions and institutional interests of recipient officer corps; and (4) the glaring failure of most nonsocialist Third-World regimes to even begin closing the per capita GNP gap with the nations of the North Atlantic Community.[n] Despite these and other obstacles, we shall argue that psychosocial dimensions of MAP training have been moderately effective in both making foreign military elites more responsive to U.S. definition of mutual interest and more disposed to accept the advice of American military personnel and diplomats. The transmission of factual images which reinforce or inculcate simplistic free-enterprise/anti-Communist values has increased rightism or conservativism among some officers and diffused such attitudinal orientations to military colleagues with weakly held or nonexistent political frames of reference.[o] This process has affected the

[n]As a form of imperialism, neocolonialism can be distinguished from colonialism not only on *de jure* grounds (i.e. legal sovereignty) but also in terms of the greater degree of indeterminacy and reciprocity in the varying patterns of dominance. While we do not devote much attention to the matter, it must be acknowledged that at times even friendly Third World officers may bargain for particular weapon systems, increase the rental costs of U.S. bases on their soil and persuade American officials that it was indeed necessary to depose a pro-Western civilian government. The scope of their relative policy autonomy is, of course, circumscribed. No cases have been mentioned in which such officers have persuaded American counterparts that military pacts were provocative, or U.S. bases functioned as "lightening rods," or that aspirations for societal development and national dignity required national ownership of basic resources! This study seeks to explain why.

[o]As used here, and particularly for the operationalization of the hypothesis in Chapter seven, *right-wing* does not necessarily imply opposition to all social change or technological "modernization." It is defined as encompassing at least one, though generally several, of the following policies: (1) high priority to the protection of domestic *and foreign* property holdings in profitable areas; (2) *de facto* refusal to redistribute existing wealth in favor of the blue-collar worker and peasant sectors; (3) opposition to publicly-owned and controlled enterprises when generous tax privileges or other subsidies will induce private investors to assume responsibility for the undertaking; (4) anti-Communist suppression; (5) a *de facto*

"internal power balance"—to use CIA terminology—in recipient countries which have had their officers trained over long periods or in particularly large numbers.[p]

The traditional justification for training has been that it is necessary to enable foreign military men to handle equipment which has been sold or given to them. In fact, however, training programs antedated the provision of equipment. During the late 1930s, training was given to Latin Americans in order to cultivate their friendship and a rejection upon their part of German, Italian, and other European military missions which had traditionally performed such activities in those countries.[q] The same objective of attitudinal manipulation has remained a priority one for recipient countries in the underdeveloped areas. Thus, no less than thirty-three such nations are currently receiving free training despite the absence of grant aid in the form of equipment.[r]

Militarism as a Systemic Imperative

Most American corporate and official policy-makers undoubtedly would prefer to rely upon the cultivation of civilians in Third World countries. Although many programs have been directed at influencing these sectors and politically relevant institutions in the underdeveloped areas,[18] for a variety of reasons their efficacy has been regarded as problematic by American policy-makers. Among the impoverished and even within the elite strata of these societies, the existing political systems are not regarded as "legitimate" or authoritative. Because of the previously mentioned expropriation of surplus by foreign capital as well as

---

pro-Western foreign policy where the former one was independent or non-aligned. Leftist or radical coups involve one or more of: (1) nationalization of profitable economic enterprises; (2) egalitarian social reforms; (3) toleration of Marxist and other radical parties which mobilize the lower class sectors; (4) adoption of a nonaligned or militantly neutralist foreign policy.

[p]The use of the MAP as a "cover" by the Central Intelligence Agency is examined in Chapter six and more briefly in seven.

[q]Concomitant with this was a drive to induce Latin American governments to close their open door to economic investments and trade from Japan, Germany, and Italy—that is, to establish an American sphere of influence. These events are detailed in Gardner, *Economic Aspects*, and Green's *Containment of Latin America*.

[r]"Of the 35 individual country military assistance programs, 22 will provide training only—U.S. and/or overseas." And of the MAP total of 46 recipient countries, "the remaining 11 will receive training in the United States only." Hence, only 13 out of 46 countries will be receiving equipment as well as training. U.S., Department of Defense, *Military Assistance and Foreign Military Sales Facts: March 1971* (Washington: Office of the Assistant Secretary of Defense for International Security Affairs, 1971), p. 3.

This distribution is consistent with our view that the political or manipulative aspects of training are of paramount importance to the MAP as a foreign policy instrument. In 1969 the commander in chief of the U.S. Southern Command (for Latin America) testified that even if those countries had no more need for American equipment in five or ten years, "we should continue our training program indefinitely." *Foreign Assistance Act of 1969*, House, p. 646.

other factors such as the decline in terms of trade, the "buy-American" restrictions upon economic aid, corruption, an unwillingness to reform, and administrative incompetence, many of these regimes have failed to improve the quality of life for their "citizens."[19] In this situation, unsuppressed leftist and radical groups are able to mobilize and lead underprivileged constituencies who demand effective access to governing elites. This, of course, materially threatens not only American corporate remissions but also local upper and middle classes who constitute the privileged sectors of these societies. And at times, democratically inclined leaders do reject their former upper-class reference groups to champion populist or nationalist demands.[s]

Although military officers occasionally "succumb" to such radical aspirations, in general they represent one of the more conservative elites to be found in underdeveloped areas. Most officers are recruited from the property-owning upper or middle classes (merchants, industrialists, bankers, landowners, professionals), from the sons of military officers, and increasingly from the socially insecure lower middle class (teachers, clerks, etc.). Many in the last category are economically and socially privileged vis-à-vis workers and peasants. They often are disdainful toward the manual classes and feel threatened by radical elites who encourage mass demands for increased welfare and greater equality of opportunity, particularly in economies with marginal per capita growth rates. Hence, middle-class consciousness in these countries is frequently associated with hostility to radicalism.[20]

There are at least two other major sources of conservative predispositions among officers. Like the middle classes with whom they identify and interact, the military often fear that populists or radical egalitarians—let alone socialists or communists—would curtail their often sizable portion of the governmental budget. Senior officers in many of these countries have resort clubs, expensive cars, servants, many services, and high salaries which are in excess of actual command or military needs. These privileges consume vitally needed development capital. Secondly, military organization itself inculcates a commitment to obedience, order, ascriptive patriotism, the use of force and an unquestioning respect for hierarchical authority. These values engender antipathetic dispositions to political demonstrations, strikes, and the uncertainties associated with redistributive political conflict.[t] Except in the most backward countries where mobilized underprivileged sectors did not constitute a threat, most military regimes have contributed little to socioeconomic development. Functionally,

[s]The distinction between civilian and military can be easily overdrawn. Considerable social interaction and ideological affinity can often be found in certain—usually but not always—convervative circles. Nevertheless, given the relative openness and fluidity of relationships among civilian parties and interest groups, it can be said that civilian politicians tend to be subject to more cross-pressure—including downward ones—than their heavily armed and disciplined military counterparts.

[t]At the perceptual level, officers with a need to rationalize their conduct in moral terms tend to associate such movements with eventual communization.

then, they have operated to conserve the status quo of dependence and underdevelopment.[21] Thus Third World officers are a particularly apt target elite for the counterrevolutionary political goals of the training program.

Scope and Organization

In chapter two we shall demonstrate that such goals are primary though not exclusive objectives of the military assistance training program. The testimony of executive officials, congressmen, American officers associated with MAP, the views of scholars and others will be excerpted and summarized. Hostility to self-determination by other countries when this process results in closing the door to corporate investment or rejecting a pro-Western foreign policy orientation will be shown to be well within the realm of the decision-making framework of American policy-makers and executors. In addition, their cognizance of the essentially manipulative nature of the training process will be made apparent.

The third chapter examines the crucial social and hospitality aspects of the program, while the fourth concentrates upon the process of ideological indoctrination which was incorporated between the late 1950s and 1965. It is subordinate to and has been designed to reinforce the responsiveness of foreign officers to American military attachés and mission or military advisory group (MILGRP or MAAE) officers assigned to their country of origin. In chapter five we describe major U.S. training installations in the United States and abroad. Chapter six examines the crucial role of MILGRP personnel and attachés in selecting and maintaining contact with foreign officer trainees. In the following chapter, an attempt is made to assess the effectiveness of training by relating the proportions of foreign officers trained in various countries to the incidence of rightist and leftist coups. The testimony of American officials and the views of experts are also examined. Brief consideration is then given to the pro-American policy changes which followed the overthrow of nationalist governments in several countries. Finally, in the concluding chapter, we explore recent trends in military assistance training and speculate on what role the program is likely to play in the future. The relevance of our findings to theoretical perspectives on civil-military relations is also discussed.

2 Foreign Officers as Instruments of American Policy

A five-square type of thinking emerged in the 1950s from the Pentagon, aided and abetted by the State Department which, in essence stated: train a nation's officer corps, arm its military and the nation belongs to the United States.[1]

L. Dupree, 1968

These governments do not tend toward measures which are 'radical' in the American view, 'expropriation' of United States mining and industrial interests is less likely under the 'system.'[2]

Harold A. Hovey, 1965

The justifications given to Congress for the program have been a maze of contradictions.[3]

Robert D. Tomasek, 1959

Before the efficacy of a program can be assessed, it is necessary to establish that the objectives it is believed to serve *are deliberate and intended* ones rather than fortuitous outcomes of an uncoordinated policy process. In the preceding chapter, reference was made to descriptive and theoretical works which tend to establish an historical pattern of American hostility to exercises of national self-determination when these involved social egalitarianism, a closed door to the siphoning off of economic surplus by foreign corporations or an "independent" (nonaligned) foreign policy stance.

The Military Assistance Program operates along with civilian propaganda, economic aid, cultural exchange, and diplomacy to further this counterrevolutionary pattern.[a] As the following quotation suggests, conflicts within and among Third World nations—rather than defense of the American mainland—constitute the overriding objectives of training.

---

[a]According to recent testimony by Ronald I. Spiers, the director of the State Department's Bureau of Politico-Military Affairs, no priority is accorded to civilian as opposed to military training programs. Whether or not there ever was such a preference, today "a distinction is rarely made between civilian and military requirements, since local needs are all embracing. Indeed, in many cases, the preferred strategy is to build effective institutions across the board; that is, in the cultural, academic, technical, and military spheres." U.S., Congress, House, Committee on Foreign Affairs, *Military Assistance Training, Hearings*, before the Subcommittee on National Security Policy and Scientific Developments of the Committee on Foreign Affairs, 91st Cong., 2d sess., October-December 1970, p. 150.

In conclusion, the study mission wishes to point out that the majority of issues which must be addressed about MAP training are political and economic in nature, rather than strictly military. This emphasis reflects our strong convictions that military assistance programs are primarily an instrument of American foreign policy and only secondarily of defense policy. So long as the concept of national sovereignty remains dominant in human affairs, the United States can never be certain, regardless of amounts given in training and equipment, that a given nation's armed forces would be prepared to assist ours in a crisis.[4]

This highlights the previously mentioned distinction between colonialism and neocolonialism. Under the former, the concept of national sovereignty is repudiated for subordinate territories whose armies are raised and directly commanded by the imperial power. In the neocolonial situation, other countries are formally accorded sovereign rights and consequentially there is always a residual risk of loss of *de facto* control over their institutions and politically active elites. In fact, the degree of control varies greatly in different countries.

The following sections of this chapter establish that most American policy-makers perceive the MAP and especially its concomitant training relationships as a means of maximizing the responsiveness of foreign officers to U.S. policy goals for their countries. From a variety of official and nonofficial sources, we have selected material which also tends to confirm that these goals include corporate penetration and a dependent pro-American (or where that is unattainable, a pro-Western) foreign policy orientation. Those who are already cognizant of these purposes may prefer to omit this chapter and proceed to succeeding ones.

Other MAP-related foreign policy goals include: Americanizing doctrine, organization, language, tactics, and equipment in other countries; ensuring that bases and communications facilities are available; strengthening regional alliance systems; obtaining intelligence; and promoting the disposal of obsolete American war materials.[b] In varying degrees, the pattern of the alliance systems and these other goals are all subordinate to and conditioned by the underlying counter-revolutionary objective.[c] Thus it is not coincidental that most alliances and defense pacts are directed against regimes which pursue socialistic economic policies even though such "free world" countries as Israel, India, Britain, France,

---

[b]And the interlinkages among these goals are considerable. Thus, equipment grants or easy credit sales may result in acceptance of training that eventuates in secret military pacts which imply larger military missions that sell more equipment and American tactics. The latter requiring English language instruction and staff college level training in the United States. It would be easy to underestimate the complexity and mutually reinforcing character of these relationships.

[c]The term *underlying* implies an historical contextual dimension, a virtually worldwide geographical extension and of course the operational definition of other goals so that they are compatible with the containment or suppression of movements of a socially radical and economically nationalistic character. Again we refer to general tendencies which do upon occasion admit exceptions often compelled by a fortuitous concatenation of events, e.g. Tito's break with the Soviet Union in 1948.

the United States, and Turkey have militarily attacked other countries or territories in the post-World War II period.[d]

Logically then, in 1959 and 1960 there was no interest in furnishing military equipment or economic aid to the Revolutionary Government of Cuba. Radical nationalistic regimes in the underdeveloped areas receive equipment only when this is viewed as a means for inducing their army officers training.[e] The goal of using the military to subvert such civil governments explains why the ubiquitous euphemism "stability" does not always mirror the "real" purposes of military aid to a particular target country. This and the equanimity with which executive officials and most congressmen have viewed the extension or continuation of military aid to dictatorial regimes are aptly illustrated by the following colloquy with the assistant secretary of defense for international security affairs.[5]

Mr. Fraser. In some of these countries we are providing assistance to the side that has seized the power.

Secretary Nutter. We are furnishing assistance in the form of military training to almost all of the countries of Latin America. It is sometimes difficult to sort out those that have elected governments from those that don't.

We feel it is extremely important to maintain our relations with the people who are in positions of influence in those countries so we can help influence the course of events in those countries.

Mr. Fraser. The record does not suggest that you have had success.

---

[d]That alliances such as SEATO have purposes which transcend mutual defense is implied by executive complaints lamenting Pakistan's improved relations with her supposed enemy China. No member of Congress has voiced any criticism of such remarks. Obviously if peace and defense against hostile neighbors were primary American Alliance objectives, such developments would have been welcomed. U.S., Congress, House, Committee on Foreign Affairs, *Foreign Assistance Act of 1963, Hearings*, before the Committee on Foreign Affairs, House of Representatives, 88th Cong., 1st sess., p. 729. U.S., Congress, Senate, Committee on Appropriations, *Foreign Assistance and Related Agencies Appropriations for 1965, Hearings*, before the Committee on Appropriations, United States Senate, 88th Cong., 2d sess., p. 224. U.S., Congress, House, Committee on Appropriations, *Foreign Assistance and Related Agencies Appropriations for 1966, Hearings*, before a subcommittee of the Committee on Appropriations, House of Representatives, 89th Cong., 1st sess., p. 267. *Military Assistance Training, Hearings*, House, p. 89.

This antagonistic disposition is probably best explained by the fact that the SEATO alliance had been "directed against [sic] China." Admission of its hostile nature is by a military aid expert whose classified research was funded by the Defense Department. S.P. Gibert, "Soviet-American Military Aid Competition in the Third World," *Orbis* 13 (Winter 1970), p. 1136. Joshua Wynfred and Stephen P. Gibert, *Arms for the Third World: Soviet Military Aid Diplomacy* (Baltimore: Johns Hopkins Press, 1969, p. ix. Cf., *Economic Issues in Military Assistance, Hearings*, Joint Economic Committee, pp. 133-34, 179.

[e]While our information does not permit us to document a large number of cases, it is clear that even such American "allies" as Jordan, Pakistan, Thailand and Japan initially accepted training programs only in order to obtain U.S. war materials. *Military Assistance Training, Hearings*, House, pp. 89, 102. U.S., Congress, House, Committee on Foreign Affairs, *Military Assistance Training, Report*, of the Subcommittee on National Security Policy and Scientific Developments to the Committee on Foreign Affairs, April 2, 1971, pp. 14-15.

Secretary Nutter. We would have to know what would have been the case if we had not had the influence.

Mr. Fraser. Don't we find that the trend in Latin America is not a very favorable one at the moment?

Secretary Nutter. Yes, sir; we are very much concerned about what is happening.

Mr. Fraser. I gather your office, at least doesn't worry about what kind of government you are helping.

Secretary Nutter. I do worry about what kind of government I am helping; yes, sir.

Mr. Fraser. Then it does not have any effect on your decisions whether or not to continue the training of some of the military personnel.

Secretary Nutter. It has an effect if such a course also runs in favor of our national security. My concern is to promote the national security and–

Mr. Fraser. Of the United States?

Secretary Nutter. Of the United States; yes, sir.

Mr. Fraser. In your judgment, that means internal stability in these countries, is that right?

Secretary Nutter. Not always. Sometimes it does, and sometimes it does not. It means maintaining our influence in some areas of the world that are critical to our security. It means helping to promote, as best we can, the developments that are most in our national interest, but that does not necessarily mean providing for the internal security of those countries. Sometimes it does. Sometimes the stability of these countries is important in order to prevent something worse from happening.[f]

The unfavorable trend referred to in Latin America was one of raising nationalism particularly evident in such countries as Peru, Bolivia, and Chile.[g]

Nationalism, Nonalignment and Neutralism

At the congressional level, there is abundant testimony that the United States has set itself against those countries whose governing elites have sought to pursue independent foreign policies. Thus the Libyan military regime which overthrew that country's autocratic monarchy in 1969 was characterized as "a more pro-Arab Government" and "a less pro-U.S. Government" to the satisfaction of

[f]The Nixon administration's under secretary of state has defined "security" as involving considerably more than defense of the American mainland, i.e.; it extends to the internal policies of Third World countries. U.S., Congress, Joint Economic Committee, *Economic Issues in Military Assistance, Hearings*, before the Subcommittee on Economy in Government of the Joint Economic Committee, 92d Cong., 1st sess., January-February, 1971, p. 300.

[g]The neocolonial mind, however, is incapable of admitting the legitimacy of such self-determination. Hence, Chairman of the Joint Chiefs of Staff Admiral Thomas H. Moorer unself-consciously reiterates President Nixon's paternalistic assertion that "the destiny of every nation within our [sic] inter-American system remains of foremost concern to the United States." U.S., Congress, House, Committee on Foreign Affairs, *Foreign Assistance Act of 1972, Hearings*, before the Committee on Foreign Affairs, House of Representatives, 92d Cong., 2d sess., March 1972, p. 58.

committee members who shared the executive attitude that American aid had somehow not achieved its political ends in that country.[6] Senator Stuart Symington, a former secretary of the Air Force who has of late emerged as a moderate critic of the Pentagon, has gone so far as to assert that Washington consistently opposed nationalists and supports oligarchies in the Third World.[7] A systematic review of congressional hearings on military aid provides us with the following list of military or executive depictions of MAP political targets:[8]

neutralism
leftist revolution
forces of disruption
nationalism
radical African states
home-grown insurgents
preventing or eliminating insurgencies inimical to U.S. interests
political instability
extremist elements
radical elements
extreme nationalists
political dissidents
insurgents and their allies
other extremists
radical elements
militant radicals
revolutions
Arab nationalism
revolutionary ideas
leftist, ultranationalist, anti-American, Nasser-type group

If one excludes actual Communists who are obviously also within the enemy category, what types of groups remain in these countries to be supported by the United States?

One might of course argue that this is an oversimplified portrayal of the essence of neocolonial involvement in other peoples' affairs since Communists frequently infiltrate and support such radical nationalist movements. According to Pentagon data which has been presented to Congress, however, most insurgencies are divorced from Communist involvement—let alone leadership.[h] And these studies indicate that Communists—when they do join a movement—do

---

[h]Between 1958 and 1966, "only 38 percent of the 149 insurgencies involved communism in any degree." Hughes, "The Myth of Military Coups," p. 6.

so *after* it is close to victory.[i] Nevertheless, it seems that all insurgencies which are not engendered by "pro-American" elements are encompassed within the definition of the enemy.[j]

Although a few official statements have declared that MAP's goal is to offset Soviet aid and to "at least" enable nonaligned nations to maintain their independence of the Soviet Union, it would seem that this is a short-term minimal objective.[9] Thus American officers associated with MAP in Brazil have argued the essentiality of continued training which was functional to "exerting U.S. influence and retaining the current pro-U.S. attitude of the Brazilian Armed Forces."[k] Executive justifications using almost identical terminology are ubiquitous.[l] The dedication to having foreign officers "on our side with their minds and hearts" explains former Secretary of Defense MacNamara's categorical refusal to consider phasing out military aid to a number of underdeveloped countries which did not border upon any Communist nation: "Were we to take our assistance away entirely I am certain they would cut their forces and as a result within a matter of 2 or 3 years they would become neutralists."[10]

Several implications of this posture should be mentioned. First, in the case of some nonaligned regimes such as Indonesia and Laos in the past, the United States draws a *clear distinction* between building up or cultivating the friendship of an army on one hand, and supporting that army's government.[11] Second, when it is believed that an army will force its government to decline Eastern offers of assistance, no attempt will be made to "offset" that aid in order to enable the recipient country to maintain its independence of both blocs. This

[i]Thus, "we are not able to directly link this insurgency with Communist leadership or direction in most instances. The general pattern has been for Communist nations and domestic groups to lend their support to insurgent movements once the latter have shown some staying power." U.S., Congress, House, Committee on Appropriations, *Foreign Assistance and Related Agencies Appropriations for 1970, Hearings*, before the Subcommittee on Foreign Operations and Related Agencies, House of Representatives, 91st Cong., 1st sess., p. 716.

[j]Although stability and counterinsurgency are commonly mentioned MAP goals in military literature and congressional hearings, there are apparent exceptions as suggested by the previously quoted exchange between Congressman Fraser and Assistant Secretary of Defense G. Warren Nutter. For another example stipulating that there should be no "doubts that insurgent elements are pro-United States oriented" prior to U.S. withdrawal of aid to an incumbent military regime, see Col. A.H. Victor, Jr., USA, "Military Aid and Comfort to Dictatorships," *U.S. Naval Institute Proceedings*, 95 (March 1969), p. 47.

[k]It was stressed that Brazil was the only Latin American country to provide troops in an abortive effort to legitimize the U.S. invasion of the Dominican Republic during 1965. *Reports of the Special Study Mission*, House, p. 5.

[l]The reasoning behind this is generally of a rough and ready type. Thus, one investigator reported: "Thousands of foreign military officers . . . are being molded by American military education. The Department of Defense accepts the inevitable risk, confident that the stamp left by Fort Levenworth or Fort Benning, the shared experience of a tactical exercise, the friendships formed in wardrooms or mess halls will sometime, somehow pay a dividend to the interests of the United States, despite national political differences that may exist. In classrooms, training exercises, and cocktail parties . . . ." Drew Middleton, "Thousands of Foreign Military Men Studying in U.S.,"*New York Times*, November 1, 1970, p. 18.

strategy contributed to the overthrow of Arbenz's government in Guatemala (1954) and Lumumba's short-lived Cabinet in the Congo (1960), but it seems to have failed in Cuba and Somali. The latter country "sought Western aid during the 1960-63 period, but the response was not acceptable, primarily because the Western offer was conditioned upon the newly independent state not accepting aid from other sources."[12] Third, in contradistinction to Latin America and a small number of other countries, most justifications for states which were formerly territories or mandates of Britain and France use the term "pro-Western" rather than "pro-American."[m] In some cases this represents a cooperative approach which has been followed in traditional Anglo-French spheres of influence.[n] Elsewhere—particularly in the case of neutrals—it is the best that can be hoped for from a small training program. Occasionally, "Western" functions as an ideological symbol and is used interchangeably or in conjunction with "pro-American." Thus, in the oil rich monarchy of Saudi Arabia whose export economy is still largely dominated by U.S. corporations, a recent justification read:

---

[m]Hence, in Morocco "our specific objectives include the continuing evolution of a pro-free world attitude within the Moroccan armed forces." Similarly, "U.S. military assistance has contributed to the maintenance of Senegal's free world oriented government and established a link between United States and Senegalese military forces." For these and many other "free world" dictatorships with relatively small armies, the number trained need not be large:

Mr. Zablocki. As it concerns military aid to the 18 countries listed on page 15, do you think we are overextending ourselves? . . .

General Wood. The amounts, you see, are not large. As I said . . . the program is valuable because even in some cases [censored] and train a few people, we think that their influence after their training may accord us a degree of free world orientation which makes it worthwhile. *Foreign Assistance Appropriations for 1966, Hearings*, House, pp. 278, 732. *Foreign Assistance Act of 1965, Hearings*, House, p. 672.

[n]This cooperation has been dictated as much by limited American power as it has by tradition or any commitment to NATO partnership. Hence, a double standard—what one might call a revisionist interpretation of the Monroe Doctrine—is applied to Latin America where the United States has with decreasing success endeavored to maintain an exclusive sphere of influence vis-á-vis not only the USSR but also Britain and France. Advocating easier credit terms and a $150 million increase in the authorized ceiling for equipment sales to Latin America, Secretary of Defense Laird recently warned "that during the past few years we have seen the French move in with hundreds of millions of dollars worth of sales in that area of the world. We have seen the British move in and the Soviets move in." In this instance, the stipulation of losses referred to American industrialists, bankers and trade unions. His failure to explicitly mention influence may, however, be discounted in light of other evidence presented in this chapter as well as the testimony of Lt. General George M. Seignious (deputy assistant secretary of defense for international security affairs and director of the Defense Security Assistance Agency) at the same hearings: " . . . our military assistance is provided to nations for purposes of enhancing their self-defense capabilities against either external or internal threat. Some assistance however is provided for other purposes as well. These purposes do not always lend themselves to precise definition of a sole purpose. They include arrangements for the maintenance of important U.S. base and other rights such as in the case of Spain, Portugal and [censored] or the maintenance of U.S. influence and friendship as in the case of the Latin American program. [censored]." *Foreign Assistance Act of 1972, Hearings*, House, pp. 61, 161.

Our training program in this country has provided a source of contact with Saudi military leaders on all levels and has thus afforded an opportunity to influence programs and decisions affecting United States objectives in Saudi Arabia. Not only has the effectiveness of the Saudi forces been increased by our training efforts, but United States policy objectives and Western ideas have been disseminated by the continued personal contact and cooperation.[o]

Saudi Arabia is interesting because there were no militarily threatening Communist states and no need existed to offset Communist aid. The only danger was Arab nationalism. It is no wonder then that experts who have at one time or another been associated with the Pentagon or State Department agree that MAP training aims at using "professional ties" and comraderie to ensure recipient officers will be "on our side."[13]

Corporate Investment and Access to Resources

Although experts who have studied military aid have also linked it to the maintenance of capitalism with its concomitant access to raw materials by U.S. corporations, few have related this objective to the subversion of neutralism in these countries.[14] The following statement by a Defense Department associated scholar is exceptional in that it does tend to associate both of these goals with an overall commitment to the status quo:

The potential resources of Africa are needed on the side of the free world to aid in the preservation of U.S. security. . . . A U.S. policy for Africa and the Africans must be designed and implemented promptly or we shall lose [sic] Africa—to obstructive neutralism, to the communists, or to a polarization on a basis of colored vs. white peoples of the world.[15]

Since Communist and nonaligned countries have never refused to engage in trade with the free world, the reference to resources can only be a euphemism for corporate exploitation of them. When the latter occurs, underdeveloped countries receive a fraction of what they can earn by exploiting the resources themselves.[16]

Few officers have been as explicit concerning MAP objectives in this area as has General Robert Porter, the pre-1969 commander-in-chief of the U.S. Southern Command. The following remarks were made while he was still exercising command responsibility over Latin America.

---

[o]*Foreign Assistance Act of 1965, Hearings*, House, p. 732. Thus the United States "quietly encouraged" the Saudi Arabian regime to supply arms to Yemeni Royalists in their attempt to depose the neutralist Republican government. George Thayer, *The War Business* (New York: Simon & Schuster, 1969), p. 236.

Many of you gentlemen are leaders and policy makers in the businesses and industries that account for the huge American private investment in Latin America. . . . Some misguided [sic] personalities and groups in our own country and abroad call you capitalists who seek profit. Of course you do. . . . You can help produce a climate conducive to more investment and more progressive American involvement in the hemisphere. . . . The Alliance envisages some $300 million a year in U.S. private investment.

. . . .

As a final thought, consider the small amount of U.S. public funds that have gone for military assistance and for AID public safety projects as a very modest insurance policy protecting our vast private investment in an area of tremendous trade and strategic value to our country.[17]

The institutionalized nature of this role orientation is attested to by the following exchange concerning the 1968 expropriation of the International Petroleum Corporation which "has touched virtually every aspect of United States–Peruvian relations."[18] House Foreign Affairs Committee Chairman Dante Fascell is questioning Porter's successor who exercises direct command authority over military missions in twenty Western Hemisphere republics.

Mr. Fascell. In understanding that factor and in understanding also the political necessity to stay in power, do you agree that each one of these nationalistic regimes, with a change in its political bases, will be forced to take a stronger anti-American stand in order to justify their position?

General Mather. I think that is very logical to expect, Mr. Chairman, because really, I think the true profile of America in Latin America is the multibillion dollar investment we have here. This is what these economic materialists are after, and causing a lot of our problems. It is $12 billion. About a fifth of our total foreign investment.[19]

A similar preoccupation was exhibited by the Rockefeller Report on Latin America. In addition to urging *expanded* military assistance training as an instrumentality for ideological dominance and the maintenance of a sphere of influence, it warned that:

The forces of nationalism are creating pressures against foreign investment. The impetus for independence from the United States is leading. . . .

. . . .

The central problem is the failure of governments throughout the hemisphere to recognize fully the importance of private investment.[20]

Other military officers in addition to Generals Porter and Mather have regarded *corporate penetration* of underdeveloped economies to be within the purview of the MAP.[21]

This perspective is shared by high level executive officials. Rationalizing the continuance of military aid to the moderately nationalistic Libyan military regime, Assistant Secretary of State David D. Newsom recently testified that "[t]he primary U.S. objectives remain the protection of U.S. citizens and American investments in Libya and the reestablishment of a basis for mutual confidence between the United States and Libya. [That country should] . . . not become a destabilizing force in the surrounding area."[22] The extension of military aid to Morocco has also been viewed as serving to maintain an open door for American investors in that country along with base rights and a desire to keep King Hassan on his throne. Like Haile Selassi, the Moroccan dictator has exercised a "moderating" role vis-à-vis the "radical" Middle Eastern and African states.[23] Finally, a similar justification was recently proferred by the ambassador to Taiwan who viewed the U.S. invasion of Vietnam as integral to protecting general investment and trade interests *in that area* of the world.[24] Most members of the American congresses which have authorized and appropriated funds have acquiesced in or overtly approved of corporate expansion as a MAP goal. No member participating in hearings has suggested that the United States should be indifferent to–let alone endorse–public ownership of profitable economic enterprises by Third World countries which would prefer such a developmental strategy.[p] On the other hand, some have been extremely vocal in their advocacy of the so-called private responsibility system.[25] In general, this objective is taken for granted as it infuses most foreign activities of the United States in such areas as propaganda, economic aid conditions, leader grants, and so forth.

Parliamentary Democracy

Executive implementation and congressional oversight of military programs share one other significant characteristic with economic aid. Very few congressmen and virtually no administrators express the view that these instrumentalities should be utilized to promote free elections and civil liberties in recipient countries.[26] The most recent tendency is to claim that nothing can or should be attempted in the civil liberties area. In the first section of this chapter, a typically blasé statement by the Defense Department's assistant secretary for international security affairs was quoted. His counterpart in the State Department shares the same perspective, although the position is stated with greater clarity.

---

[p]Based on the hearings cited in this chapter. Yet because of the overextension and *failure* of American policy in Vietnam, several Congressmen such as Fulbright, Church, Symington, Fraser and Culver have in recent years espoused opposition to intervention in general terms. See, for example, the hearings referred to in note 7 and *Reports of the Special Study Mission*, House, pp. 9-13, 18-20. Cf., *Economic Issues in Military Assistance*, Hearings, Joint Economic Committee, p. 48.

The Department of State, therefore, vigorously endorses military assistance training from both a foreign policy and a political standpoint as one of the most productive forms of assistance investment. Training enables our friends and allies to provide for their own defense with their own people. It also affords us a means of entree with a vitally important segment of society in many countries. Whether we like it or not, the military holds political power in many countries, and exerts a great deal of influence on national policies in others. Our training program affords military leaders from all over the world an introduction to the United States and to Americans, to our free and open society, and to the achievements of our form of government.

I do not think anyone can claim that such exposure will result in a democratic outlook or the adoption of our democratic ways. . . .

I think it very hard to estimate the extent to which military training either inculcates values of a democratic character in the people that go through the training or otherwise. I think this cannot be the primary criterion. It really has to be our bilateral security relationship with a given country.[q]

If the existence of parliamentary democracy is operationally peripheral to official definitions of "security," the same cannot be said for such objectives as subverting neutralistic and socialistic manifestations of self-determination. Like political freedom, the latter *high priority* goals cannot always be attained. That

---

[q]Hence, when the question of providing arms to dictatorial regimes is raised, civilian policy-makers at the highest levels evidence an historic indifference to democracy which might just as well have been articulated in the eras of Machado and Batista:

"Mr. Fraser. I understand also it is U.S. policy to furnish military aid for authoritarian governments in Latin America?

Secretary Rogers. I didn't say it that way.

Mr. Fraser. I know, but we have some authoritarian governments in Latin America. It is the object of U.S. policy to provide military machines for authoritarian governments?

Secretary Rogers. No.

Mr. Fraser. What is the policy?

Secretary Rogers. Our policy is to assist other nations in this hemisphere to maintain their stability. We are dealing with the governments as they are, as I say, in a very modest way. We are not passing moral judgments on the nature of their government. We found it didn't work.

So the policy of the Government is to assist, in very modest ways, Latin American countries when they request material which will help with the internal security of that country.

Mr. Fraser. Suppose a dictator controlled a country and you decided the dictator should be overthrown. Would you welcome and aid that dictator or would you see this aid as contradicting democratic values?

Secretary Rogers. I don't think that kind of hypothesis helps.

Mr. Fraser. Imagine yourself as a citizen of a Latin American country with an authoritarian regime. You see the United States giving aid to a government which denies your people their freedom, what would your opinion be?

Secretary Rogers. I can't answer that.

Mr. Fraser. In neither case can you put yourself in the position of these people?

Secretary Rogers. I don't think these people put themselves in that position."

Military Assistance Training, Hearings, House, pp. 151, 155. *Foreign Assistance Act of 1972, Hearings*, House, pp. 33-34.

they are more often achieved merely reflects the fact that for several decades the United States has allocated both military and nonmilitary resources to their furtherance. With a few exceptions such as the Dominican Republic in 1963, this cannot be said of the largely rhetorical goal of promoting free elections abroad. In fact, as we shall see in Chapter Four, the indoctrinational materials used in the course of training have *never placed great stress* upon the suitability of parliamentary democratic institutions for political systems in underdeveloped areas.

Responsiveness and Neocolonialism

Our selection of corporate penetration and opposition to nonalignment was determined by the belief that these objectives are themselves closely intertwined and have historically conditioned America's need for bases, alliance systems, anticommunism, vast military establishment, and the MAP. De-emphasis of other military aid goals is not meant to imply that they are not also presented as justifications for military aid and training. Thus, the evils of a reified and largely nonexistent "international Communism" are reiterated ad nauseum.[r] While unsophisticated and socially insecure audiences tend to be catalyzed by such diabolical symbolism—and these undoubtedly include some congressmen—many if not most decision-makers in government are aware that communism and national independence are reconcilable and actually converge in such countries as Rumania, Korea, Vietnam, Mongolia, and Cuba. And it has been many years since policy-makers imagined China to be a Soviet satellite.[s]

---

[r]At the ideological level, stereotyped symbols such as the "Free World" function as did "Progress" and "Civilization" in the nineteenth century. Of motivational significance for some policy-makers and especially among those who implement imperialist goals at lower echelons, they serve a cultural need that domination and intervention be rationalized in moral or normative terms. While some liberal congressional critics have realized that alleged motives are not in fact of causal significance, their own ideological bias against systemic perspectives and particularly being accused of an "anti-business" attitude inhibits sedate analysis of what we have called more fundamental sources of neocolonialism. See, e.g., Suzannah Lessard, "What's Wrong with McGovern," *Washington Monthly*, June 1972, pp. 6-16.

[s]That it is the socioeconomic order and its normative values which constitute the primary enemy, rather than militarily aggressive designs of particular governments is implied by the following exchange:

Mr. Passman. When we speak of Communists, we are thinking primarily of Russia, are we not?

General Wood. Russia and Red China. Both of them are Communist. They are arguing with each other, just now, but they are still both Communist. *Foreign Assistance Appropriations for 1966, Hearings*, House, p. 386. Cf.: David Horowitz, *Empire and Revolution* (New York: Random House, 1969), pp. 71ff; *Foreign Assistance Appropriations for 1970, Hearings*, House, p. 872; *Economic Issues in Military Assistance, Hearings*, Joint Economic Committee, pp. 46-7, 195; *Foreign Assistance Act of 1972, Hearings*, House, pp. 52, 176-79, 220, 231; and footnote d.

Like anticommunism, most other goals are also consistent with those which are central to this study. Thus MAP is viewed as promoting the sale of "excess" or obsolete American equipment with consequential standardization.[27] It naturally strengthens such military alliances as NATO, SEATO, ANZUS, the Rio Treaty, etc. Diplomatic support in the United Nations and such regional groupings as the Organization of African Unity are likewise regarded as MAP objectives.

Mr. Cohelan. General, on Ethiopia. I am disturbed about the levels. . . .
General Wood. We feel Ethiopia is one of the more stable governments in Dark Africa. Haile Selassie is a force for leadership in Africa. He is an influence for good with the African states. . . . [censored] All of this makes it worthwhile to keep the country stable and to give it some military potential. . . .[t]

A few moments later, General Wood reinforced the justification by citing the availability of Ethiopian forces in Korea and the Congo.[28] Thus the MAP is conceived of as a means of increasing the potential military forces available to the United States for fulfillment of its objectives in various countries.[u] As the British withdraw most of their forces from the Persian Gulf in the early seventies, Iran is expected to provide its American trained and equipped troops for the purpose of suppressing Arab nationalism and social revolution in that area.

The Shah has said in effect to President Nixon: "Would you rather have American troops take up where the British left off, or would you like us to do the job?"

. . . .

Their military leaders have an excellent relationship with our men. The Shah understands that he has a role larger than simply maintaining the independence of his own country, and he has expressed a willingness to help achieve stability in the vital Persian Gulf. In such a situation, a reasonably large U.S. military mission and training program seems to me justified.[29]

---

[t]Thus, in the case of Jordan, it was argued that the MAP has strengthened the "loyalty" of the armed forces to the pro-American government and has induced such countries to maintain an active alliance status. Similarly, according to General Wood, "in return for our military assistance, Turkey has been a staunch and loyal ally and has become a pivotal member of NATO and CENTO." *Foreign Assistance Appropriations for 1970, Hearings*, House, pp. 812, 845. *Foreign Assistance Appropriations for 1966, Hearings*, House, p. 266.

[u]We do not exclude the high probability or virtual certainty that as the program has developed and gathered momentum, "power for its own sake" has also functioned as a motivational factor both with respect to what is now called "total force planning" as well as domestically. Thus, from the standpoint of social status and organizational privileges for American officers, the MAP is functional for purely institutional–if never articulated–military interests. Although initiated and partially sustained to further civilian defined foreign policy goals, military prestige, comfort and career prospects contribute to explaining the almost universal endorsement of the program by Pentagon senior officers.

The objectives of MAP also include "third country training." Israel, a MAP recipient, has provided training and advice to such African countries as the Congo, Uganda, and Ghana.[v] Between 1966 and 1971, the latter two experienced right-wing military coups. Their military aid donors had also included Britain and the United States.

As the preceding discussion suggests, a panoply of goals are associated with the extension of military aid. To the extent that they are attained, a recipient state falls increasingly within an American sphere of influence.[w] This is epitomized by the four nations which have received the largest proportionate amounts of MAP aid: Turkey, Greece, South Korea, and Taiwan. Occasionally, Iran is added as a "forward defense" country. Because of such massive aid injections, these corrupt capitalist regimes remain as dependent as ever upon American "aid." And Washington's commitment–despite the touted Nixon Doctrine–has been to keep them that way.[x]

Essential to all of these MAP goals is a reservoir of pro-American bias among the officer corps. Although traditionally depicted as a collateral benefit resulting from technical training in the United States, the cultivation of pro-American sentiments is equally if not more important.[30] According to the then Secretary of Defense Robert S. MacNamara, it was "beyond price to the United States to make friends of these men."

---

[v]In a similar fashion, the German Federal Republic has provided aid to Greece and Turkey while Indonesia has been training Cambodians. *Foreign Assistance Act of 1972, Hearings*, House, pp. 31-36.

[w]As noted previously that this is an ultimate goal beyond a common Western policy of suppressing radical nationalism and social revolution is suggested by two factors: (1) proposals to curtail military credit sales are typically answered by the warning that disappointed African and Latin American purchasers will turn to Britain or France for the desired equipment and (2) "it seems without question that the training of foreign military forces influences the weapons chosen by the force trained. The most direct link between training and acquisition appears to be in those areas of weaponry that are relevant to the military missions the United States feels are priority goals for the recipient." When considered in conjunction with "instructions directing that training must insure standardization (by foreign armed forces) with U.S. doctrine, tactics, techniques, and methods of operation," it would seem that the essential distinction between imperial rule and American neocolonialism pertains to the existence of the formal regalia of sovereignty. Naturally, complete *de facto* control is rarely attained due to residual nationalist antagonism and the illegitimacy of such domination. With regard to the factors cited above, see: *Foreign Assistance Appropriations for 1966, Hearings*, House, pp. 391-92; *Foreign Assistance Appropriations for 1969, Hearings*, House, p. 263; *Foreign Assistance Appropriations for 1970, Hearings*, House, pp. 604, 621; *Military Assistance Training, Hearings*, House, pp. 165, 171. Cf., *Military Assistance Training, Report*, House, p. 15.

[x]*Military Assistance Training in East Asia, Report*, House, pp. 6-10, 14-17, 27-29. We are not suggesting that self-sufficiency is in fact an operational MAP objective. See, e.g., pp. 2-5, and the last source referred to in the preceding note. Both sources express apprehension that Japan's self-sufficiency will result in an independent foreign policy which would deleteriously affect American economic and political goals or its security interests. Cf., John Dower, "The Eye of the Beholder: Background Notes on the U.S.-Japan Military Relationship," *Bulletin of Concerned Asian Scholars* 2 (October 1969), 16-28.

> I have said before, and I think it bears repeating, that in all probability the greatest return on any portion of our military assistance investment—dollar for dollar—comes from the training of selected officers and key specialists in U.S. schools and installations. These students are handpicked; they are the coming leaders of their nations. It is of inestimable value to the United States to have the friendship of such men. Accordingly, appropriate training is strongly stressed. . . .[31]

Or, in the words of General Lemnitzer, training is the "most important portion, dollar for dollar, of the entire program," because it "promotes long-term mutual understanding and respect for the United States."[32] The strategy involves little in the way of mutuality, since foreign officers are to equate American foreign policy goals with their own national interest.[y] Referring to British programs, one specialist has described this manipulative process in terms of structuring new external military reference groups which may outweight identification with and deference to one's own civil government.[z]

The following chapters examine how and to what extent the United States succeeds in transforming foreign military elites into increasingly responsive promoters of American national interests.

---

[y]We do not know whether American officialdom is more ambitious in this respect with regard to small recipient states than to middle-sized powers which tend to manifest greater military pride and often have memories of past neoimperialist glories. In the latter case, the task is a formidable one to say the least.

[z]"The individual trainee is isolated from his past reference and membership groups, that is, from the sources of social and psychological support for his previous beliefs. He is enmeshed in a network of new membership groups, characterized by high levels of cohesiveness and a reinforcing ideological homogeneity. . . . The training process undergone by the officer corps of many of the new states is such as to produce reference-group identifications with the officer corps of the ex-colonial power and concomitant commitments to its set of traditions, symbols, and values." Robert M. Price, "A Theoretical Approach to Military Rule in New States: Reference-Group Theory and the Ghanaian Case," *World Politics* 23 (April 1971), pp. 405, 407.

3 Social Aspects of Training

> Where changes of attitude are sought, it often happens that the more personal channels of communication are the really effective ones.[1]
>
> Frederick S. Dunn, 1950

> We must study and teach all the subtle nuances, all the Carnegie-like techniques of winning friends, and we must learn to be patient.[2]
>
> Capt. R.A. Jones, January 1965

> U.S. training assistance, accordingly, is aimed less at hard military expertise than it is at the cultivation of internal political attitudes favorable to the United States.[3]
>
> Dr. R.K. Baker, October 1970

The foreign military trainee may be exposed to the most ingeniously fabricated propaganda, but this will have little effect upon his attitudes if his personal relations with American officers have been marred by indifference or overt unpleasantries. At the nexus of the training program then are the "interface" contacts which condition the trainee's receptivity to communications during and after his one or more tours of MAP training.

Although a varied array of techniques have been brought to bear on the problem of structuring friendships between American officers and trainees, they will be subsumed here under the following: orientation tours; visits as adjuncts to technical instruction; and social hospitality at training installations and nearby communities.

Orientation Tours

Since their initial use more than three decades ago, orientation tours (i.e., "guided vacations") have been designed to affect the political attitudes of recipient officers.[a] The indifference or outright hostility manifested by the armed forces of some Latin American nations towards the establishment of regular training relationships occasioned the initiation of hospitality tours during the late 1930s. While in 1938 there were four German and eleven Italian

[a]"The Department of Defense considers these tours one of the most effective ways of creating understanding and friendship between U.S. and foreign officer corps." Institute of International Education, *Military Assistance Training Programs of the U.S. Government* (New York: IIE, Committee on Educational Interchange Policy Statement no. 18, July 1964), p. 15.

missions in Latin America, within four years all had been expelled and replaced by U.S. military missions. Renunciation of the "right" to intervene, technical assistance, Eximbank loans, and largely unfulfilled promises of modern military equipment accompanied the offering of these tours. During that period and in subsequent decades, "orientation training" along with technical aid or P.L. 480 grants have been used to induce Third World elites to accept military equipment and associated training which the United States had previously determined was in its own interest to extend. The primary objective has been to create "friendly" officers who are both amenable to influence and reliable sources of intelligence. When foreign armed forces already receive U.S. training, these tours serve to reinforce and thus further institutionalize established patterns of dependence.[b]

During the late 1930s and the following decade when Latin America and Europe provided the bulk of such officers, hospitality tours were generally limited to those of high rank such as generals and naval captains—prominent military figures who might induce their government to sign a military staff agreement. This may be what Senator Fulbright had in mind when he recently ridiculed a Pentagon rationalization by declaring "you can always generate an invitation to really anything."[4] In any case, the policy of excluding low-ranking officers has been retained although exceptions are made for "new" nations which have small military establishments or few senior officers.[c] The following colloquy is instructive on this priority as it is on the manipulative nature of the tours.

> Mr. Passman. 'Orientation' covers the cost of bringing VIP's to the country and exposing them to the Nation in the way of tours, and whatever is considered necessary to indoctrinate them in whatever we are trying to indoctrinate them to.
>
> Admiral Heinz. It includes only military personnel.
>
> Mr. Passman. You have VIP military personnel, too, do you not?
>
> Admiral Heinz. Yes, sir.
>
> (Discussion censored)[5]

Although during the first decade or two, little in the way of alternative tours were available, by the early sixties three or four special tailored visits were being programmed.

[b]Japan is a recent case in point. Orientation tours were maintained despite the completion of equipment deliveries in 1965. U.S., Congress, House, Committee on Foreign Affairs, *Foreign Assistance Act of 1965, Hearings*, before the Committee on Foreign Affairs, House of Representatives, 89th Cong., 1st sess., p. 368. Cf., footnote x to Chapter Two, and Lt. Col. Margaret Brewster, "Foreign Military Training Bolsters Free World," *Army Digest* 20 (May 1965), 17.

[c]A recent Army Regulation simply refers to "specially selected key foreign officer personnel" who "are considered to be senior officers and in a travel status during the entire period while in CONUS, Alaska, or Hawaii." U.S. Army, *Foreign Countries and Nationals: Training of Foreign Personnel by the U.S. Army* (Washington, D.C.: Headquarters, Dept. of the Army, Army Regulation AR 550-50, October 1970, Effective 1 December 1970), pp. 4-1, 4-2. Hereafter cited as: U.S. Army, *Foreign Nationals*.

Thus, "distinguished visitor groups" of five or more officers generally spend two weeks visiting U.S. sites of interest and various bases.[d] This is the most common type of military tourism and was probably what the Special Study Mission had in mind when it reported:

> Orientation tours are visits to the United States during which participants are taken to military installations, important governmental centers, scenic areas and tourist attractions across the country over a period of from several weeks to more than a month. Among locations regularly visited are Washington, D.C., New York City, Los Angeles, and Disneyland. . . .
>
> While not denying that orientation tours are more in the nature of guided vacations than actual training courses, members of the mil-group staff supported them strongly as an important method of exposing a significant leadership group to American life and society at a crucial point in their intellectual development, thereby affecting their future philosophy and convictions.[6]

The Air Force, Navy, and to some extent the Army have three additional types of orientation tours. Unlike the previously mentioned "distinguished visitor" category, the second includes five officers but no American escort. One of these five—presumably an enthusiastically pro-American repeater—must be fluent in English. Another type of tour is for an individual who is "assigned to an American unit for an extended period of time in a job commensurate with his rank." Although the rationale is that this increases the "foreign officer's professional skill in an occupational speciality," the creation of personal rapport and impressions are also of major concern.

> Person-to-person contacts with his American counterparts, as well as with civilians, are encouraged and facilitated by U.S. military personnel assigned to assist him. . . . The last of the orientation visits is a MAP program under which a small number of U.S. and foreign officers are exchanged (not necessarily on a numerically reciprocal basis) and assigned for several years to a foreign military organization."[7]

The group tours for senior officers are regarded by the Pentagon as being particularly important. The "military departments state that brief visits to the United States afford 'the biggest return on the American training dollar.' "

> This is the only . . . crack we get at the people in whom we are most interested. . . . Naturally, it is good to impress a lot of company grade officers with our sincere desire to be friends, but in the last analysis . . . it is *the division commander or his superiors who make the important decisions in time of crisis.* (Emphasis added.)[8]

The orientation tours are not limited to sight-seeing and travel, but also include social events (cocktail parties, dinners, receptions, etc.) and "entertain-

---

[d]Recently the Army and possibly other services seem to have lengthened many of these visits to three weeks. U.S. Army, *Foreign Nationals*, p. 4-1.

ment at various bases."[9] This, in fact, may well be the most important aspect of the visit, although one cannot doubt their being deeply impressed by modern American equipment and organizational efficiency. As we noted earlier, the emergence of many "new nations" during the 1950s and 1960s has made it necessary to provide such tours for field grade officiers. In Africa, for example, during the initial postindependence years, most senior positions continued to be filled by British and French officers. Some racist Congressmen have openly voiced resentment over the VIP treatment accorded to African visitors who were expected to occupy senior positions and possibly oust governments in the near future. The following colloquy is instructive on this point as well as Defense Department objectives:

Mr. Passman. Who are these distinguished people?

Admiral Heinz. These are chiefs or staff, other military leaders and potential military leaders.

Mr. Passman. What is a potential military leader?

Admiral Heinz. A potential military leader is one who shows promise, shows ability.

Mr. Passman. Does he come out of that distinguished category if he flunks your orientation course?

Admiral Heinz. He is expected to become one of the senior men in the military establishment.

Mr. Passman. Why don't we say it?

Admiral Heinz. These people come here on less than the full protocol basis that would apply in the case of a visit by the Chief of Staff of the Army of a large country.

Mr. Passman. I know. You said a potential leader. You bring him over clothed as a distinguished visitor, then when he flunks this course, do you return him as a distinguished person or does he go back a little lower down on the totem pole? When does he lose that distinction?

Admiral Heinz. He doesn't flunk out of this course.

. . . .

Mr. Passman. Mr. Secretary, I am not quarreling, even if you put gold chains on them. I am just indicating for the record that we never mean what we say. . . . How long do you keep them here?

Admiral Heinz. They are exposed to our schools. They meet many people of importance in the United States. They are here from 2 to 4 weeks.

Mr. Passman. That's not my question at all. I asked you what you exposed them to—is it only the military, or is it our civilian free enterprise system?

Admiral Heinz. We try to expose them to a broad segment of U.S. life, not just the military. It centers around the military, but their exposure is not restricted to military. . . . These personnel are selected by our Ambassador, whose choice is usually approved by the President of the country. They are people with whom it is important to us to have relationships [censored] . . . . It is prospective chiefs of staff, the [censored] . . . . It is a program . . . to familiarize them with the United States and to establish relations with them. It gives them at least a small chance to understand what the United States is doing, and what America is all about. It takes about 2 to 4 weeks to tour this country; and, during that time, relationships are established with them which we feel are helpful.

. . . .

Mr. Shriver. It is public relations?
Admiral Heinz. It is public relations, yes.

In the mid-sixties, field grade officers on tour received a modest spending allowance of one dollar per day and nine dollars for each military installation visited, but the total was not to exceed eighteen dollars per day. Higher allowances were authorized for senior officers.[e]

The intensification of cold war competition in the underdeveloped areas along with such "threats" as the Cuban were responsible for greater emphasis upon this type of informal ideological exposure in the 1960s. As early as 1958 a National Security Council directive authorized the Pentagon to institute cold war indoctrination throughout the armed forces. A year or so later, the unified commands which administer military aid in various areas of the world were instructed to put cold war planning "on the same basis as hot war" preparation.[10]

Reflecting this preoccupation, the secretary of defense in 1959 directed a 15 percent increase in orientation tours. Although systematic data is unavailable, we know that in FY 1961, "175 group tours of high ranking foreign officers [were] scheduled by the Department of the Army alone."[11] "During fiscal year 1963, 1,430 individuals were given these orientation training tours"[12] either singularly or in groups. In the following fiscal year, three senior officers from the Nasserite Republic of Yemen received a tour followed by English language instruction and a Basic Infantry Course at Fort Benning. Through social contact and exposure to propaganda, it was hoped that these officers would become pro-Western (U.S.).[13] The same goal was operative for Africa area orientation tours which brought 287 officers to the United States in FY 1968 and 473 were proposed for FY 1969.[14]

On a global basis, "orientation influence training" reached a numerical zenith in FY 1967 when approximately 4,254 individuals (about 80 percent officers, the remainder enlisted men) were brought to the United States and $3,887,000 was budgeted for this program. Because of appropriation reductions due to Vietnam expenditures and the weakening of the American economy, the FY 1968 program was curtailed to an estimated $2,587,000 for about 3,200 individuals.

It is immediately evident that these prestigious entertainment tours must have

---

[e]For senior officers in such high level positions as vice chief of staff the basic allowance is doubled. Given commissary privileges and various free services, one could easily take issue with Defense Department contentions that the milieu is "austere." Opulent hospitality is of course reserved for Chiefs of Staff who are explicitly precluded from participating in the MAP orientation tour program. They are invited as honored guests of the service secretaries where protocol officers who specialize in catering to such foreign dignitaries ensure appropriate hospitality, sightseeing, interviews, and of course entertainment. With respect to MAP orientation tour allowances, see: U.S., Congress, House, Committee on Appropriations, *Foreign Assistance and Related Agencies Appropriations for 1969, Hearings*, before a subcommittee of the Committee on Appropriations, House of Representatives, 90th Cong., 2d sess., pp. 459, 540-41, 568. U.S. Army, *Foreign Nationals*, pp. 2-8, 6-7.

included a large contingent of majors and colonels as well as generals despite the surfeit of the latter in many Third World military establishments. Few if any enlisted men were offered such excursions from most countries except for the (Canal Zone) Republic of Panama which accounted for about 90 percent of the 739 brought in FY 1966 and the 784 in FY 1967.[15] Since the country of origin for about 50 percent of all officers given these trips was censored, Table 3-1 should only be regarded as suggestive. In some cases, governments may even be unaware of such visits by their officers, and in others the regime desires to conceal the existence of these "guided vacations." This may be the situation for such countries as Mexico, Tanzania, the Sudan, and Cambodia prior to the 1970 coup. In FY 1969, for example, the MAP incorporated no less than twenty countries with which the United States did not have a defense treaty. Five received both equipment and training, four sent officers for training in the United States and at overseas installations, while eleven received training only in the United States.[16]

Table 3-1
Orientation Tours for Officers: FY 1966-FY 1967

Country	Fiscal Year 1966	Fiscal Year 1967
Far East		
Taiwan	52	52
Indonesia	0	39
Japan	27	43
Korea	68	42
Malaysia	5	5
Philippines	8	6
Thailand	30	34
Vietnam	46	0
Near East and South Asia		
Greece	82	44
Iran	112	51
Iraq	0	4
Jordan	3	3
Lebanon	2	3
Nepal	Censored	Censored
Saudi Arabia	16	15
Turkey	75	61
Europe		
Austria	0	4
Belgium	4	9
Denmark	13	3
Censored	Censored	Censored

Table 3-1 (cont.)

Country	Fiscal Year 1966	Fiscal Year 1967
Germany	43	29
Italy	12	22
Luxembourg	3	0
Netherlands	8	16
Norway	3	4
Spain	6	0
Africa		
Congo	10	5
Dahomey	0	4
Ethiopia	6	7
Ivory Coast	0	4
Mali	2	0
Morocco	9	10
Nigeria	0	3
Sudan	2	4
Tunisia	2	0
Upper Volta	4	4
Latin America		
Argentina	154	209
Bolivia	62	68
Brazil	310	300
Chile	39	79
Colombia	84	32
Costa Rica	19	0
Dominican Republic	18	26
Ecuador	25	43
El Salvador	12	36
Guatemala	21	30
Honduras	17	20
Censored		
Nicaragua	32	37
Panama	20	22
Paraguay	52	34
Peru	100	200
Uruguay	22	46
Venezuela	88	78
Nonregional	1,040	1,560
Total	2,800	3,380

Source: U.S., Congress, House, Committee on Appropriations, *Foreign Assistance and Related Agencies Appropriations for 1968, Hearings*, before a subcommittee of the Committee on Appropriations, House of Representatives, 90th Cong., 1st sess., pp. 567-69.

The recipients of orientation tours are of course varied. Even firm allies are targets for this form of psychological warfare. Thus in FY 1970, 85 percent of all Brazilian military trainees were slated for orientation tours as against 15 percent who would enroll in technical courses when they arrived in the United States.[17] This is probably explained by that country's recent nationalistic heritage which was manifested by expropriations and an independent foreign policy until the U.S. backed overthrow of her elected president in 1964. Hence both Brazil and Argentina–which also boasted an independent nationalistic government until 1955–have been given orientation tours for *all* graduates of their national war colleges. Apparently preoccupied with what are viewed as long-term national interests and friendship with these two countries, a congressional subcommittee recently recommended that "this practice of furnishing mass postgraduate vacations should be stopped."[f]

---

[f]U.S., Congress, House, Committee on Foreign Affairs, Military Assistance Training, Report, of the Subcommittee on National Security Policy and Scientific Developments to the Committee on Foreign Affairs, April 2, 1971, pp. 14-15.) And within the Senate, several members of the Foreign Relations Committee were equally perturbed as the following UPI dispatch makes clear: "The Pentagon has spent more than $1 million flying future Brazilian military officers to the United States for two-week 'orientation tours,' including nightclub shows in Las Vegas, Nev., trips to Disneyland, poolside luaus and accommodations in plush hotels.

The Senate Foreign Relations subcommittee on Latin America released a hearing transcript containing a Defense Department accounting of the trips. The bills for the tours in 1968, 1969 and 1970 added up to $1,163,902.

Sen. Frank Church (D-Ida.), the subcommittee chairman, said the trips were "junkets."

But Maj. Gen. George S. Beatty, who runs the orientation program, testified "orientation visits provide a unique opportunity to acquaint selected groups of current and future leaders with U.S. culture, technology and government. To the staff and command schools, these visits mean knowledge of the United States at a formulative time of the officers' development."

Church questioned whether the United States has become too closely identified with the "repressive Brazilian government." He said Brazilian military and police, which the United States help to train, operate without political or judicial restraints.

The orientation tours for Brazil's future military leaders consisted mainly of visits to U.S. military bases, briefings by American officers and examination of U.S. military equipment.

The Brazilian delegations did not stay in barracks or officers quarters, but in such hotels as the Sheraton Park in Washington, the Sheraton-Palace in San Francisco, the Edgewater Hyatt House in Los Angeles and the Flamingo in Las Vegas.

One National War College tour of 108 students and seven staff members cost $138,160 and took the participants to Oklahoma, Kansas, North Carolina, Kentucky and Georgia.

The 1970 visit for the Brazilian Air Command and Staff College visit, which cost $101,600 included this itinerary:

Monday, May 25: 1400 (2 p.m.), arrive McCarren Airport, Las Vegas, and proceed to Flamingo Hotel; evening, at leisure; 2330, midnight dinner/show at the Stardust Hotel.

The 59 visitors from the Brazilian Naval War College, at a cost to American taxpayers of $67,090, got a 15-gun salute in San Diego at the naval training center, a tour of Universal Studios in Hollywood, a trip to Disneyland and "lunch at the Blue Bayou."

In New London, Conn., there was a beach party on July 27, 1968, for 74 visitors from the Brazilian Naval War College, who were here for two weeks at a cost of $90,900.

Other groups got a poolside luau in Charleston, S.C., a visit to Radio City Music Hall as guests of the Coca-Cola Co., and, in San Francisco, a reception on Treasure Island *Chicago Sun Times*, July 25, 1971.

Extracurricular Tours

Brief orientation-type tours are welcome diversions for foreign officers who attend classes in one of more than a hundred technical training installations open to them in the United States. In recent years, each installation commander has received $250 per MAP trainee to implement a Defense Department Information Program which was inaugurated at various bases between 1963 and 1967. Because the program is also directed at non-MAP foreign trainees, the actual per capita figure is close to $150.[g] It is used for guided tours as well as media presentations and social events. Although we lack comparable financial data, there is no doubt that such manipulative activities were funded during the late fifties and perhaps earlier at some locations.

Well before Secretary of Defense McNamara ordered the introduction of a systematic and enlarged information program exclusively for foreign military trainees in September 1963,[18] foreign officers were being escorted to *carefully selected* tour sites calculated to impress them with the vitality and dynamism of American society. Illustrative is a 1958 report on a tour by senior foreign officers enrolled at the Naval Mine Warfare School at Yorktown, Virginia:

Each of the officers devoted considerable time to becoming familiar with the American way of life. Industrial and educational tours included visits to the United States Naval Academy at Annapolis, the American Oil Company Refinery at Yorktown, . . . historical tours of Yorktown, Williamsburg, and Jamestown, the Newport News Shipbuilding and Dry Dock Company, the College of William and Mary . . . plus others.[19]

During the 1960-61 ten-month Command Course for Senior Foreign Naval Officers at the Naval War College, *no less than six* three-to-ten-day orientation tours were scheduled:[20]

Washington Orientation Visit (7 day).

Purpose: To acquaint students with some of the civil, political and military aspects of the national capital of the United States. . . . this orientation visit supports the study of the United States system of government.

Method: Briefing and orientation visits to the Pentagon, Naval Weapons Plant, Naval Observatory, Naval Medical Center, and United States Naval Academy. . . . Tour of Jefferson Memorial, Lincoln Memorial, Washington Monument, the White House, the Supreme Court of the United States (remarks by Justice John Marshall Harlan), the Department of Justice, the Inter-American Defense Board, (Reception, host, Lieutenant General L.C. Mathewson USA, Chairman, Inter-American Defense Board), individually arranged visits as desired to the Library of Congress, National Gallery of Art, etc.

---

[g]Army "representation allowances" or entertainment funds were fixed at one dollar per day for each FMT allocated to the installation commander. It is unclear whether this supplants or is complementary to the $2.50 mentioned in the text. *Military Assistance Training, Hearings*, House, p. 103. U.S. Army, *Foreign Nationals*, p. 6-7.

New London-United Nations Visit (3 days)
Observation of underway submarine operations and briefing by the staff of Deputy Commander, Submarine Force, U.S. Atlantic Fleet. . . . Observation of the United Nations in session . . . meet with country representatives, briefing on 'Developing Role of the United Nations' and 'Disarmament' . . . Luncheon in Delegates Dining Room (host, Vice Admiral Charles J. Wellborn USN, Chairman U.S. Delegation, U.N. Military Staff Committee),–attend Council/Committee meetings. . . .
Norfolk, Key West, Charleston Orientation Visit (7 days).
Visits to the schools of instruction . . . observation of underway anti-submarine and mine warfare operations . . . tours and briefings at selected activities of the shore establishments.
San Juan, Puerto Rico and Guantanamo Bay, Cuba, Orientation Visit (8 days)
Tours of San Juan and briefing by Puerto Rican officials, call on the Honorable Dona Felisa Rincon de Gautier, Mayoress of the city of San Juan, call on the Honorable Luis Munoz Marin, Governor of the Commonwealth of Puerto Rico, briefing on Puerto Rican economic development . . . Tours and briefings at selected Naval activities in Guantanamo Bay. . . .
Navy-Marine Corps Amphibious Assault Demonstration Orientation Visit (2 days)
Demonstration of Marine Corps Amphibious Tactical Concepts . . . to provide an understanding of the latest concepts and operational techniques. . . .
Final Orientation Visit (10 days).
Purpose: To acquire an increased appreciation and understanding of the industrial, economic, and social structure of the United States. . . . This orientation visit complements the . . . International Affairs Study. Method: Visits to selected major industrial cities. . . . Tours of significant industrial and economic activities.

"The final orientation visit of the class of 1960 included visits to ALCOA, Westinghouse Electric Corporation, J.J. Heinz Company, and the Duquesne Atomic Power Plant in Pittsburgh; the Chrysler Corporation, General Motors, and Ford Plants in Detroit; International Harvester and the Chicago Tribune in Chicago; and the Fairless Works of the United States Steel Corporation, and the Stock Exchange in New York. A tour of the University of Pittsburgh and the Museum of Science and Industry in Chicago also were on the itinerary." This tour included a large number of social gatherings and free evenings.[21]

Tours for FMTs attending Army and Air Force schools are similar, although military exercises and weapons systems of the particular service organizing the tour are emphasized. Thus at Colorado's Lowry Air Force Base, visits are organized to the Air Force Academy in Denver. High priority is also assigned to gatherings with local business and civic leaders.[22] Or in the case of the U.S. Army's Ordinance School at Ft. Aberdeen, Md., a group of FMTs during 1964 were escorted through Chrysler Corporation's tank assembly line at its immense River Rouge plant in Michigan. They also saw modern agricultural enterprises in Pennsylvania Dutch country and were entertained in the homes of Leacock Presbyterian Church members there. According to the Chief Foreign Liaison Officer at Ft. Aberdeen, these tours constituted "the primary means of implementing the Informational Objective Program."[23]

Per diem allowances are modest and trip distances are restricted in the Navy and probably by other services to 500 miles or less. For short distances, escort officers are urged to use surface transport in order to "provide foreign trainees a close look at the American countryside."[24] A September 1963 Information Program memorandum of the secretary of defense established the following guidelines for trips and visits by FMTs: (1) meetings should be arranged with "responsible" leaders of minority groups and labor unions only when foreign officers are interested and it is deemed appropriate; (2) priority should be given to arranging visits to stock brokerage firms and major corporations; (3) exposure to political parties should be used to highlight the importance of a "loyal opposition;" and (4) emphasis upon visits to model housing projects, "scientific farms," and local governments. When visiting Washington, FMTs should always be introduced to officials.[25] According to a recent report, a majority of foreign officers do in fact visit the nation's political capital before returning to their countries of origin. In the course of a five-day sojourn,

> they enjoy a guided tour of the landmarks and are guests at a luncheon usually attended by a Congressman; take part in a military-sponsored reception; have an opportunity to visit their respective embassies; are dinner guests of local families; and attend lectures arranged by the Washington International Center, a non-profit organization headed by the former Ambassador to Ethiopia, Arthur Richards. The center provides an understanding of the United States for approximately 6,000 foreign trainees a year, over 1,000 of them military. The program consists of a week-long series of seminars, tours, and discussions designed to assist the students to put into perspective what they have seen in this country.[26]

What does the euphemism "perspective" connote?

The manipulative design of the tours and visits is suggested not merely by what is carefully arranged for inclusion in the itinerary, but even more by aspects of American society which are excluded. Of special importance are fifty-million impoverished Americans residing in rural areas and urban slums. Not one description of these tours mentions the most infamous black ghetto in the world—Harlem—let alone hundreds of others populated by visibly oppressed and discontented "citizens." Nor are there "arranged meetings" with those who have led the civil rights and peace movements. Alienated workers, idealistic students, racially discriminated against Puerto Ricans, Chicanos, and Indians are all nonexistent as far as this "balanced" portrayal of American society is concerned![27] Nevertheless, it is certain that a small though indeterminant number of FMTs are inadvertantly exposed to such failures of the American system or at least to situations which are ambiguous in their normative connotations. When this occurs, the role of the escort officer is crucial since he is "in the best position to influence their [the FMTs] impressions of what they see and hear."[h] After completion of visits and tours by foreign officers, the U.S.

[h]Here, of course, not merely ideological self-consciousness but personal traits themselves may be quite determinative. According to a political socialization specialist, "we would

escort officer files a report describing and evaluating how each FMT reacted.[28] Presumably, derivative summary data are channeled through one of the military intelligence branches to the attaché or MILGRP in the foreign officer's home country.[i] Generally there are between one and three escort officers for each ten FMTs on tour.[29]

The "Low Profile"

In Pentagon jargonese, this term refers to *avoiding any appearance* of manipulation or domination. It is impossible to exaggerate the sensitive nature of indoctrinating foreign officers for once they perceive that they are being subjected to such a process, the entire effort becomes nugatory. This has been noted in connection with Eastern military aid programs for several African countries.

> It seems likely that a few well indoctrinated individuals may acquire key positions, but in general direct political training is as likely to wear off as quickly as that which is indirect and informal. Perhaps, indeed, the resentment which intense political instruction generates will create contrary prejudices.[30]

For this reason the Defense Department has limited political indoctrination for regular technical trainees to between ten and twenty percent of available hours. Second, the DOD Information Program "stresses informality. There are no formal courses–no lectures." Third, participation "in all activities is voluntary," although "some 60-75 percent of the trainees usually take advantage" of the tours and other organized activities to be described in the next section of this chapter.[31]

A preoccupation with the delicacy of this manipulative endeavor is also reflected by the following guidelines for liaison, escort, and other American officers entrusted with its administration: (1) emphasis should be placed upon "learning by seeing," especially via "field trips," films, and "business

---

expect agents with attractive personalities, or greater self-esteem, or better capabilities for manipulation to be more *effective* in transmitting political values. . . . " U.S., Dept. of the Navy, *Informational Program for FMT*, p. 32f. T.G. Harvey, "Political Socialization Research: problems and Prospects," (paper prepared for presentation at the annual meeting of the Canadian Political Science Association, McGill University, Montreal, June 1972), p. 28.

[i]This is suggested by Dunn, "Military Aid," (dissertation), pp. 206-08. And while the 1970 Army Regulation previously referred to specifies intelligence involvement–thirty days prior to departure the MAAG is to file a biographical report on each FMT to Army Intelligence, the Defense Intelligence Agency and the installation he will attend–elsewhere in the directive it is stated that "in order to give the foreign trainee recognition for his contribution to the program, a brief report of the foreign student's participation in extracurricular and community affairs should be provided on DA Form 3288 (fig 2-1)." This form and several chapters from Regulation 550-50 are reproduced in our Appendix 11 U.S. Army, *Foreign Nationals*.

luncheons;" (2) the foregoing and discussions should be favored over lectures; (3) U.S. officers should not *appear* to act as experts; (4) at all costs they must avoid giving the "impression of 'forced feeding' " or "indoctrination;" (5) briefing material should be carefully studied by training officers but it "is not suitable for direct use by the foreign trainee;" (6) the atmosphere should be as informal as possible and questions as well as discussions are to be encouraged; and (7) the overriding objective of "exposure to American institutions, ideals and society to create understanding" is to be borne in mind at all times.[32]

The tour/visit itineraries previously referred to and the propaganda materials examined in the next chapter leave no doubt as to the meaning of "understanding"–admiration and possibly a sense of geopolitical fatalism. While MAP administrators are brazen enough to deny they are engaged in manipulative political indoctrination, they rationalize such assertions by narrowing the definition of indoctrination to the salesman's hard sell–something that only an utter fool would ever attempt. Similarly, although they claim to portray "the good and bad sides of American life,"[33] their own perceptions of this reveal how optimistically and righteously biased the term "balanced" can become.[j] The desire to engender affection (or subversion abroad) is carried to such extremes that foreign officer trainees "are designated as 'Allies' whether they come from Norway or Yugoslavia."[34] One can only conjecture the effects of such modes of address upon officers from nonaligned nations which at times espouse "anti-American" policies!

Sponsors and Social Hospitality

Supervision of the political indoctrination of foreign trainees enrolled in technical courses at U.S. and overseas installations is assigned to foreign training officers. These liaison personnel first received rudimentary instruction on non-Western cultures and "inter-cultural communication" when cold war indoctrination was intensified throughout the U.S. armed forces during the late fifties.[35] By 1970, they were administering the DOD Informational Program "at 95 military installations." According to one Pentagon authority,

[j]"The image we seek to present is that of a powerful nation, with a government based on law, a nation whose moral principles exercise a strong influence on its day-to-day actions, a nation actively and continuously in a climate of social change, one constantly modifying its political structure and systems of values to meet its aspirations and to satisfy the goals of all of its people.... Our nation is built on the fundamental principles of respect for the individual and respect for the law–the one complementing the other. If the foreign visitor understands that our values are based on the dignity of the individual, and if he also understands our concept of the law, then the workings of our institutions will be clearer to him. If both the growth of the concept of individual rights and the evolution of the law can be put in perspective of our historical struggle to achieve what has been and now is the reality of our society, then the foreign military trainee will know us for what we are. He may have no desire to adopt our concepts, social values, or political institutions as his own, but he will understand us. It is this understanding that leads to co-operation between peoples. . . . " Grand Pre, "Window on America," p. 93.

much of the success of the program can be attributed to the group of young, dynamic, energetic, and capable U.S. military officers who guide it. These young 'foreign training officers' work 7 days a week at their jobs, for it is a program of involvement requiring diverse skills–educational, social, psychological, philosophical, and managerial. It calls for a knowledge of human nature, political science, international relations, and, not least, a broad knowledge of America. It also requires empathy, patience, an understanding wife, and a group of friends, military and civilian, who will provide assistance and hospitality. They must be willing to act as sponsors for the foreign military trainee, and for the members of his family when they accompany him to America.[k]

Their crucially important role then is to assure that the foreign officer returns to his home country with a favorable appraisal of American society which optimally should border on affection.

Hence, upon the foreign officer's arrival, carefully structured orientation sessions are scheduled for from one to six weeks.[l] These involve an introduction to American customs and colloquial English among other things. With the exception of the Canal Zone base complex, the special Naval War College Command Course for Foreign Naval Officers and a few other instances, foreign officers are *always intermixed* with American officers following the orientation sessions.[m] Thus, at Fort Leavenworth, fifty-man classes are broken down into small twelve to fourteen man working groups in order to foster "social integration" and an ability to use the English language.[36] The FMTs are assured that they will have a "pleasant" experience with "plenty of time" to do "other things."[n] They are urged to participate in "sports" and "social activities" and to "otherwise take part in the American Way of Life."[37] As we have noted earlier, no one flunks the course and one of the duties of the liaison officers from the Office of the Director of Allied Personnel at Ft. Leavenworth is to aid foreign officers with their academic problems. Because during the 1950s some high ranking FMTs nevertheless did poorly in their courses and became resentful because of their loss of face, a decision was made to restrict such courses to field grade officers.[38] In fact, resentment, a failure to apply oneself, an inability to

[k]Foreign officers are in fact discouraged from bringing their families with them. Their presence would in many cases interfere with the goal of maximum social integration within a comparatively brief period. This explanation is not explicitly stated in the sources: Grand Pre, "Window on America," pp. 90-91; U.S. Army, *Foreign Nationals*, p. 2-9.

[l]Even before this process commences, great care is taken to ensure a favorable impression. Thus, "particular emphasis is laid on initial reception at the port of entry/departure. . . . Personnel assigned to meet foreign visitors must be acquainted with the Informational Program (app J) and be prepared to take advantage locally of any opportunities to contribute to its attainment while the foreign trainee or visitor is in their charge. The following points will be stressed. . . . " U.S. Army, *Foreign Nationals*, p. 2-6.

[m]This also applies to residential accommodations for those unaccompanied by families. U.S. Army, *Foreign Nationals*, p. 2-15.

[n]According to Dunn's interviewees at the Naval War College, recreational activities, social invitations, and speaking engagements consumed so much of the foreign officers' time that it was difficult for them to see how much serious study could be accomplished. "Military Aid," pp. 226-31.

enjoy the programmed social interaction, an unwillingness to conform to disciplinary rules and similar negative behavioral attributes, along with ill health are the primary criteria for returning a FMT to his homeland prematurely. Highlighting the preeminent social/rapport objectives of training, even these flunk-out grounds must be cleared with Army Headquarters in Washington before a foreign officer is dispatched. Similar concerns are reflected by the practice of recycling FMTs who lack academic ability and the inclusion of extracurricular participation on the academic report. Although students are graded, these are not permitted to occasion resentment through release to the student. Rather they are reported to the MAAG concerned, the Defense Intelligence Agency, the U.S. Defense Attache concerned, Southern Command U.S. Army Commander and the Commanding General, U.S. Army Intelligence Command.[39]

Planned social interaction is given high priority in towns surrounding the base. At some installations, civilian sponsors have been induced to volunteer their enthusiastic assistance and hospitality through "Community Relations Programs." These in time, were developed through military exploitation of cold war fears reinforced by special orientation tours which brought Chamber of Commerce and other civic as well as a few Labor leaders for up to eight days of hospitality and sightseeing at one or more military installations. A portion of the Pentagon's public relations budget—which increased from less than three million to more than forty-four million dollars between 1959 and 1969—is allocated to propagandizing American civilians for such purposes.[40]

> These civilian sponsors, according to the Israel colonel, who did not wish to be identified are 'really nice.' They have given him, he said, 'a close view of day-to-day life in the United States.
>
> A Chilean lieutenant colonel agreed. He found that life in America was 'much different from the movies and television.'[41]

Of special impact upon foreign officers' pride are invitations to address local civilian clubs and organizations on the subject of their home countries.[42] This appeal to their self-perceptions as authorities on particular countries, the social activities and previously described tours which are also supervised by the foreign training officers play an important role in shaping trainee attitudes toward American society and Washington's self-arrogated "free world" leadership role.

And within a milieu characterized by incessant social and sports activities, it is probable that in many cases the intended "lasting friendship between the student and his sponsor" develops as an authentic bond between the two men. Under Information Program guidelines, these relationships are to be maintained through exchanges of correspondence and occasional gifts.[43] When former American military sponsors are subsequently assigned abroad as attachés or MILGRP officers, it is reasonable to assume that they are instructed to look up old acquaintances.[o] We are not of course implying that such orders are always necessary or that all such friendships are partially coerced due to the American officer's desire for a favorable rating by his promotion review board. Their existence is however fundamentally a matter of duty and obedience.[44]

The gains resulting from such relationships transcend responsiveness by the foreign officer to American policy goals for his home country. As noted with some emphasis earlier, there are also immediate intelligence dividends.[45] In fact, all American officers are expected to act in such a capacity.[46] The interesting question here—for which no uncensored information is available—pertains to whether and how frequently military intelligence and CIA operatives are assigned as escort officers or sponsors to promising trainees from nonaligned or currently socialistic nations.[p]

Given the very substantial budget of the Central Intelligence Agency and the fact that it has worked closely with military factions in such countries as Iran, Cambodia, Laos, Indonesia, and the Congo in the past, it is reasonable to assume that sponsors in some cases are used as "covers."[q]

Only one dimension of the indoctrinational process is represented by the web of social interaction and guided tours in which foreign officer trainees are immersed. In this chapter we have only superficially touched upon the ideological content of the political values and supporting propaganda knowledge that are intended to affect trainee attitudes both directly and indirectly as a consequence of indoctrination of American officers themselves. Because of the significant contribution which a common ideology and admiration for American capitalism make in reinforcing cross-cultural friendships and reference group identification with a long-range potential, Chapter Four is devoted to dissecting some major themes of such progaganda.

---

[o]This probably occurs even more frequently at higher levels. A notorious case in Brazil was the assignment of Brig. General Vernon Walters—a close World War II friend of Costelo Branco—to the embassy in that country about six weeks before the latter launched the U.S. backed 1964 coup which ousted that country's elected left-leaning government.

[p]Explicitly referring to the CIA function of penetrating Third World institutions including the military, former Director Richard Bissell noted several years ago that "many of the 'penetrations' don't take the form of 'hiring' but of establishing a close or friendly relationship (which may or may not be furthered by the provision of money from time to time. . . . The essence of such intervention in the internal power balance is the identification of allies who can be rendered more effective, more powerful, and perhaps wiser through covert assistance. . . . But for the larger and more sensitive interventions, the allies must have their own motivation." *Intelligence and Foreign Policy* (Cambridge, Mass.: Africa Research Group, 1971), pp. 12, 16.

[q]On CIA involvement with military groups in Laos, Vietnam, Indonesia and Iran, see David Wise and Thomas B. Ross, *The Invisible Government* (New York: Random House, 1964). A former member of the CIA director's staff estimates that more than 12,000 of the agency's 18,000 regular personnel are engaged in clandestine political manipulation abroad—as opposed to simple intelligence gathering which he states is a secondary function of the agency. Similarly, more than two-thirds of its $700,000,000 annual budget is allocated to affecting internal power balances in foreign—primarily Third World—nations. Victor Marchetti, "CIA: The President's Loyal Tool," *Nation*, April 3, 1971, pp. 430-33.

4 Ideological Stereotypes in Military Education

The whole Army structure is pervaded with fear of reprisals that stifles any whisper of dissent. Reprisal can come in many forms–the 'efficiency report,' secret 'security' investigation, character assassination, professional ostracism and humiliation. . . . One self-deception is that the end justifies the means. . . .[1]

Col. Edward L. King, Ret., May 1970

. . . present to the student the theory of totalitarianism, showing the inherent failures of totalitarianism as contrasted with the advantages of democratic governments.[2]

Fort Gulick, January 1969

The nation [Brazil] needed it [a military coup] in order to free itself of a corrupt [elected] government which was about to sell us [sic] out to international communism.[3]

Lt. Gen. Andrew P. O'Meara, March 1965

MAP training makes no attempt to mold the political philosophy of the student, much less to tamper with his political loyalty. The program does not attempt to control his political beliefs or behavior.[4]

Dr. Ernest W. Lefever, Brookings Inst., 1970

Foreign officers are subjected to both direct and secondary political indoctrination. The former is effectuated through preplanned tours, language training, orientation materials, films, pamphlets, and speakers. Course readings and discussions at "professional" or higher training levels also have increasingly assumed such a function. With the exception of previously examined orientation tours, these mechanisms will be discussed in the latter portion of this chapter. First, however, we shall briefly treat the dominant Weltanschanung of many, if not all American military officers. This, for two reasons: (1) foreign officers who attend courses in U.S. training institutions are often exposed to progaganda designed for American nationals; and (2) to the degree that U.S. officers have themselves internalized such perspectives, it is quite likely that they color the presentation (e.g., answers to questions) of direct indoctrinational materials and [illegible]n carry over at times into personal discussions and friendships. As we shall [illegible] civic organizations and civilian sponsors in communities near training [illegible]lations have been deliberately propagandized with simplistic cold war [illegible]anda.

Institutional and Personality Factors

Although we don't know a great deal about the determinants of successful socialization, it is quite probable that these include the milieu in which it is carried out, personality traits of those being indoctrinated, and the authenticity of the entire process.[5] To the extent that officers communicating beliefs are not only conscious of their role but are also themselves committed to them, they will carry on with greater enthusiasm and hence be more likely to inspire conviction. Similarly, appropriate beliefs and cognitive supports will be structured more readily if they have also been internalized by American officer peer groups who study and socially interact with the FMTs.[a] Hence, before scrutinizing relevant thematic tendencies of propaganda beamed at foreign officers, we shall devote several pages to the intellectual milieu into which their American colleagues and mentors are themselves socialized.

While it would be undoubtedly naive to imagine that all U.S. officers display identical world views and share similar fears, it would be even more imprudent to ignore the fact of general consensus and perspective with regard to certain issue areas and goals. These are affected, in turn, by the nature of the military profession itself. Success in warfare requires not only technical proficiency but unquestioning obedience to hierarchic superiors and subordination of the individual to his unit and service. As a perhaps too harsh rendition of the careerist syndrome would have it:

> Organization-man conformity has become the pattern for success in the US Army Officer Corps. This is not the zippered lip of the professional soldier, rather it is the cover-up of the 'get on the team' boot-lickers. It has produced unthinking subservient yes-men who practice deception and concealment in order to cover up any error or failure which, if revealed, could harm either the system or the individual career. The conformity and deception are enforced by adverse 'efficiency ratings' against officers who dare speak out. A single unfavorable rating can be an effective and absolute bar to promotion. The alternative to concealing, suppressing or falsely swearing may be professional ruin for an officer. Such a system inevitably produces fearful, overcautious men, oftentimes dishonest men. Thus the Army has become a night-mare for the intelligent and honest man, and a haven for the mediocre and callous.

The author of these lines and a substantial number of other officers had been forced out of the U.S. Army because of their criticism of U.S. policies in Vietnam.[6]

---

[a]This does not exhaust the list which includes awareness of the fact that one is being manipulated and the consonance of new beliefs with one's preexisting values, especially insofar as the latter are strongly held components of the officers sense of identity [illegible] self-image. We have already noted the deliberate policy of screening out the few [illegible] conscious radical nationalists who may have chosen to apply for CONUS training. F[illegible] vast number who arrive with vague or nonexistent beliefs about the cold war, the int[illegible] social activities and "soft-sell," low-profile approach minimize the likelihood th[illegible] foreign officer will perceive himself as a psychological warfare target.

But it is quite likely that most officers are perturbed neither by the war in Indochina nor by the censorship from military periodicals of critical perspectives on U.S. foreign policies.[7] The personality types most commonly attracted by military careers tend to be those who are aggressive, threat- and force-oriented, project inner conflicts, are attracted to dogmatism and authoritarianism, and manifest an aversion for ambiguity.[8] Hence, it is unsurprising that most American officers *do not really believe in tolerance* for the liberty of dissent on major policy issues. Senator Fulbright has noted that:

> The view Admiral McCain expressed is shared by many military men who hold that critics of the governments's Vietnam policy, whatever their credentials, play into the hands of the enemy and prolong the war. Military speeches on the subject are larded with invidious references to long-haired young men, the press, the academic community, 'so-called experts,' and, usually in veiled form, members of the Congress.[9]

While exceptions undoubtedly exist even within the Pentagon, it is implausible to assume that they set the tone in the war or command and staff college classrooms where to insist in expressing strongly dissenting views contributes neither to peer group admiration nor promotions.

Communism: A Paranoid Response

While American officers are comfortable with ambiguity when evaluating the merits of alternative weapons systems, their statements and written materials on Communism almost consistently betray a simplistic crusader approach. Communists are dehumanized enemies driven by an ambition to conquer the world. Communist governed countries do not exhibit institutional differentiation worthy of study, nor is there acknowledgment that their peoples in many instances support their governments and have received great social benefits.[10] Attention is focused upon Marxian theory with exclusive emphasis upon exposing its fallacies. This unbalanced and antiintellectual orientation is epitomized by General Fitch when replying some years ago to a question by South Carolina's Senator Strom Thurmond:

> Certainly, there are relatively few people in the service who are experts on dialectical materialism. On the other hand, I think the vast majority of the people in the service have a sound and healthy hate and fear of Communism.

The inculcation of hatred in military service schools is properly labeled by both senior officers and senators as "indoctrination."[11]

As a counterpart for such vitriolic antipathy toward the ideological "enemy," "balanced" portrayals of the United States systematically *define* social crises or problems either as signs of health or they are claimed to be of minor importance.

When people with erroneous impressions ask you to explain what they regard as serious flaws in our application of democracy, you can tell them that the United States Government and the vast majority of American people are trying to correct inequities. Racial developments, for example, are *signs of progress*, not indications of internal weakness.

Although this was published in 1963, it had not been superseded six years later despite the nationwide urban rebellions.[12]

One of the consequences of this demoniacal model of world struggle is a tendency to depreciate or subordinate all other values, rights, obligations and goals to the *extirpation of pure evil*—and one that is mortally threatening due to its dynamic qualities.[b] Thus aid to dictatorial regimes is acceptable whereas popular support for insurgents is intolerable:

In the less developed countries . . . the military is able to increase popular support, reduce sources of civilian discontent, separate dissidents from the people, and secure greater political stability.[13]

A second and very logical corollary to this dogmatic orientation is the refusal to acknowledge inconsistent facts. In some cases this can lead to flights of fantasy. Thus General Wood, when director of military assistance, rationalized continued military intervention in Brazil by citing an imaginary "vacuum in the civilian political structure so far as running the country is concerned, and the only force for stability and orderliness is the military."[14] Similarly, former Secretary of Defense Robert S. McNamara explained the Sino-Indian border conflict of 1962 not in terms of documented Indian provocations or disputed territorial boundaries but rather as a manifestation of the "Communist Threat"—one that did not cause border wars for Burma, Afghanistan, and Pakistan.[15]

This reductionism predictably manifests itself in a tendency *to define* significant changes or developments as being unimportant. For many years after World War II and to some extent even today the major threat has been "international communism" initially equated with a Soviet dominated world movement. In the course of 1969 congressional hearings, Admiral John S. McCain, Jr., USN, the commander in chief in the Pacific recognized the existence of polycentrism. Nevertheless, the inherent quality of aggressiveness—formerly attributed to Russia—was now unreservedly imputed to each and every Communist nation.[16] More metaphysically, the national Communist regimes may upon occasion be wholly ignored in favor of a more intangible and insidious "force." Thus, according to the 1969 Defense Department justification for MAP funds, it was imperative to

[b]Significantly, recent investigations of American civilian groups reveal a close association between militarism (defined as a force/punishment orientation), and nationalism, conservativism, authoritarianism, anti-Semitism, anti-Communism and low cognitive levels with regard to international affairs. W. Eckhardt, "The Factor of Militarism," *Journal of Peace Research* 2 (1969), 123-32. Cf., Michael Maccoby, "Emotional Attitudes and Political Choices," *Politics and Society* 2 (Winter 1972), 209-39.

contain the forces of international communism which seek to disrupt and enslave the nations and peoples the United States is trying to help toward orderly development and freedom from fear of external attack or subversion.

. . . .

Military assistance, by strengthening the armed forces of friendly countries . . . reduces the probability of and need for U.S. intervention.

. . . .

[It] provides the means of influencing key elements of the recipient nations to resist subversive Communist influence.[17]

This despite the existence of DOD studies indicating that between 1958 and 1966, "only 38 percent of the 149 insurgencies involved Communism in any degree."[18]

Identical dogmatic constraints influence portrayals of Latin American political conflict. Ignoring intra-Communist conflict and rivalry along with Castro's charismatic appeal, General Porter announced in 1968 that Fidel Castro was very "dangerous because he is part of the world Communist movement." This from the commander in chief of the U.S. Southern Command–a man who presumably should know better. He also charged that attempts to belittle the military in Latin America drew their inspiration from Communists, dupes, and fellow travelers. On the other hand, the armed forces in such countries were "in the main progressive and . . . the single strongest cohesive force to guarantee stability and thwart Communist aggression."[c]

That the foregoing oversimplifications and dogmatic constraints are not untypical is suggested by the following report from the chairman of the Senate Foreign Relations Committee who has himself emerged as a leader of a small number of congressional critics of rampant militarism.

The entrance of the military establishment into the field of foreign affairs over the past twenty years has resulted in a requirement by the Department of State that speeches and articles by the military involving questions of foreign policy be cleared 'to ensure that when the information in this material is given to the public it is accurate as to facts, it correctly expresses United States policy, and it does not include statements which might be damaging to the foreign relations of the United States.' These criteria would seem to be explicit enough to keep the military in line. The review process, too, is a thorough one, extending as necessary through the department's bureau of public affairs, to the public affairs advisers of the geographic and functional bureaus, to specific area experts as

---

[c]Gen. Robert W. Porter, "Look South to Latin America," *Military Review* 48 (June 1968), 82-83. Yet according to a former State Department specialist on Latin America, "while it is true that some radicals may look to Peking and Havana, the leaders are essentially radical nationalists and socialists. . . . The insurgency threat is not from the Communists." U.S., Congress, Senate, Committee on Foreign Relations, *United States Military Policies and Programs in Latin America, Hearings*, before the Subcommittee on Western Hemisphere Affairs of the Committee on Foreign Relations, United States Senate, 91st Cong., 1st sess., June 24 and July 8, 1969, p. 27. Cf., pp. 32-33. Porter's observations linking military aid to foreign investment which were quoted at length in Chapter Two suggest the ease with which anti-Communist and materialistic goals can be reconciled.

necessary. That review itself is necessary seems to be demonstrated by State's reactions to 124 manuscripts submitted to it by flag officers during a recent eight-month period. Two of the speeches were found to be objectionable 'in their entirety,' in sixteen cases 'mandatory' changes were made, and in twenty-eight cases, for reasons of accuracy or clarity, changes were proposed. Only thirty of the manuscripts, fewer than one-quarter of the total, needed no changes at all.[19]

While it would be unwarranted to assume that all of these modifications pertained to ideological stereotypes, it is also well to bear in mind that such demoniacal images are not the exclusive property of the American military establishment. A perusal of the *Department of State Bulletin* reveals impressive convergence in the area of ideological stereotypes.[d]

This suggests an extremely significant though difficult to document consequence of the demoniacal model. It is an extremely broad operative definition of Communism which encompasses almost any organization that supports a policy goal also espoused by Communists. Similarly, any group which seriously opposes U.S. foreign policies or criticizes American society tends to be defined as aiding Communism and therefore as an "enemy." Hence, in 1963 the American Military Assistance Advisory Group in the Dominican Republic regarded that country's elected President, Juan Bosch, as a "Communist." the United States regarded him as a tool of the Communists in 1965 just as Mexico's Lazaro Cardenas had been denigrated as a Communist in 1938.[20]

It is unsurprising then, that Admiral Heinz would testify that MAP is directed against non-Communist "Home-Grown" insurgents as well as those who are "Communist-inspired."[21] Nor should we be disconcerted at the fact that during the 1950s, the United States suspected "clandestine Soviet participation in every demonstration, riot, and uprising that attended the bedeviling process of transferring sovereignty from Britain and France to the regional successor regimes. . . ."[22]

Military training institutions have diffused a paranoid devil model of "the enemy" not only during the early cold war period of the 1950s, but also in the post-McCarthy era of the early 1960s.[23] And the same pattern of indoctrination has been perpetuated and even intensified during the late 1960s and under the Nixon administration.[24] It is reflected in virtually all articles on Military Assistance Program goals which have appeared in military journals.[25] Indeed, only a few writers have taken issue with the stereotyped portrayals and their common failure to distinguish between socialism and communism.[26] The widespread incorporation of such empirical distinctions and a balanced view of communism would, we believe, be dysfunctional to the maintenance of the open

---

[d]So does recent testimony by Assistant Secretary of State Marshall Green, who replaced Howard Jones–a genial American diplomat who got on quite well with Bung Sukarno–several months before the successful coup that brought the pro-American and rightist General Suharto to power. *Foreign Assistance Act of 1972, Hearings*, House, pp. 220, 231.

door—and that explains why the demoniacal model remains as the dominant one. It is consistent with Washington's "action policy" in the underdeveloped areas. Hence the factual distortions and affective content are both rational and functional!

Indoctrinating American Officers

Systematic dissemination of anti-Communist propaganda in the services was initiated after the Korean War because a number of captured Americans had readily "succumbed" to North Korean "brainwashing." In 1955 the Defense Department released a suggested training and education guide urging greater emphasis upon the "fallacies of Communism" as well as the "basic truths" of "our democratic institutions."[27] This fantastic dichotomy was portrayed that same year by an Armed Forces Information and Education Service pamphlet entitled *Know Your Communist Enemy: Communism in Red China.* Mao Tse-Tung was caricatured on the frontispiece surrounded by four dragons.[28] In 1956 the Army distributed the following assessment of Soviet aspirations:

If the U.S.S.R. were one of the weaker military powers, we might regard her ambitions for world conquest as more of an eccentricity than a threat to world peace or to our security. But the truth is that military power always has been recognized by Soviet leaders as an essential and even a leading [sic] factor in the pursuit of their plans for Communizing the world.[29]

During 1957-58, *Communist Propaganda—A Fact Book* was released by the Pentagon. This tract listed as "pro-Communist" various publications as well as "Ban the Bomb groups."[30]

Following the adoption by the National Security Council of a secret cold war propaganda directive in 1958, 37,400 copies of Army Regulation 515-1 were distributed. The directive advised all "information" officers to place greater emphasis upon the evil "nature" of Communism.[31] During the following year, a *Democracy vs. Communism* series of pamphlets was instituted. These were to adopt a more subtle approach than the initial *Know Your Communist Enemy* series.[32]

In the course of subsequent congressional hearings on the adequacy of cold war indoctrination in the Armed Forces, exGeneral Walker complained that anti-Communism was still not accorded high-priority, was not sufficiently hard-hitting and was poorly coordinated. Other military testimony was more optimistic. Capt. Cunha reported "Cold War indoctrination" of troops had "been going on at a constant level for many years" and was "beginning to sink in and penetrate deeper and deeper." Referring to Troop Information Officers, he concluded: "I think we have a hard core of Cold War specialists in the military." In the United States Navy, for example, there were 398 cold war propaganda officers in key staff assignments by 1962.[33]

The heightened preoccupation with such indoctrination that became noticeable in the late fifties was followed by the adoption of a new long-range internal information plan. Following Defense Department approval in 1961, the U.S. Air Force demonstrated readiness to make the changes called for by the program. Exclusive responsibility for "moral leadership" training was withdrawn from chaplains and reassigned to various commands while a new emphasis in the information program was hereafter to be placed upon *American ideals and beliefs.*[34] Here was a positive counterpart to the denigration of "the enemy"—though both aspects of the "hard sell" often were presented simultaneously. Rather than record the milestones in the Pentagon drive toward thematic parity or balance during the 1960s, we shall sample the treatment of two prominent issues: American racism and the "American way of life."

A 1961 DOD pamphlet claimed that "in most states Negroes and white students attend the same schools." At about the same time, an Army leaflet on Africa avoided criticism of Apartheid in South Africa while depicting neutralist Ghana as a "concern to the Free World." Eight years later a DOD publication on the Republic of South Africa described the Apartheid system without professing any moral criticism. Concern existed, but this was based solely upon the opportunistic criterion that eventually the situation would lead to disaster! With regard to America's own urban disaster, a 1968 training manual for urban repression alleged that black ghetto riots were inspired *primarily* by subversives and other malcontents who led mobs.[35]

The categorical refusal to acknowledge the possibility of some relationship between continued racial oppression and rising black aspirations for equality was not a unique or deviant case of military obfuscation. Conscious distortion and a blind spot for such chronic problems as monopoly pricing, unemployment, mass poverty, and planned obsolescence contribute to a grotesquely *unbalanced* portrayal of American capitalism. Of equal import, however, is the calculated equation of policies often advocated by economic nationalists, radicals, and socialists in the underdeveloped areas with totalitarianism or Communism.

What choice, then, have the governments of the underdeveloped countries? They can pursue the path of authoritarianism: regiment labor, *expropriate property*, stifle initiative, reduce consumption, and build up their economy from the forced savings of their people. *This is the way of Communism.* Alternatively these governments can pursue the path of freedom. This means that they must build up their capital, at the same time allowing for increased consumption and *economic liberties.* Since they have such limited resources, they can only achieve these goals with outside aid (i.e., corporate investment). That is *the crucial element* in the formula. (Emphasis added.)[36]

Often *leaders* of less developed countries have turned to *socialism* and socialist practices not so much to place the means of production in the hands of government as to assure that production facilities will be used with some measure of social justice. Very often these leaders *fail* to see that we have produced in our free, competitive system forms of social justice not generally associated with capitalism.[37]

In *totalitarian regimes, including* Communism, there is no economic competition. The government, in control of an elite party, has a monopoly of all production and distribution. Such systems lack the many benefits derived from competition in democratic regimes. (Emphasis added.)[38]

Similar themes appear in counterinsurgency manuals for American officers.[e]

Although we have quoted only from printed propaganda, the anti-Socialist line has also infused a fair number of DOD films and to a lesser degree base newspapers and electronic media broadcasting.[39] Books recommended for military personnel do not include those critical of either the military or the counterrevolutionary goals of American foreign policy.[40] While the exposure of American officers to this propaganda may be marginal in many instances, it seems reasonable to conclude that the overall effect is to condition an intolerant and rigid ideological atmosphere in the services. This is particularly ironical in view of the substantially higher educational levels which prevail in the armed forces. While 7.7 percent of the American population over twenty-five were college graduates during the early 1960s, 69 percent of the officer corps in the same age bracket held such degrees.[41]

Many of the degrees, however, were probably awarded by third rate civilian institutions. Intellectual and aesthetic norms are of ornamental significance at these "colleges," as they clearly are within the military's own educational system for senior officers.[f] Most colonels or naval captains who will be promoted to the highest positions must pass through not only the Command and Staff Schools, but will also attend joint or service War Colleges.

These institutions which collectively process several thousand officers per year "set the tone for the intellectual life of the armed services."[42] Their purpose is to *broaden* the officers' perspectives so that the role of a particular service or the armed forces can be related to general but unquestioned national

---

[e]A 1967 counterinsurgency handbook claimed: (1) underdeveloped countries were committed to "development;" (2) stability was always a prerequisite for "development," and military rule was a means of attaining the necessary stability; (3) rebellion and insurgency were condemned; (4) only three types of government were acknowledged: autocracy, communism and democracy; (5) groups or parties which encouraged the lower classes to make welfare demands upon the political system were viewed perjoratively; (6) military and other types of regimes should tolerate a "loyal" opposition to avert insurgency; (7) the "political culture" of a country represented its "real" Constitution; (8) the middle classes should be favored and strengthened; (9) in industrialized countries, *both* conservative and liberal parties represented the same class–impliedly a long-range goal for the Third World. U.S., Department of the Army, *Stability Operations–U.S. Army Doctrine* (Washington, D.C.: Field Manual 31-23, December 1967), pp. 5-11.

[f]In fairness to the military institutions which are about to be described, it should be admitted that perhaps 90 percent of civilian colleges qualify as little more than authoritarian "diploma-mills." Neither a commitment to excellence or academic freedom are honored at these occupational schools. Indeed, many of the students value intellectuality only when it can be directly related to their materialistic career goals. Some of the forementioned patterns are examined in: Edgar Litt, *The Public Vocational University* (New York: Holt, Rinehart & Winston, 1969); H. Bahr, "Violations of Academic Freedom: Official Statistics and Personal Reports," *Social Problems* 14 (Winter 1967), 310-20.

security and foreign policy goals. This is especially true of the Army War College, the Naval War College, and the Air Force War College as well as such joint institutions as the National War College and the Industrial College of the Armed Forces. All are on approximately the same educational level except for the Industrial College which is at the apex of the system. The joint institutions also emphasize interservice coordination and the key role of civilian corporations. Courses are attended by officials from the CIA, State Department, USIA, IDA, etc.

With the qualified exception of the Air War College, these "colleges" are in fact pseudoacademic schools whose standards rank well below major civilian institutions of higher learning. Their treatment of international affairs has been described as "superficial" even by their own graduates. Except for some of the civilians, the faculty is not professionally trained in political science and in fact is little more knowledgeable than the students.[g] In selecting the latter, "the services make no special effort to determine which officers are most likely to do distinguished academic work." There are no grades, no flunkouts and little systematic research, although students are expected to write a "thesis" on some problem of current interest to the armed forces or their service. The average four-hour working day is more often than not devoted to group discussions of a problem or a lecture by a representative from government, academia, or a major corporation. It is to be expected then that with such academic fare, these War Colleges have been the target of criticism as places where officers spend several months without doing much serious work. The Air War College is rather more demanding intellectually, and naturally there are some highly committed and talented individual officers at all of these institutions.

Although War College administrators assert that academic freedom exists and is encouraged, this principle does not extend to questioning the basic counterrevolutionary foreign policy imperatives nor does it sanction invitations to "controversial" speakers. Even on such matters as new means for effectuating existing policy, social pressures and career sanctions severely limit the willingness of an officer to persist in the articulation of dissenting views. The common outcome is "agreeable conversation." Similarly, there is pervasive obeisance to the demonic model of Communism and the inevitability of conflict with the

---

[g]This may not hold for State Department political officers. According to its director of politico-military affairs, "there are certain training programs where we have a much greater role and involvement, particularly in the service colleges, the Naval War College, the Armed Forces Staff College, the Air War College. The State Department has faculty advisers who are resident there and who help shape the curriculums both for the U.S. elements of the course and in those cases where foreigners attend. Some of these courses have a much higher content of a political character.

We also had a key role in the institution of the information program which you have been informed about. We participate in the briefing of these groups when they come to Washington. Our desk officers are in frequent contact with foreign military trainees from their areas. On the social side, my office hosts an annual luncheon for foreign trainees from the Naval Command School at Newport." *Military Assistance Training*, Hearings, House, p. 156. Cf., *United States Military Policies, Hearings*, Senate, pp. 31, 51, 58-59, 65.

Soviet Union, China, Cuba, etc.[43] If it is true that "war College graduates are extremely well informed concerning the Communist threat,"[44] we may properly regard them as indoctrinated rather than educated through a balanced and objective appraisal of the parametric variations of different Communist societies.[h] Exemplary of this designed socialization is the assessment of China by an Industrial College National Security Seminar for officers and civilian leaders. China is defined in purely aggressive terms and internal social measures are ignored as is the notorious corruption of the Kuomintang regime on Taiwan.[45] Similarly, an Army *Area Handbook for Venezuela* uncritically extolls the virtues of foreign investment while categorically denigrating not only Fidelismo but also "extreme nationalism."[46] Although these treatments are somewhat more sophisticated than the crude distortions which characterize the films and pamphlets discussed earlier in this chapter, they remain unbalanced and confined within a unidimensional "hard-line," cold-war-policy consensus. Perhaps this is best explained by the fact that "military schools are not relatively autonomous institutions. They are part and parcel of the armed services which have many responsibilities of which military education is only one, and not a primary one at that."[47]

If the strengths of War College instruction are in the field of "policy scientism"–stressing nonmilitary factors which obstruct or facilitate the attainment of strategic goals–their empirical and theoretical deficiencies are only partially compensated for by the more academic service academies. At these widely esteemed halls of elite socialization, the "curriculum and educational levels attained are equivalent to those of our leading undergraduate colleges. . . ."[48] Indeed, over the past decade, liberal arts and social sciences offerings at the military academies have been strengthened. In 1964 eighteen credits in social sciences and history were required at the U.S. Military academy (USMA), while slightly more was mandatory at the U.S. Air Force Academy (USAFA). The minimum course load in these areas at the U.S. Naval Academy (USNA) was somewhat lighter. "Insofar as prescribed courses in liberal studies

---

[h]Of special significance is the fact that the Brazilian War College "was established twenty years ago by U.S. military advisors and Brazilian staff officers. It was patterned after the U.S. National War College and to this day receives generous financial support from the Pentagon. In addition, at least one U.S. military advisor is assigned full time to maintain liaison between the ESG and the Pentagon. The Story of the Escola Superior de Guerra–told only very briefly here–is a good example of how a relatively small investment by the United States can have tremendous leverage in training a governing elite and shaping the ideology and destiny of a major country such as Brazil . . . all this with a very low U.S. profile." In general, most "Brazilian Service schools are patterned after U.S. schools. . . ." The first quoted passage appears in "Where the Soldiers Learn to Rule," *Brazilian Information Bulletin*, May 1971, p. 5. The second is from: U.S., Congress, House, committee on Foreign Affairs, *Reports of the Special Study Mission to Latin America on Military Assistance Training and Developmental Television*, submitted by Hon. Clement J. Zablocki, Hon. James G. Fulton and Hon. Paul Findley to the Subcommittee on National Security and Scientific Developments of the Committee on Foreign Affairs, 91st Cong., May 7, 1970, p. 5.

are concerned, there are only minor distinctions among the schools."[49] The increasing attention devoted to liberal arts during the 1960s was necessitated by the progressive militarization of American foreign policy in the Third World.

> Over the past years, the Armed Forces have been phasing into a role in world affairs quite different from the one they had a generation ago. . . . Victory on the battlefield of the future requires political sophistication and an understanding of the indigenous populations.[50]

By the late 1960s, the Air Force Academy featured no less than twenty political science courses whose catalog titles and descriptions indicated a conventional academic approach.[51] This was an obvious consequence of the fact that a growing proportion of faculty members had received some graduate education at civilian institutions. In FY 1955, for example, forty-nine officers from all the services were doing graduate work in international relations and foreign area studies.[52] The number has assuredly increased in subsequent years, and presumably some of these officers returned to teach at one of the service academies.

Despite the rigorous demands and academic orientation of the academies, it is doubtful that cadets were exposed to much in the way of positive aspects of social revolution. Nor was there significant opportunity to criticize the hard-line cold war consensual basis of American foreign policy. For academic goals are clearly secondary to the major function of these *character molding* institutions. Every possible "contrivance" is utilized to instill the service ethos—"flags, songs, mottoes, and customs." Obedience rather than sceptical questioning is inculcated as the cadet soon "discovers that it is not his to reason why. . . ."[53]

This indoctrinational approach is reflected in the admission that courses on history, government, international relations and economics at the USMA "provide a fundamental understanding of the general objectives of our troop information program." Communism is not analyzed with even a semblance of balance—emphasis is deliberately placed upon its contradictions and weaknesses."[54] At the Air Force Academy in Colorado "the fundamentals of our democratic way of life and the world-wide threat communism poses, including its techniques of infiltration and subversion" are detailed, while a Harvard Ph.D. teaching at the USNA has reported:

> Through electives in the English, History, and Government Department, a midshipman may broaden his education in the nontechnical fields, which were never more important to a naval officer than in today's Cold War conflicts between Western humanism and the crass materialism of Communism.[55]

Thus in 1967, eight out of thirteen political science and international relations course descriptions at the Naval Academy clearly indicate ideologically biased course materials.[56] And there is similar evidence for the USMA at West Point.[57] The special impact of such an unbalanced approach is underlined by the fact that those matriculating in a military academy "might well be a little more conservative in its political orientation than the average young American male."[58] Furthermore, the treatment of political subjects has been reputed for its superficiality.[i] Invited outside lecturers are given security checks and students naturally must be concerned with their own clearances. This and continuous evaluation of each cadet by a group of his peers tends to discourage expression of controversial political views and contributes to a pattern of stifling if functional conformity.[j]

Considerable attention has been devoted to the military academies not merely because they are the most "liberal" and "academic" military schools. More important, their graduates constitute the elite officer caste of the services and are very disproportionately represented in senior positions. In addition, the service academies "define the professional standards that the ROTC, OCS, and other programs seek to attain."[59] Their curricula are essentially self-established as civilian politicians in the Defense Department and congress have chosen not to exercise oversight on these matters because for the most part they share similar strategic objectives and assumptions. These facts explain the general adoption of a demoniacal approach by other military training installations and the widespread inability of American officers to evaluate their own society from a balanced perspective.

Indoctrinating Foreign Officer Trainees

Initial exposure to America's virtues by the foreign officers selected for training in the United States generally occurs before departure from his native land. As noted earlier, most prospective trainees are already pro-American or at least neutral toward the United States. This is ensured by a security check prior to the proffering of an invitation by the MILGRP or attaché. Before departure, American military representatives furnish Defense Department and USIA publications which describe the trainee's future program of courses and provide necessary information on American customs. Brief, superficial, and highly favorable descriptions of American society are also bestowed upon him.[60]

---

[i]"For example, there is much discussion of public policy, especially foreign policy, but less discussion of the underlying processes and mechanisms of social change. The values or ideals for which men strive are explored, but less thoroughly than more tangible matters such as geography, resources, or governmental structure. Something important was lost when the old nineteenth century courses in moral philosophy fell into disrepute at the academies. An appropriate substitute has not yet been found. Again, while students are exposed to many of the great ideas and doctrines of Western man, they rarely have opportunity to wrestle with them as they should." Masland and Radway, *Soldiers and Scholars*, p. 239.

[j]"The cumulative effect of the methods and attitudes recounted above is a greater tendency than we think desirable toward conformity, rejection of the unorthodox, and acceptance of the status quo. This is probably the most formidable obstacle the academies will have to fight. . . . " Ibid., p. 244. Cf., Sanford M. Dornbusch, "The Military Academy as an Assimilating Institution," *Social Forces* 33 (May 1955), 316-21.

Following arrival in the United States, the Foreign Military Trainee (FMT) is given a *Guidebook for Visiting Military Students*.[61]

Their English language competence is in many cases quite limited. Since the late 1950s American attachés and MILGRP officers have been instituting predeparture training in the English language under the haughty assumption that "any officer in the armed forces of any of the Free World countries who doesn't learn English is a fool."[62] In fact, imperialistic aspirations seem as important than superpower chauvinism. According to a MAP administrator in the Defense Department's Office of International Security Affairs, "English as the lingua franca cuts across political and ideological lines and cements the bonds formed between foreign military trainees and the United States."[63] Or put more euphemistically even when such linguistic competence is not "fully achieved," it nevertheless "opens the door to a wider world of information and communication far beyond the trainee's immediate military needs.[64] In other words, it maximizes the FMT's "understanding . . . of American society, its institutions, and its people." By 1970 language training programs had been instituted "in most of the countries receiving military assistance," and in more than a few English was the new "second language."[65]

In general English language proficiency requirements vary according to the training course in which a FMT will enroll—and one suspects that what is regarded as adequate is also governed by how badly the Defense Department wants officers or individuals from a particular country to be trained in the United States. If they do poorly on the MILGRP administered English language proficiency test, the FMTs may be assigned to the Defense Language Institute at the Lackland Air Force Base.[66]

Those slated for the Command and General Staff Course receive special orientation and preparation at Fort Leavenworth. This often means a five-week English language refresher course that propagandizes as it instructs.[k] Thus the readings—according to our rough and ready nonquantitative content analysis—offer the following insights: [67]

Chapters One and Two: The Armed Forces of the United States; and U.S. Defense Organization. This is essentially descriptive and notes military subordination to "civil" government which is administered by either civilians or military men. The claim is made that the officer corps is representative of all social groups in the population.

Chapter Three: Local and State Government. Harmony is stressed between people and governors who are portrayed as "servants" of the citizens. Emphasis is upon formal structure and responsibilities rather than upon behavior.

Chapter Four: Federal Government, The approach is similar to the preceding

[k]According to a recent U.S. Army Regulation, all FMTs must attend "appropriate phases" of the "Allied Officer Preparatory Course." In addition to the five-week English segment, it may include three weeks on U.S. military terminology, organization and tactical doctrine. U.S. Army, *Foreign Countries and Nationals: Training of Foreign Personnel by the U.S. Army* (Washington, D.C.: Headquarters, Dept. of the Army, Army Regulation AR 550-50, October 1970, Effective 1 December 1970), pp. 2-3-4.

chapter. Justice and popular control prevail, and the government functions efficiently.

Chapter Five: The Judicial System. A descriptive treatment with a clear implication of no problems. Government is defined as one of laws, not of men.

Chapter Six: Structure of U.S. Courts, Highly descriptive stressing the complexity of government. Emphasis upon desegregation decisions, but no attention to lack of enforcement or acceptance by the white population.

Chapter Seven: American Political Parties. No criticism. All groups adequately represented through coalitional parties where compromise is the essence of the democratic tradition. Democracy held tantamount to party competition, but a two party system functions best.

Chapter Eight: Electoral Procedures. Descriptive. Nothing on financing, disenfranchisement of blacks or others, obstacles to voting, electoral manipulation, etc.

Chapter Nine: Popular Factors in American Politics, Part 1. No criticism, a just system. Emphasis upon the dispersion of authority, many access points, and a plurality of groups representing all interests.

Chapter Ten: Popular Factors in American Politics, Part 2. More emphasis upon pressure groups—all impliedly having equal access.

Chapter Eleven: Agriculture in the United States. A success story. No references to income distribution, rural poverty, migrant laborers, etc.

Chapter Twelve: American Economy, Part 1, and Chapter Thirteen: American Economy, Part 2. No major economic problems discussed. Stress is on the role of *private enterprise promoted by the government* in a mixed economy. *Corporate investments held to be a major importance for economic developments in the underdeveloped areas.* Denies the exploitive theory. Accepts government planning in the "Third World," and urges trade liberalization. Emphasizes the U.S. role in countering *Communist aggression*.

The references to "Communist aggression" in a chapter largely devoted to extolling the virtues of government-subsidized corporate investment may be as coincidental as the ubiquitous term "civil" rather than "civilian" government. Nevertheless, since Communists are not the only radical elements who espouse an "exploitive theory" of capitalism in underdeveloping areas, the juxtaposition of these themes tends to be functional for the Open Door policy goal. This conclusion is reinforced by the advocacy of trade "liberalization" and tacit acquiescence for military rule in Third World countries as it is by the emphasis upon interest groups. Rightist military regimes and interest groups representing business sectors are frequent collaborators in the suppression of economic nationalists and radical leaders of unions or other lower class mobilizational entities.[1]

We have summarized the banalities infusing these English language readings at length because systematic examination of other indoctrinational materials on

[1]It is noteworthy that the orientation tours discussed in the preceding chapter either omit organized labor or direct that interviews be arranged with responsible types when requested by the FMTs and deemed appropriate. And the American two-party system is presented as epitomizing the virtues of a "loyal" opposition and a party in power which "are in fact more united than divided on most of the basic problems facing [sic] American society." Cf., U.S. Army, *Foreign Nationals*, Appendix K.

American life and Institutions suggests the typicality of such contrived portrayals. Thus in the regular course curriculum at Fort Leavenworth, there is a fair amount of hard-sell anti-Communism which stresses such themes as: limited Marxist success in creating an industrial base–and this at the cost of *depressing* popular living standards; terror; dictatorship; atheism; tactical Machiavellianism; aggressiveness; etc. Communist "doctrine" as well as Communist governed states are defined as the enemy.[68] The Soviet Union is incessantly denigrated, although more fervent antipathy is reserved for a satanic China. Hence, one recent assessment of Chinese policies emphasized the following:

1. Attempts to counter and surmount Western civilization underlie current Communist Chinese policies.
2. The United States, as the epitome of Western civilization, is a natural enemy of Communist Chinese culture.
3. The 'cultural revolution' had deep roots in the basic question of how to build a new China capable of coping with Western civilization.
4. The Chinese Communists have attempted to destroy the ancient Chinese beliefs.
5. The political organization growing out of the Ninth Party Congress reflects a concentration of power in an 'inner court.'
6. To achieve World Domination through a protracted struggle and thereby return China to her rightful place as the center of the world.

These vitriolic absurdities–despite a clear acknowledgment that the Chinese armed forces are equipped and organized primarily for mainland defense.[69]

In general, the process of inculcating counterrevolutionary hatred at the U.S. Army School of the Americas relies upon identical themes. But they are conveyed with much more grotesque stereotypes in an even more crude and oversimplified fashion with a considerable reliance upon cartoon booklets and cinematic media.[70] The second major difference is that there is relatively little propaganda idealizing U.S. society. This is undoubtedly due to the Canal Zone's colonial states which is tantamount to negative reinforcement. Trainees are, however, given brief exposure to the "democratic theory" as redefined to conform with an empirically distorted pluralist/brokerage model of government. Military intervention is wholly ignored as a hemispheric problem.[m]

Both at Leavenworth and in the Canal Zone, FMTs receive a limited amount of direct indoctrination in the virtues of capitalism and foreign investment. While a few pamphlets used at the CGSC do condone "socialism," this is defined either as European "Christian socialism," or identified with European social democracy–a movement noted for its refusal to nationalize the means of production and for its acquiesence in growing American corporate control to

[m]Yet, somewhat amazingly, the "legitimacy" of democratic government is held to have derived from God as much as or more than from the consent of "a rational people" and their laws! Escuela de las Americas, Ejercito de los EE. UU., *Teorias de los Gobiernos Democraticos, Seccion I* (Fuerte Gulick, Zona del Canal: MI 4600, Febrero 1965), pp. 2-15.

West European economies.[n] References to such "socialism" are rare and occasionally there are even implications that it is to be distinguished from Communism principally in terms of methods.[71]

In accord with a 1963 decision by the Department of Defense to adopt a "soft-sell" emphasis upon the virtues of North America civilization for trainees in the United States, the Armed Forces Institute in Madison, Wisconsin prepared and published in 1965 a series of instructors' manuals entitled *Informational Program on American Life and Institutions*. They are utilized at the CGSC and other installations.[o] Their message and approach mirrors that contained in *U.S. Army English for Allied Students*–summarized earlier in this chapter. The new multivolume series, in a superficially liberal but systemically conservative manner, conveys the image of a system and way of life in the United States which works well, is effective and affirmatively evaluated by most people. The portrayal is "balanced" only in terms of: (1) presenting traditional conservative and liberal viewpoints (more often in seriatum than as conflicting perspectives); and (2) by briefly mentioning selected "problems" and shortcomings. The existence of widespread political alienation and criticism of manipulative/repressive aspects of American institutions by radical critics and some liberals is ignored. Similarly, the seriousness of racial polarization is minimized and the growth of antiwar sentiment is ignored. The antisocialistic bias is pervasive while the anti-Communism remains simplistic! Capitalism, euphemistically described as "private" or "free" enterprise is celebrated within a "mixed" economy framework. Military subordination to civilians rather than to civil government is a nonsubject and there are not any references to Western military dictatorships. Instead, and within a militant cold war context, there are the self-assumed "responsibilities" of the military to "society" and the tasks which confront "the nation." Attention is focused upon the importance of political knowledge for officers and the legitimacy of their aspirations for national leadership roles to solve socioeconomic development problems and to combat the world wide menance of communism.[p] The United States is the empirical

[n]And while these movements and other proponents of "mixed" economies have been quite content to tax the wage and salary earning majority in order to socialize risks and to create external economies for firms, they have been more than content to safeguard the private distribution of profits. Similarly, the technocrats who regulate tend to be linked to and dependent upon the industries in question. Ralph Miliband, *The State in Capitalist Society* (New York: Basic Books, 1969). Cf. Grant McConnell, *Private Power and American Democracy* (New York: Knopf, 1966).

[o]See: U.S., Department of Defense, *Directive ASD(M) No. 5410. 17*, January 15, 1965. Previously FMTs were indoctrinated on American institutions during periods they had been excluded from regular courses when classified material was being presented. Dunn, "Military Aid," (dissertation), 97-98, With regard to the soft-sell approach designed "to avoid seeming [sic] to propagandize our friends," see: *Winning the Cold War, Hearings*, House, pp. 1022, 1038-42.

[p]And with remarkable candor, one widely used course guide argues that because political parties in underdeveloped countries are usually ideological (i.e., "committed to nationalism or socialism or some religious doctrine," a competitive party system is undesirable since

refutation of Marx, because of its widely shared prosperity, open political system, and growing social equality![q]

Since in most instances military intelligence agencies or the CIA screen out "anti-American" trainee applicants, it is likely that such blatant distortions inspire little classroom scepticism. On the contrary, it is probable that student discussions are even more militantly counterrevolutionary than the content of the published course materials cited in preceding paragraphs.[r] Such a prognosis is based upon the structured authoritarian personalities and limited education of officers from underdeveloped areas. Aside from military instruction, most "middle-class" officers have attended only secondary schools.[72] Moreover, their socially insecure petty-bourgeois backgrounds are generally correlated with political conservatism and anti-Communism.[73] Hence, the effect of such nympholeptic propaganda—even when little is actually absorbed—is to heighten their fears of all radical nationalists while inculcating junior-partner symbolic identification with the United States and prospective responsiveness to MILGRP advice or encouragement.

But the effectiveness of such manipulative indoctrination is as much or more determined by the propaganda *environment* than it is by the FMT's social origins or educational limitations. When this is structured so as to reinforce the trainee's predispositions, an optimum outcome is likely. For this reason considerable space was devoted in the preceding chapter to the socio-personal milieu which so profoundly affects the foreign officers willingness to learn on tours as well as within the classroom.[74] We shall now turn to some of the major base complexes where ideological indoctrination and personal cultivation have been integrated into MAP training curricula.

---

there "can be no 'loyal opposition' as in the United States." U.S., Department of Defense, *Informational Program of American Life and Institutions for Foreign Military Trainees: Vol II, Popular Participation in Government* (Madison, Wisconsin: United States Armed Forces Institute, 1965), p. 9.

One "explanation" of the lack of concern for political democracy—but not its denigration—strikes the author as hilarious. According to Col. Amos A. Jordan, Professor and Head of the Department of Social Sciences, U.S. Military Academy at West Point, "certainly, if respect for democratic values and institutions and appreciation of the American way of life were to become primary objectives of MAP training, then the U.S. armed services would be committed to a form of political warfare far beyond their traditional role." *Military Assistance Training, Hearings*, House, p. 100.

[q]See vols. 1-4 of the source in note p for the propaganda themes summarized in this paragraph. Some recent course guides have even linked antiwar protests in the United States to Communism. See, e.g., U.S. Army Command and General Staff College, *Strategic Subjects Handbook RB100-1* (Fort Leavenworth: 1 August 1969), A-10, Lecture R244. Cf. A-12, A-14, A-34.

[r]Statement by Martin Needler to the author, April 16, 1970. Based upon his conversations with MAP associated officers.

5 Major Foreign Officer–Training Installations

> Some courses of instruction, particularly at the Special Warfare School and the Civil Affairs School, have been selectively slanted for the benefit of foreign officers.[1]
>
> Lt. Col. Harry F. Walterhouse, USA, Ret., 1964

> The Canal Zone is said to provide a congenial atmosphere for the trainees, and their exposure to specially trained U.S. instructors forges bonds of friendship and affection for Americans generally.[2]
>
> Cong. C. Zablocki et al, May 1970

> Second, SOUTHCOM is responsible for contingency planning for crisis situations in countries of Latin America which might require a military response from the United States. According to SOUTHCOM's general staff, the U.S. military presence in the Canal Zone serves as a credible deterrent to adventurism by radical elements who would be more active in the hemisphere if SOUTHCOM did not exist.[3]
>
> Cong. C. Zablocki et al, May 1970

With the qualified exception of those visiting the United States on orientation tour "guided vacations," the preponderance of foreign officer trainees are subjected to ideological indoctrination while being given technical instruction at particular bases. After a brief overview, the discussion in this chapter will focus upon several of those training schools or installations which have played a significant role in socializing foreign officers over the years.

Overseas Training Programs

Since the Second World War, close to a third of foreign military trainees have received their instruction outside the continental United States (CONUS).[4] While those individuals who did not subsequently visit U.S. based installations would have been in general less exposed to "the American way of life," they nevertheless were affected in varying degrees by some ideological indoctrination. This was due to their personal contact with American officers and exposure to the *formal* information program that was instituted during the early 1960s.

Although in 1969 the United States operated 428 major and nearly 3,000 minor overseas installations, relatively few of these were centers for such training.[5] The most important complex of training installations is located in the Panama Canal Zone. Because of the scale of this Spanish/Portuguese language undertaking, it is described in a separate section of this chapter. Other significant overseas training bases are: Clark Air Force Base (Philippines); Wheelus Air Force Base (Libya—prior to 1970); U.S. Army Pacific Intelligence School (Okinawa); U.S. Army (Germany; Thailand). Except for the bases in Thailand and Libya which had processed foreign trainees for less than a decade, the remaining installations have been training foreign officers since the late 1940s. These and others which perform such roles on a smaller scale are administered by the unified area commands or the particular services.

While "some 12,000 Americans were engaged in military training operations overseas (excluding Vietnam)" during 1967,[6] many were not in fact assigned to the forementioned installations. In fiscal year 1969, for example, there were 101 mobile training teams which administered tactical or logistical training in twenty-six countries.[7] The scale of this effort is indicated by the assignment overseas between 1950 and 1958 of 2,536 such teams.[8] These teams tend to be small and often instruct foreign personnel in the use of U.S. supplied equipment. During the 1960s, those assigned to seventeen Latin American countries were generally bilingual.[9] In some instances this instruction has been performed by Contract Technical Services Personnel who are provided by the weapons manufacturer.

Installations in the United States

One of the difficulties encountered in assessing the training program is the lack of systematic data or full disclosure. Hence, "the Military Assistance Program brings about 18,000 foreign military officers and men to this country a year. And an additional 10,000 are trained here by Defense under other programs."[10]

Some may be joint undertakings with such civilian agencies as the CIA or AID. Thus, during the early 1960s the Inter-American [sic] Police Academy was moved from the Panama Canal Zone to Washington and renamed the International Police Academy. Operated by the AID Office of Public Safety, the IPA trains police from about fifty Third World countries. Well over 2,000 graduates have completed courses, of which more than 60 percent returned to Latin America. Their training includes a period at the U.S. Army's Special Warfare Center in Fort Bragg, N.C., where they are indoctrinated in anti-Communism and given some exposure to counterinsurgency tactics.[11]

Most military officers not directly financed by the MAP grant aid appropriation were probably trained in conjunction with Defense Department credit, guarantee, or cash sales programs. Thus in fiscal year 1971, it was reported that the

largest group, about 8,000, are going through a crash training program financed by the Department of Defense. Another group, numbering about 5,300 is training or studying in the United States at the expense of the governments concerned. About 2,800 officers are being educated at various Army, Navy, Air Force and Marine Corps institutions under the grant aid program.[12]

The expenses for the 5,300 may in a good number of instances have been financed indirectly by the U.S. Agency for International Development.[a] As noted in Chapter One, since the "Food for Peace" program was initiated in 1954, more than $1.3 billion in counterpart funds have been used to support military budgets of recipient countries.[13] This facilitates the diversion of other funds for the purchase of foreign training in the U.S. In any case, a very substantial though indeterminate percentage of the non-MAP grant aid officer trainees were exposed to ideological indoctrination.[14]

A relatively small number of foreign officers have in recent years been sent to U.S. universities. From FY 1949 through FY 1969, a total of 416 attended institutions of higher education (169 Army; 199 Air Force; and 48 Navy). For FY 1970 it was proposed that seven Naval and seven Air Force officers be accepted.[15] These students from underdeveloped areas divided about equally between technical fields such as medicine and engineering and the social sciences including management.[b]

Some training–about 25 percent of that provided by the U.S. Navy–is classified as "Unit Training." This naturally includes enlistees as well as officers and is quite limited in its ideological/personal effects. Yet, contacts at the officer level are encouraged and informational program activities such as tours have been programmed for some foreign units.[c]

---

[a]A recent U.S. Army Regulation identifies the following FMT categories: (1) MAP grant aid; (2) Foreign Military Sales; (3) Military Assistance Service Funded; (4) Free World Military Assistance Forces (FWMAF); (5) Non-MAP training. All of these involve some degree of U.S. subsidy, although for the last one it is limited to course costs. U.S. Army, Foreign Nationals, p. 1-1.

[b]"Normally the subjects available in United States institutions denote a degree of academic sophistication far beyond that achieved in the schools of the less developed countries. Yet in these countries . . . the need for training in management, economics, public administration, the social sciences, and related fields is most critical. The requirement is across the board–in the local military establishments as well as in the indigenous political and social institutions. . . . In many of these countries where the military plays a predominant role in national development, the collateral benefits accruing from the training of senior officers . . . are obvious." U.S., Department of Defense, "Collateral Benefits Derived from Military Assistance and Other DOD Training Programs," (unpublished memorandum, Office of the Assistant Secretary of Defense for International Security Affairs, 1960), p. 7, quoted in John M. Dunn, "Military Aid and Military Elites: The Political Potential of American Training and Technical Assistance Programs," (unpublished Ph.D. dissertation, Princeton University, 1961), pp. 81-82. With regard to the recent data referred to in our text, see U.S., Congress, House, Committee on *Foreign Affairs, Foreign Assistance Act of 1969, Hearings*, before the Committee on Foreign Affairs, House of Representatives, 91st Cong., 1st sess., p. 457.

[c]At least since 1966, *all* foreign visiting naval personnel were encompassed by the U.S. Navy's ideological indoctrination effort designed to inculcate an "active appreciation of American values and ideals, strong confidence in American power and purpose and an

Since the early sixties, a few senior officers from allied countries have attended the Army and Air Force War Colleges. The Naval War College instituted a special course for foreign officers in 1957. While neither the National War College nor the Industrial College of the Armed Forces (ICAF) accepted nonnationals, the latter paralleled the NWC in providing correspondence courses for them. These courses which include a prespective on international affairs that is heavily larded with anti-Communism, have *been used as curricula in the military staff or war colleges of at least five Latin American nations–Argentina, Brazil, Ecuador, Peru, and Venezuela.* In the late 1960s, annual ICAF foreign correspondence course enrollment was in the vicinity of 1,500.[16]

For those "hand-picked" to attend regular CONUS (U.S. Continental Command) institutions, special English preparation is given before departure and if necessary in the United States. During the late 1960s, approximately 600 per year attended the Defense Language Institute English Language School at Lackland Air Force Base.[17] Some of the language learning materials were highly ideological in nature.

As noted in Chapter Three, personal contact and exposure to selected aspects of American society are stressed for foreign trainees who attend classes with U.S. officer students. The proportion of American to foreign students varied considerably–depending upon the particular course, installation and program. Between FY 1964 and FY 1970, approximately 2,000 two-week to two-year courses were open to foreign officers at 189 U.S. bases.[18] They were actually enrolled, however, in only 142 such installations located in 31 states.

This listing is necessarily partial as another 1964 report on the military assistance program noted the existence of "some 350 different installations in the United States where foreign military trainees take format courses."[19] Since "a key objective of the MAP training program is to promote a pro-U.S. attitude,"[20] it is unsurprising that:

> Only a small number of trainees are unable to complete their assigned courses, either because of health problems or because of insufficient previous training. Care is taken in the latter case to assure that no one is returned home without the opportunity to receive instruction in another field more in line with his abilities.[21]

In other words and unlike American student colleagues, foreign officers are incapable of "failing" their courses.[d]

understanding of the role of the armed forces in a democratic society." This was to be furthered by "a warm reception in the United States and home hospitality." Only extremely high ranking foreign officers personally invited by the Chief of Naval Operations, etc., were exempted from this propagandization. Institute of International Education, *Military Assistance*, pp. 14-15. U.S., Department of the Navy, *Information Program for FMT (Foreign Military Trainees) and Visitors in the United States* (Chapter VI, OPNAVINIST 4950 ID CH-1, 14 December 1966, p. 31.

[d]The nonacademic grounds for dismissal were related in Chapter Three. For reasons which should be quite obvious by now, every effort is made to avoid resort to them. These efforts

Table 5-1
CONUS Installations Enrolling Foreign Trainees

Alabama
- U.S. Army Ordnance Guided Missile School, Huntsville
- U.S. Army Chemical Corps School, Ft. McClellan
- U.S. Women's Army Corps School, Ft. McClellan
- Craig Air Force Base
- Grunter Air Force Base
- Maxwell Air Force Base
- U.S. Army Aviation School, Ft. Rucker
- Brookley Air Force Base

Arizona
- Williams Air Force Base
- Luke Air Force Base

California
- Mather Air Force Base, Sacramento
- Hamilton Air Force Base, Sacramento
- Net School, Tiburon
- Harbor Defense School, San Francisco
- Electronics Material School, San Francisco
- Electronics Technician School, San Francisco
- Radiological Defense School, San Francisco
- Letterman Army Hospital, The Presidio, San Francisco
- U.S.N. Postgraduate School, Monterey
- U.S. Army Language School, Presidio of Monterey
- Vandenburg Air Force Base
- Edwards Air Force Base
- Norton Air Force Base
- McClellan Air Force Base
- Fleet Training Center
- Fleet Sonar School, San Diego
- Fleet Gunnery and Torpedo School, San Diego
- Engineman School, San Diego
- Electricians Mate School, San Diego
- Interior Communications Electricians Mate School, San Diego
- Metalsmith School, San Diego
- Metalwork School, San Diego
- Patternmakers School, San Diego
- Radioman School, San Diego
- Storekeepers School, San Diego
- Welding School, San Diego

Colorado
- USAF Academy, Colorado Springs
- Fitzsimmons Army Hospital, Denver
- Lowry Air Force Base

Connecticut
- U.S. Submarine Base, New London

District of Columbia
- Bolling Air Force Base, Washington, D.C.
- Deep Sea Diving School, Washington, D.C.
- Walter Reed Army Institute of Research, Washington, D.C.
- Walter Reed Army Hospital, Washington, D.C.

Table 5-1 (cont.)

Florida
- Fleet Sonar School, Key West
- Bartow Air Base
- Tyndall Air Force Base
- Graham Air Force Base
- Eglin Air Force Base
- Naval Air Station, Pensacola

Georgia
- Bainbridge Air Force Base
- Sperice Air Base
- Robbins Air Force Base
- U.S. Army Infantry School, Ft. Benning
- Supply Corps Officers School, Athens
- U.S. Army Southeastern Signal School, Ft. Gordon
- The Provost Marshal Generals School, U.S. Army, Ft. Gordon
- U.S. Army Civil Affairs and Military Government School, Ft. Gordon

Hawaii
- Fleet Training Center, Pearl Harbor

Illinois
- Scott Air Force Base
- U.S. Army Granite City Engineer Depot, Granit City
- Chanute Air Force Base
- U.S. Army Medical Service Meat & Dairy Hygiene School, Chicago
- U.S. Army Quartermaster Subsistence School, Chicago
- Electronics Technician School, Great Lakes
- Electronics Maintenance School, Great Lakes
- Engineman School, Great Lakes
- Electricians Mate School, Great Lakes
- Machinists Mate School, Great Lakes

Indiana
- The Adjutant General's School, U.S. Army, Ft. Benjamin Harrison
- Finance School, U.S. Army, Ft. Benjamin Harrison

Kansas
- U.S. Army Command and General Staff College, Ft. Leavenworth
- McConnell Air Force Base

Kentucky
- U.S. Army Armor School, Ft. Knox

Maryland
- U.S. Army Intelligence School, Ft. Holabird
- U.S. Army Ordnance School, Aberdeen Proving Ground
- Andrews Air Force Base

Massachusetts
- Laurence G. Hanscom Field
- Lincoln Laboratory
- U.S. Naval Hospital, Chelsea

Mississippi
- Kessler Air Force Base
- Greenville Air Force Base

Table 5-1 (cont.)

Missouri
- U.S. Army Medical Optical and Maintenance Activity, St. Louis
- Malden Air Base

Nevada
- Nellis Air Force Base
- Stead Air Force Base

New Jersey
- Salvage School, Bayonne
- McGuire Air Force Base
- U.S. Army Signal School, Ft. Monmouth

New York
- U.S. Army Chaplain School, Ft. Slocum
- U.S. Army Information School, Ft. Slocum
- U.S.N.R. Training Center, Ft. Schuyler
- U.S. Naval Hospital, St. Albans
- Griffiss Air Force Base

North Carolina
- U.S. Army Special Warfare School, Ft. Bragg

Ohio
- Wright-Patterson Air Force Base, Dayton

Oklahoma
- Vance Air Force Base
- Tinker Air Force Base
- U.S. Army Artillery and Missile School, Ft. Sill

Pennsylvania
- U.S. Naval Hospital, Philadelphia
- Damage Control Training Center, Philadelphia
- Boilerman School, Philadelphia
- U.S. Naval Station, Willow Grove
- Olmstead Air Force Base

Rhode Island
- Fleet Training Center, Newport

South Carolina
- Fleet Training Center, Charleston
- Mine Warfare School, Charleston

Tennessee
- Sewart Air Force Base

Texas
- Amarillo Air Force Base
- Sheppard Air Force Base
- Perrin Air Force Base
- U.S. Army Aviation School, Camp Walters
- Reese Air Force Base
- Webb Air Force Base

Table 5-1 (cont.)

James Connolly Air Force Base
U.S. Army Aviation School
Brooks Air Force Base
Kelly Air Force Base
Lackland Air Force Base
Randolph Air Force Base
Army Medical Services School, Ft. Sam Houston
Brooke Army Hospital, Ft. Sam Houston
Naval Air Station, Corpus Christi
Naval Air Station, Kingsville
Harlingen Air Force Base
Moore Air Force Base
Laredo Air Force Base
Laughlin Air Force Base
U.S. Army Air Defense School, Ft. Bliss

Utah
Hill Air Force Base

Virginia
U.S. Army Engineer School, Ft. Belvoir
Judge Advocate General School, U.S. Army, Charlottesville
U.S. Army Quartermaster School, Ft. Lee
Fleet Training Center, Norfolk
Radarman School, Norfolk
CIC Watch Officer's School, Norfolk
Radioman School, Norfolk
Air Conditioning and Refrigeration School, Norfolk
Naval Air Reserve Training Unit, Norfolk
Fleet Air Defense Training Center, Virginia Beach, Norfolk

Wyoming
FE Warren Air Force Base

Source: Institute of International Education, *Military Assistance Training Programs of the U.S. Government* (New York: IIE, Committee on Educational Interchange Policy Statement no. 18, July 1964), pp. 26-8.

The priority assigned to the neocolonial political influence objective over that of technical military instruction is also suggested by the Americanizing effects of the curricula used in CONUS courses. They are designed for the tactical, logistical, and material needs of American military services, and these are often—though not always—at variance with the particular circumstances of armed forces in the underdeveloped areas. And despite the fact that the cost per man exceeds that for overseas training—where courses are directly geared to foreign as well as American needs—by 100 percent on the average, the trend in

and the recent directive to award identical diplomas to foreign and American officers constitute a thinly veiled discriminatory pattern which is bound to engender resentment among American officers who enroll in the same courses. How much the latter will inhibit interface rapport—the key long-range objective of training—is an interesting question.

recent years has been to bring an increasing proportion of foreign officers to CONUS installations. Courses in the United States which are specially adapted to Third World needs have been uncommon because of fears that they would prompt "resentment" by foreign officers. No less important is the certainty that such treatment would severely reduce opportunities to socialize with Americans.[22] Although there are probably others, the only exception to the author's knowledge was for a contingent of Africans being trained to assume officer responsibilities during the early 1960s. The decision to segregate them at Fort Knox was probably occasioned in part by the racism which permeates the American military. Given the manipulative goals of the MAP, this was an exceedingly delicate undertaking. According to a social scientist who has done considerable research for the U.S. Army,

> the entire effect of bringing Africans to the United States to participate in military training is not only nullified but reversed by such incidents. I have, in fact, been told of such incidents occurring which involved not only Africans but Latin American officers of the Negro race. . . . With any degree of cultural sensitivity, our MAP personnel can maximize the political impact of their relations with African military personnel.[23]

The "incidents" being referred to were occasioned by civilian discrimination at public facilities which adjoined bases and had *not been placed "off limits."* Why?

Army Officer Programs

Before attempting to evaluate the political impact of MAP propaganda, it may be instructive to glimpse at its diffusion in the field. The installations treated in the following paragraphs include all of those about which information was publicly available.

Panama Canal Zone Complex

U.S. military aid to Latin America since World War II has exceeded $1 billion. During that period, more than 200,000 Latin American officers and men have been brought to the United States while an additional 30,000 have been instructed in Panama. Testifying in 1965, Military Assistance Director General Robert Wood boasted that "nearly all of Latin America's commissioned and non-commissioned officers have been trained either in the States or Panama."[24] During the same fiscal year, the $14.2 million requested for training would finance "2,998 students in Panama and 2,155 students in the United States, as well as for the cost of U.S. mobile training teams for in-country training." For

FY 1970, only $10.8 million was requested for such purposes, while total MAP grant aid for Latin America reached $21.4 million.[25]

Latin American air force pilots and maintenance crews were first brought to the Canal Zone in the 1940s. During the latter part of the Second World War, army officers and possibly some naval personnel visited such U.S. bases in the area as Fort Gulick, Fort Sherman, Fort Clayton, and Albrook Air Force Base. These initial contacts were integral to the short-run U.S. objective of making Latin American military establishments dependent exclusively upon the United States. Washington policy-makers assigned coastal patrol and antisubmarine roles to Latin American navies, while the armies were to maintain internal security. During the critical 1940-42 period, these armed forces lacked both the equipment and organization to repel an invasion.[26] Only the Brazilians and Mexicans were disposed to contribute combat units during World War II, while a number of Latin American countries firmly resisted U.S. pressure to sever diplomatic and trade relations with the Axis powers until the last phase of the war.[e]

Visits and training activities in the Canal Zone were terminated after the 1945 Allied victory. They were reinstituted in 1949 with renewed emphasis upon the creation of friendly ties with officers and internal security training. These cold war goals were incorporated into the Mutual Security Act of 1951, and subsequently embodied in the Military Assistance Agreements which most Latin American countries were induced to sign during the 1952-55 period.[27]

The revolutionary challenge symbolized by the Cuban Revolution prompted a renewed emphasis upon internal order and riot control training in 1960.[28] By 1961, intensified anti-Communist indoctrination was being programmed, and a year later the commander in chief of the Panama-based U.S. Southern Command reported that the traditional political influence goals of the training would be strengthened by assigning *highest* priority to anti-Communism. General O'Meara designated the enemy as "neutralism" or "leftist revolution" which were *defined* a priori as Communist-led or supported. The role assigned to Latin American officers in furtherance of U.S. national interests was the maintenance of "law, order and stability."[29] Neither O'Meara nor the training materials stress the goal of civilian supremacy.

During 1962 the high priority assigned to "order" was underlined by the dispatch of the 800 man U.S. 8th Special Forces Group to Fort Gulick. Within four years its more than eighty mobile training teams had put in more than 55,000 man-hours in the course of assignments to seventeen Latin American nations.[30] By 1964 plans for joint counterinsurgency maneuvers involving land,

---

[e]Although one might surmise that U.S. neocolonial objectives had been completely frustrated, any assessment would have to take account of: (1) the institutional importance of national traditions in some of the larger countries; and (2) the close previous ties between Latin officers in a number of states and military establishments in Germany and Italy. Furthermore, the missions of the last mentioned countries had in fact been expelled.

sea, and air forces from the U.S., Venezuela, Colombia, Argentina, Paraguay, and Peru had been devised. In December of that year they were carried out by 7,000 men in Peru.[31]

As of 1969, a new chief of the Southern Command was publicly defending military dictatorship where necessary to maintain stability due to the threat of Fidel Castro's "revolutionary ideas." General Mather did, however, recognize the desirability of socioeconomic progress as a means of obviating political discontent. There was little, in his opinion, that the United States could do to curtail political intervention by the Latin American military. On the other hand, he voiced a paranoid apprehension of social revolution similar to the one used by Lyndon Johnson to rationalize the American invasion of Santo Domingo:

> What worries me about revolutions, is the attendant risk of chaos and anarchy, because I know that there is a Communist presence in all of these countries, which can take advantage of those conditions.[32]

And so, one might add, can non-Communist nationalists and social revolutionaries.[f] But the meaning of his statement is that the United States and its dependent elites would have lost assured control of the client state–and herein lies the threat to Washington's informal empire.[g] Hence, it was only *after some Constitutionalists began distributing arms to the people* that U.S. Marines were ordered into the Dominican Republic–followed by an OAS "legitimization" and the dispatch of small contingents from Brazil and Costa Rica.[h]

The most well known U.S. counterinsurgency-training installation in the Canal Zone is Fort Gulick–referred to throughout Latin America as the "escuela de golpes." Officially opened in 1949, the school prescribed Spanish and Portuguese as the exclusive instructional languages in 1956. By the late sixties, most of the faculty were Cuban exiles, Mexican-Americans or Puerto Ricans.[33] Other U.S. officer instructors have attended the Defense Language Institute. The training program's Latin profile has been enhanced through extensive utilization of guest instructors. In 1964 there were fifteen officers and six enlisted "guest instructors" from eleven Latin countries, while four years later fourteen nations

[f]According to a journalist who recently interviewed Southern Command officers, the actual U.S. policy is to rely upon the Latin American military establishments to suppress, "radical elements" as well as "leftist revolutionary movements." Tad Szulc, "Influence of the U.S. Military is Expected to Wane in Latin America," *New York Times*, November 1, 1970, p. 18. Cf., Irving Louis Horowitz, "The Military Elites," in *Elites in Latin America*, Seymour M. Lipset and Aldo Solari, eds. (London: Oxford University Press, 1967), pp. 175-77.

[g]See, for example, Gen. O'Meara's testimony in the *Foreign Assistance Act of 1965, Hearings, House*, p. 346. Hence, one Special Forces training exercise at Fort Gulick involves the organization of a guerrilla force to restore a deposed "pro-Western" government "when it turned out the coup leaders leaned toward Colonia, a nearby communist country." Zalin B. Grant, "Training, Equipping the Latin American Military," *New Republic*, December 16, 1967, p. 13.

[h]State Department interviews revealing this as the operational meaning of "Communist" in the Dominican case are referred to in James Petras, *Politics and Social Structure in Latin America* (New York: Monthly Review Press, 1970), pp. 346-47.

were represented by twenty-one officers and twelve senior noncommissioned officers. All of these guest instructors were themselves "graduates" of the school. By 1971, in excess of 300 Latins had served as guest instructors.[34]

Fort Gulick trained 16,000 Latin Americans between 1949 and 1964 and enrolled an average class of 1,600 in 1970. It is organized on the basis of its two functional specialties. Although there is a technical department, close to seventy percent of course hours are allocated to internal security instruction. The forty-four courses (sixty-seven by 1970) offered vary in duration from two weeks to a forty-week command and general staff course which combines both military and civic action roles.[35] While engaged in study and recreation, "the students from the different countries get together and have an unsurpassed opportunity to discuss their common problems."[36] One such "problem" is intimated by the following pamphlets:

Asi es El Comunismo
Secretary of War, Army Information Service
no date or place indicated

Los Agentes Enemigos y Usted: Temas para Tropas
no publisher or date; 25 pp.

Como Funciona El Partido Comunista
Folleto No. 5, last date cited in text: 1959

Conquista y Colonizacion Comunista
Folleto No. 10, 24 pp., last date 1960

El Dominio del Partido Comunista en Rusia
Folleto No. 7, 14 pp., last date cited in text: 1960

Que Es El Comunismo?
Folleto No. 2, 20 pp., last date cited 1960

J. Edgar Hoover, *La Respuesta de una Nacion al Comunismo* Enero 1961, no publisher

Como Logran y Retienen El Poder Los Comunistas
Folleto No. 6, 22 pp., last cited dates are 1961

Como Controla El Comunismo La Vida Económica del Pueblo
Folleto No. 8, 15 pp., last cited dates are 1961

La Democracia contra el Comunismo
Folleto No. 1, 17 pp., last cited date is 1961

Como Viven los Obreros y Campesinos Bajo Comunismo
Folleto No. 9, 14 pp., last cited dates are 1961

Que Hacen Los Comunistas de la Libertad?
Folleto No. 3, 35 pp., last cited date is 1962

Como Controla el Comunismo las Ideas del Pueblo
Folleto No. 4, 24 pp., last cited date is 1962

El Atractivo del Comunismo
Fuerte Gulick, Zona del Canal, Abril 1964, MI4514, 5 pp.

Tactica y Tecnica del Comunismo Internacional
Fuerte Gulick, Zona del Canal, Marzo 1964,
MI4530, 3 pp.
Plan de Conquista del Comunismo
Fuerte Gulick, Zona del Canal, Marzo 1964,
MI4541, 3 pp.
Expansion en Europa: Seccion III
Fuerte Gulick, Mayo 1964, HA4525, 9 pp.
La Politica Exterior de la Union Sovietica: Seccion I
Fuerte Gulick, Zona del Canal, Junio 1964,
MI4535, 40 pp.
Lecutra Suplementaria: Principios y Practicas
Del Gobierno Totalitario
Fuerte Gulick, Zona del Canal, Febrero 1965,
HA4600, 25 pp.
Teorias de los Gobiernos Democraticos, Seccion I
Fuerte Gulick, Zona del Canal, Febrero 1965,
MI4600, 17 pp.
Ideologia Comunista II
Fuerte Gulick, Zona del Canal, Junio 1965,
MI4501, 40 pp.
El Desarrollo de Tacticas Revolucionarias–Primera Parte
Fuerte Gulick, Zona del Canal, Septiembre 1965,
HA4526
El Desarrollo de las Tacticas Revolucionarias I
Fuerte Gulick, Zona del Canal, Octubre 1965,
MI4526, 47 pp.
Partido Comunista de la Union Sovietica
Fuerte Gulick, Zona del Canal, Octubre 1965,
MI4516, 56 pp.
Ideologia Comunista I
Fuerte Gulick, Zona del Canal, Octubre 1965,
MI4500
Desarrollo de las Tacticas Revolucionarias, II Parte
Fuerte Gulick, Zona del Canal, Noviembre 1965,
MI4527, 26 pp.
J. Edgar Hoover, *Ilusion Communista y Realidad Democratic*
Fuerte Gulick, Zona del Canal, Marzo 1966, 17 pp.
La Teoria del Comunismo
Fuerte Gulick, Zona del Canal, Abril 1966,
HA4500-2, 23 pp.
Expansion del Comunismo en America Latina
Fuerte Gulick, Zona del Canal, Mayo 1968,
MI4531A, 71 pp.

Expansion Comunista en Europa Oriental
Fuerte Gulick, Zona del Canal, Agosto 1968,
MI4525
Teoria, Practica e Instituciones de los Gobiernos Totalitarios
Fuerte Gulick, Zona del Canal, Enero 1969,
MI4606, 26 pp.

Although two "special courses" were "instituted to combat the Communist threat" in 1962, the devil image was also incorporated into many of the regular courses. By the end of the decade, ideological indoctrination accounted for from 2-4 percent of enlisted trainees' time, and 15-20 percent for many of the officers.[37] Approximately half of the forty-one courses listed in the 1969 Fort Gulick Catalog dealt with the "menace," as the following enumeration reveals:[38]

Information Officer (0-4)
The "Introduction to Information Activities" includes "Communism vs. Democracy." Forty students, 3 weeks, for commissioned officers of all services and civilians in comparable positions.
Irregular Warfare Operations (0-6)
The "Introduction to Special Warfare" includes "Communist Doctrines." Forty students, 10 weeks, for lieutenants and captains.
Irregular Warfare Orientation (0-6A)
The "Intelligence and Military Police Subjects" include "National Objectives, Communist Ideology." Thirty students, 2 weeks, for majors, senior officers and comparably ranked civilians.
Military Police Officer (0-9)
The "Communist Threat" includes "nature of the communist world insurgency, communist ideology and democracy." Thirty-five students, 13 weeks, for those with 12th grade education, basic military training and a general knowledge of military police activities–potential instructors.
Military Intelligence Officer (0-11)
The "Purpose" is "to examine communism, the threat it poses, and military intelligence countermeasures against this threat." Forty students, 16 weeks, for selected commanding officers–cleared for confidential material–who have been or expect to be assigned to intelligence or counterintelligence.
Officer General Supply (0-26)
"Irregular Warfare" includes the "Causes and background of insurgency movements, the nature of the communist threat." "The Communist Threat" includes "Communist ideology, democracy." Thirty students, 12 weeks, for company grade officers with high school education.
Automotive Maintenance Officer (0-40)
The "Communist Threat" includes "Theory of Communism; fallacies of the communist theory; communist front organizations in Latin America; and communism vs. democracy." Forty students, 14 weeks. For potential instructors who anticipate assignment to automotive repair or maintenance.
Cadet (C-2)
About 190 hours are devoted to such "special subjects" as: "A study of

communism; theory and practices of a democratic government; fundamentals of military instruction; English vocabulary; speech and writing techniques; language laboratory exercises; governmental organization of the USA and its social and economic organization plus its military system; tour of the United States; seminar." Eighty students, 40 weeks, for cadets who have completed all but their final year at a recognized military academy in their country.

Cadet Orientation (C-3)

More than 100 hours are devoted to "General Subjects" which include "Communist Policy and Theory; Democratic Government." Eighty students, 12 weeks, for cadets enrolled at a recognized Latin American military academy.

Cadet Special (C-4)

About 120 hours are devoted to "Special Subjects," which embrace–among others–"Communist Policy and Theory; Democratic Governments." "Tactics" includes "Chinese Communist Doctrine." Eighty students, 18 weeks, for cadets enrolled at a recognized Latin American military academy.

Command and General Staff Officer (0-3)

"Command" includes the "Nature of Communism; Communism in Practice; Patterns of Communist Aggression; the Communist Powers; Development of U.S. Strategy; U.S. Foreign Policy." The topics presented by guest speakers include: "Concepts of International Relations; International Economic Trends; Nature of Contemporary Communism; Communist Eastern Europe; Communist China." The "General Subjects" include a "U.S. tour; the military system of Canada; the military system of the Federal Republic of Germany; the military system of Great Britain." Forty-two students, 40 weeks, for majors and senior officers who have completed any branch officer advanced course or have equivalent credit.

Military Police Criminal Investigation (OE-12)

The "Communist Threat" includes: "Sabotage and Countersabotage; Nature of the Communist World Insurgency; Communist Ideology; Democracy." "Irregular Warfare" includes the "Presentation of the Communist Threat–particularly in Latin America." Thirty-five students, 13 weeks, for officers and NCO's who are potential instructors and have both basic military training and a knowledge of police subjects.

Military Police Enlisted (E-10)

"Irregular Warfare" includes the "Presentation of the Communist Threat, particularly in Latin America." Thirty-five students, 8 weeks, for military police with 6 years of civilian schooling.

Combat Intelligence Staff Noncommissioned Officer (E-13)

Its "Purpose" encompasses "an orientation on the threat posed by communism." Forty students, 7 weeks, for career noncoms or civilians with a sixth grade education who are or expect to be assigned to combat intelligence. Clearance for "confidential" material.

Counterintelligence Noncommissioned Officer (E-14)

"Purpose" includes "to provide an orientation on the threat posed by communism, and counter-measures against this threat." Forty students, 9 weeks, with the same prerequisites as E-12.

Radio Operator (E-23)

"Irregular Warfare" encompasses the "causes and backgrounds of insurgent movements, nature of the communist threat in Latin America, the military, political, sociological and community development programs that should be instituted by the government to control an insurgent movement at any stage of development." Forty students, 14 weeks, for enlistees who pass the Morse Code Aptitude Test.

NCO General Supply (E-26)

"Irregular Warfare" is identical to E-23. Thirty students, 9 weeks, for noncoms and selected students with a secondary level education.

Engineer Equipment Mechanic (E-29)

"Irregular Warfare" includes "Concept and orientation on irregular warfare; nature of the communist threat." Thirty students, 16 weeks, for those with an elementary school education and a present or anticipated assignment to an engineer unit as a mechanic.

Basic Medical Technician (E-30)

"Intelligence and Security" includes "Nature of the Communist world insurgency threat; countering the insurgency threat." Forty students, 20 weeks, for noncoms with an elementary school education.

Weapons Repair (E-44)

"Irregular Warfare Operations" includes "Background of insurgency movements; . . . psychological operations and basic information programs necessary to support counterinsurgency actions." Ten weeks, 20 students, for enlistees with present or anticipated assignments in weapons maintenance and repair.

The relationships promoted by indoctrinating students with inversely distorted images of Communist and North American societies have contributed to requests by Bolivia and Paraguay that U.S. instructors be assigned to their military schools.[39] This may be a consequence of the fact that in the Panamanian base complex as elsewhere, the forging of personal friendships is considered to be as or more important than instruction in military subjects.[40] Transport to the Canal Zone is funded by the U.S. Agency for International Development, and students received a per diem spending allowance of $1.20 during the mid-1960s.[41] In addition to providing free lodging and meals, the Southern Command sponsored scenic trips, festive parties, outings, and athletic events for the Latin Americans. Each trainee was assigned an American "social sponsor." In assessing the long-term objectives of these manipulative efforts, one North American general referred to the *security of U.S. business investments* in Latin America and the goals of the now defunct Alliance for Progress. His optimism was premised upon the assumption that:

As the thousands of the schools alumni rise through the ranks of their various armed forces, we will see increasingly the direct impact of the school's efforts in promoting Inter-American understanding.[42]

It is clear, however, that the Latin Americans were not politically indoctrinating their neighboring officers from the North.[i] The possibility that the fox and the

[i]"According to the prevailing view at SOUTHCOM, Latin American military men play a major role in national political life, whether or not the United States approves, and their

chickens may have divergent interests is simply defined as unthinkable and systematically excluded from consideration.[j]

A much smaller Canal Zone installation is the Inter-American Geodetic Survey School at Fort Clayton. Opened in 1952, it has equipped more than one thousand Latin American officers with mapping skills. In 1969 the United States was forced to reduce the school's budget because of the higher priority assigned to the war in Vietnam. The *New York Times* reported that the "30 percent cut . . . is forcing the Inter-American Geodetic Survey here to close its cartographic school and severely curtail its mapping operations throughout the hemisphere."[43] Fort Clayton's relationship to counterinsurgency tactics is almost as obvious as that of the U.S. Army Jungle Warfare School at Fort Sherman, also in the Canal Zone.[k] While both of these installations trained Latin Americans, little has been published concerning their activities. Presumably, their courses incorporate anti-Communism, and the instructors promote affective interpersonal relationships. This is probably also true of Fort Davis, where riot control training was given to Latin American police officers. In many countries the national or federal police are semimilitarized.

Inter-American Defense College

Under the nominal jurisdiction of the Inter-American Defense Board, this war college level institution was established by the United States in 1962—about the same time that anti-Communist indoctrination of Latin American officers was stepped-up at the Canal Zone complex. Both events were, of course, direct responses to the antiimperialist and Socialist challenge of the Cuban Revolution. Opened in October 1962, the first five-month course enrolled twenty-two officers of the rank of Lt. Colonel and above. Since then the course has been lengthened to nine months. Up to five officers and one civilian may be nominated by each country.

The United States funded the school with a million dollars from MAP

views have direct effects on the potential for attaining U.S. objectives in each country. Hence, the SOUTHCOM commander and his staff claim they are in a position to exert 'maximum constructive influence' on Latin American armed forces not only in military matters, but in support of political, social and economic modernization." *Reports of the Special Study Mission*, House, p. 22.

[j]Nevertheless, there *are* dysfunctional environmental factors which have not as yet been neutralized. One sensitive observer has noted that "the atmosphere and life style of U.S. citizens within the Canal Zone, and especially the sense of their different social and economic status vis-à-vis Panamanians, leaves one with an overall impression of U.S. colonialism. Is this the best environment in which to expose young nationalist Latin American military men to United States values"? *Military Assistance Training, Hearings*, House, p. 129.

[k]Without accurate maps, it is extremely difficult to locate and plan tactics against insurgent rebels in rural areas.

appropriations. Located at Fort Lesley J. McNair in Washington, D.C., it adjoins the U.S. National War College. There is a similarity of some segments of curricula at both institutions. Approximately 50 percent of the courses at the IADF are devoted to anticommunism and international politics:[44]

The curriculum is equivalent to that of the U.S. National War College, the NATO Defense College and the Imperial Defense College of Great Britain. Beginning with a study of the international situation, the first portion covers the Communist bloc and the Free World.[45]

According to Edwin Lieuwen, there is great stress upon the evils of Communism, "but little about democracy."[46]

Always directed by a North American general, this "Inter-American" institution's curricula are premised upon the unquestioned *assumption* that a unity of interest defines relationships between the United States and nations of the southern hemisphere. Although the Pentagon recruited and organized its faculty, Spanish is the official language. The morning sessions feature a guest speaker followed by an "excursion to some nearby point of interest. . . . The cultural aspects of American life are not neglected." During the afternoon, small "working committees" chaired by students discuss some "problems."

There is considerable emphasis upon both sightseeing and entertainment.

During the two major trips made by each class—one to Latin America and one through the United States—the students visit important military installations, industrial centers, and places of cultural interest. Shorter trips during this year have included visits to the Norfolk Naval Base, Norfolk, Virginia; historic Williamsburg, Virginia; the United Nations Headquarters in New York; and the US Naval Academy at Annapolis, Maryland.

Activities in the Washington area typically include visits to the White House, the US Capitol, the US State Department, the Inter-American Defense Board, the Inter-American Development Bank, the Pan American Union, and the Pan American Health Organization, as well as a visit to a university.

These excursions are consistent with the primary objective of this "college"—the structuring of friendships and "understanding." By June 1968, 227 senior officers had "graduated" from the institution in which no one is graded or fails out. Since then, the average enrollment has been 35 for each class.[47]

U.S. Army Command and General Staff College (CGSC)

A French senior officer has contrasted the CGSC at Fort Leavenworth with his nation's Ecole Superieure de Guerre by noting the American institution's much greater size and its superb social program.[48] The role of this installation in promoting pro-American orientations among foreign officers *cannot be overestimated.* For unlike Fort Gulick, here they associate constantly and informally with upwardly mobile North American counterparts.

Most American field grade officers who aspire to senior rank must take the 10-month Regular Course at the CGSC. This prestigious tactical school is at the heart of the U.S. military educational system where it attracts the most capable and ambitious "potential commanders and generals of the future."[49] At least since 1950, the curricula have encompassed geopolitical subjects. The growth of nationalist and revolutionary struggles in the underdeveloped areas was reflected by the heightened priority assigned to geopolitical and anti-Communist indoctrination in the 1957-58 curriculum. Added emphasis was given to training in these areas during the 1961-63 period, and in the following year approximately 10 percent of the regular course or about 100 hours was devoted to such topics.[1] The curriculum is described as placing

> high priority on widening the visits of the professional officer.
>
> Two-thirds of the course is devoted to intelligence and operations; the gathering and assessment of information; the conduct of operations in hypothetical situations in which the student assumes various roles.
>
> Civil affairs, logistics and training in the use of personnel constitute the remainder of the course. Here the student deals with the problems of administration, civil as well as military, in a friendly or hostile nation.[50]

By the late 1960s, a large array of up-to-date materials on Communism were available for use in conjunction with both the regular course and the four-month fall and spring associate courses—these having been introduced in 1956. As we saw in the preceding chapter, these subjects are discussed within the context of an *unquestioned* U.S. counterrevolutionary perspective. The treatment is slightly more sophisticated than at Fort Gulick, but portrayals of communism and socialism are similarly distorted and hence, attitudinally manipulative.[51]

Foreign officers have attended the school at least since 1908 when an

---

[1]"While teaching its students how to use effectively this Nation's ground forces in tactical operations, the USA CGSC is, at the same time, attempting to show them that US strategy combines in proper proportion all elements of national power—political, economic, psychological and military strength. . . . It has been in response to the Army's growing worldwide responsibilities that the College has increased its instruction dealing with strategic matters. The curriculum could not ignore that the search for victory in the cold war struggle has placed additional responsibilities on the military man at all echelons. Officers in the Regular Course in 1963-64, therefore had approximately 100 hours of their classroom time devoted to subjects like politico-military theory, national strategic planning, the national military program, communism in practice, and strategic appraisals of significant world areas." Lt. Col. Selwyn P. Rogers, Jr., USA, "Unshuttered Vision," *Military Review*, 44 (September 1964), 17-21. With reference to anti-Communist course indoctrination during the 1950s at the CGSC, see: U.S., Congress, Senate, Committee on the Armed Services, *Military Cold War Education and Speech Review Policies, Hearings*, before the Subcommittee on Special Preparedness of the Committee on Armed Services, United States Senate, 87th Cong., 2d sess., p. 1092.

invitation to send students was accepted by Mexican dictator Porfirio Diaz. Yet between that date and the end of 1942, only 49 officers from ten nations had attended. Paralleling the emergence of the United States as a neocolonial world power, enrollment skyrocketed to more than 1,750 foreign officers from 1943 until June 1959.[52] The swift expansion in the school's role during the late 1950s is revealed by Table 5-2. During the eight years between 1959 and 1967, approximately 150-170 foreign officers attended the school annually. By the end of fiscal year 1970, 3,309 foreign trainees had been graduated. Only 103 foreign majors and senior officers were enrolled in FY 1971. Approximately 60 percent of them had attended command and staff colleges in their own countries. The foreign enrollment cutback was occasioned by the need to divert funds for the war in Indochina. Foreign officers constitute about 10 percent of the total student body and are systematically intermixed with American officers both within classrooms and socially. For "allied" officers and school's nonmilitary function is to create or reinforce pro-U.S. identification and interofficer friendships.[53] With regard to the increasing numbers of "friendly" officers from neutral states, the political objective clearly overshadows instruction in military tactics:

Mr. Morse. . . . I would like, however, to get some clarification from you, General Fuqua, about these modest training programs in Afghanistan, Syria, and Yemen. What is the advantage of giving these military officers military training at the command and general staff level? I don't see any particular advantage to it from our point of view.

General Fuqua. We feel if we can get an officer of the field grade into one of our courses at the Command and General Staff College, the mere association with Western people there would rub off on him during the course of his

Table 5-2
Foreign Officers Enrolled at USA Command and General Staff College: 1948-59

Year	Countries Represented	Officer Enrollees
1947-48	23	56
1948-49	26	52
1949-50	26	47
1950-51	24	38
1951-52	29	61
1952-53	29	60
1953-54	29	64
1954-55	35	72
1955-56	41	120
1956-57	41	148
1957-58	44	138
1958-59	45	170

10-month period and we feel that the expenditure of such funds might have some effect in making this man a little bit pro-Western.

If he rises . . .

Mr. Morse, Pro-Western?

General Fuqua. If he rises in his military hierarchy over there perhaps he can exercise some influence to keep the military from going completely into the Soviet camp. We don't expect [censored] to work any miracles, but we hope by expending this money we can preserve or generate a pro-Western military thinking in the various countries.

It is merely for that. We hope that this will happen.

As we demonstrated in Chapter 2, U.S. strategy seeks to inhibit not only total dependency upon the Soviet Union, but nonalignment as well. This is conveyed to the "neutralist" officers in several ways: (1) through good vs. evil subject matter indoctrination; (2) by frequent references to all participating officers as "allied"; and (3) by implying distrust for those whose governments have not wholly allied themselves with the "forces of light." The colloquy continues:

Mr. Morse. Implicit in Mr. Fraser's question was the thought that we are opening ourselves up to Soviet Intelligence by this device. Isn't this a factor?

General Fuqua. This is quite possible. However, let me say this, in all of these courses items of classified nature are denied these people, except when they are allies, such as British, people like that, who are on our side completely. But when we have people of questionable ideologies [censored] they are denied classified information of any kind. They are excluded from all courses that contain these classified items.

Mr. Talbot. If I may interject one comment. This program arises out of the deep and passionate conviction that there are advantages in a free and open society and that if some members of the armed forces of a country which is very closely related militarily to the Soviet Union can see this free and open society, that the effect will be good from our point of view.[54]

Analysis of course content was provided in Chapter 4 while social activities and aspects of the touring program were essayed in Chapter 3. Because of its scope and carefully planned nature—as well as the constant interface contact with relatively talented and indoctrinated American field grade officers—the affective impact of enrollment at Leavenworth probably exceeds that of any other CONUS training installation. As a by-product of the élan so manifest at the school, "in several countries 'Leavenworth Clubs' meet regularly. They are composed of local nationals and Americans who have been graduated from the Army's command and general staff college at Fort Leavenworth, Kansas."[55]

Another indication of this is the global readership of the CGSC journal *Military Review.* With one or two exceptions, its articles have consistently portrayed Communists as Machiavellian devil types and the United States as a global savior nation. A recent article has actually endorsed military coups aimed at neutralizing a possible "leftist takeover or conservative corruption." According to the author there has been a ubiquitous failure of civilian political leadership in the Third World"![56] Since 1952 a Portuguese language edition has been published in Brazil, while Spanish speaking officers could place subscriptions directly with the armed forces of the following: Argentina; Bolivia; Chile; Colombia; Cuba (until 1959); Ecuador; El Salvador; Mexico; Nicaragua; Peru; Uruguay; and Venezuela. Occasionally, the journal publishes contributions by extrainees. Thus, an anti-Castro article by Burmese Major Thoung Htaik appeared following his graduation from the 1960-61 Regular Course at the CGSC.[57]

Military Academies

The most prolonged aculturation and ideological indoctrination is experienced by foreign students who attend American military academies. In the past, some West Europeans and a very small number of Asians or Latin Americans were enrolled in the U.S. Military Academy (West Point) and in the U.S. Naval Academy (Annapolis). The appeal of nationalism and socialism in the underdeveloped areas during the 1950s and 60s was a significant factor in the expansion of facilities at the academies; and this, in turn has made more spaces available. During 1963 or 1964, State Department faculty advisers were assigned to all three academies.[58] By 1967 a cadet exchange program had been instituted between West Point and "selected" Latin American military academies in Argentina, Bolivia, Brazil, Chile, Colombia, Ecuador, Guatemala, San Salvador, Honduras, Nicaragua, Paraguay, Peru, Uruguay, and Venezuela. This followed "an experimental exchange program conducted in July-August 1966 in which twenty-eight U.S. Military Academy cadets visited fourteen Latin American military academies. In September, twenty-eight cadets from Central and South America returned the visit."[59] By March 1969 Army cadets from seventeen Third World nations were enrolled in the four year course at West Point. Smaller numbers attended the U.S. Naval and Air Force academies. Instruction in such areas as political science and international politics at these undergraduate institutions is more academic than that provided by the war colleges, but the indoctrinational imperative remains dominant. Character formation is the preeminent objective of military academy education.[m]

Fort Benning

Along with the CGSC at Fort Leavenworth and the USMA at West Point, the U.S. Army Infantry School at Fort Benning, Georgia was one of the first to

[m]Thus courses in such areas as history, government, international relations are intended to "provide a fundamental understanding of the general objectives of our troop information program." *Military Cold War Education, Hearings*, Senate, p. 1102.

receive foreign officers. The goal of winning them to "our side with their minds and hearts"[60] prompted invitations to Latin American armies in 1939 as it did to such countries as the neutralist Republic of Yemen in 1963.[61] The magnitude of this endeavor is indicated by that fact that during 1959 "the Allied Training Program at Fort Benning" was training "about 800 officers from 30 different countries."[62] The number of trainees markedly increased under the Kennedy administration. These FMTs have been indoctrinated by USIA propaganda and exposed in other ways to specially selected aspects of American society.[63]

The importance assigned to training foreign officers here was reflected in the late sixties by the use of specially designed curricula for the advanced infantry course. According to the school's commandant,

> the advanced course is the primary one for Allied students, but Allied officers and enlisted men also undergo ranger and airborne training, the pathfinder course and, in the case of Vietnamese students, our OCS. The lessons these Allied students learn will be reflected in years to come in the doctrine and practices of their own armed forces, and the bonds of friendship established between individuals will help our country in many instances in the future.[64]

General Talbott also noted that in 1970, "some 350 Allied officers from 50 foreign nations" would be in attendance. While the number of countries of origin had risen by over two-thirds since 1959, the total foreign enrollment had declined by more than a third. This again probably reflects the reallocation of military appropriations to the American campaign in Indochina.

Fort Bragg

As early as 1952, the U.S. Army Psychological Warfare School had instituted "a 4 week block of social science subjects on the economic background of the Soviet Union, and Communist philosophy." Within ten years, an "18-hour block of instruction entitled 'Communism and Communist Propaganda' " had been added to the curriculum. During 1962 the commandant informed a congressional committee that "we are turning our attention to the task of assisting the military forces of friendly indigenous countries to organize and conduct psychological operations in order to forestall [sic] or cope with the Communist threat."[65] As implied by the materials cited in our Chapter Four, a satanic portrayal of Communism probably infused the lectures and pamphlets that were distributed. Army regulation 515-5 requires information officers and programs to utilize perjorative language whenever reference is made to Communism.[66]

The Fort Bragg complex is also the site of the very important U.S. Army Special Warfare Center. Recognizing that "technical training would lose much of its impact if conducted in an ideological vacuum," the Defense Department directed the integration of indoctrinational materials on anti-Communism and U.S. foreign policies into the new counterinsurgency course that was inaugurated in 1962. While some officers have urged the need for even more attention to geopolitical subjects, it should also be noted that in field exercises the insurgents tend to be perceived by trainees as Communists.[67] A Spanish language manual used at the installation during the 1960s depicted *rebels* as "Guerrilla-Terrorists" in the service of "International Communism." It warned that such elements sought to take advantage of nationalistic sentiments among the populace and advocated socioeconomic improvements not for their own sake but in order to limit *popular sympathy* for the "terrorists." Officer trainees were advised to accept modest numbers of foreign (i.e., American) military and civilian counter-insurgency advisers in order to avoid the introduction of large foreign armies which would make the indigenous regime *appear* to be a "puppet" government. Naturally, the exemplary relevance of Saigon was ignored.[68] A 1967 field manual candidly announces that when a host country army is unable to contain insurgent forces, U.S. Army units "may" be sent in. Again as one might expect, the 1965 Santo Domingo analogue is omitted.[69] The Special Warfare Center counterinsurgency course—which since 1962 has benefited from the counsel of a permanent State Department faculty advisor—was praised as a "model" for the inclusion of "broader" educational [sic] objectives."[70]

Fort Bragg is also a major U.S. Special Forces base. This elite corps has "performed missions on assignment for high ranking members of the United States intelligence establishment in South Vietnam."

> Unlike normal Army intelligence units, the sources said, the Special Forces maintain close liaison with the military-intelligence headquarters within the Military Assistance Command of Vietnam and with the Central Intelligence Agency. Most regular Army intelligence units report to divisional commanders.[71]

We may surmise that the close linkage between the Special Forces and CIA is not geographically restricted. The former may be used as a CIA "cover" at Fort Bragg, Fort Gulick, and elsewhere in efforts to recruit dependable contacts among foreign officers. In assessing the plausibility of this conjecture, we should recall that the CIA and the American Military Mission coordinated their efforts when the nationalistic Iranian government was overthrown in 1953.[72]

Fort Knox

Just as the Cuban Revolution and the increasingly successful national liberation struggle of the Vietnamese NLF prompted the qualitative upgrading assigned to counterinsurgency training at Fort Bragg after 1961, so the "threat" represented by the emergence of such "radical African states" as Zanzibar and Algeria occasioned a marked shift in U.S. tactics for that continent.[73] As noted earlier, until 1963 the maintenance of pro-Western armies in the new African states was left largely to Britain and France. Their seeming inability to prevent the appearance (Ghana, Guinea) and hence, possible spread of radical nonaligned regimes resulted in a decision to sharply step up the U.S. training effort for the "Dark Continent."[n] Thus in 1963 responsibility for U.S. military activities in Africa was shifted from the European Command to the Florida based unified command for the Near East and South Asia (NESA)–one with more experience in dealing with officers from underdeveloped areas.

During July of that year, an Army Officer Leadership Training Course was inaugurated at the Armor School in Fort Knox, Kentucky. Between fifty and seventy-five carefully selected African officer candidates from such countries as Nigeria, Liberia, Senegal, Mali, Tunisia, Congo (Kinhasa), joined 20-30 officer candidates from Syria, Jordan, and other unstable capitalist or nonaligned regimes. The six-month Officer Candidate School type of course was intended among other things to provide trainees with the skills necessary "to protect their governments against subversion and domestic disorders."[74] Training objectives also encompassed the customary anti-Communist indoctrination and furthering of pro-American loyalties. In subsequent years, the Defense Department justified such aid in terms of maintaining "our commitments to our allies *and friendly armed forces.*" (Emphasis added.)[75] This distinction is crucial in the case of radical or nonaligned regimes.

Okinawa

Like Fort Gulick, the U.S. Army Pacific Intelligence School on Okinawa caters exclusively to foreign trainees. Although "many" MAP students have attended, only a portion of its funds are from appropriations for military aid grants and sales. Between 1958 and 1970, this service-operated and funded school enrolled 3,670 students from sixteen "friendly" and "allied" Asian nations. The Intelligence Staff Officer Course was given in English–often via an interpreter–to officers ranging up to the rank of colonel. Little has been disclosed concerning extracurricular activities. But given the "underfunding" of the installation and the "overcrowding" of its facilities, they were undoubtedly less than spectacular. On the other hand, one should not therefore conclude that this was an unappealing bleak site for foreign officer trainees. Every year the U.S. Army has conducted a week-long Psychological Warfare Workshop for between three and four hundred foreign military personnel on Okinawa. And the U.S. Marines have used Okinawa for orientation tours granted to officers from fifty allied nations since 1969.[76] Hence training stints on this Japanese island almost certainly had somewhat more to offer than poorly communicated technical instruction. How much more, we do not know!

[n]When "we did step in," it was with "training or training combined with 'impact' shipments of limited military equipment." Hovey, *United States Military Assistance*, p. 110.

U.S. Air Force Programs

The elaboration of a global anti-Communist alliance and base system following World War II has required the training of entire air forces in many Third World countries. This has involved many maintenance and other enlisted personnel as well as officer pilots. Between 1950 and 1968 more than 45,000 such students have been brought to the United States, while "many thousands more have been trained by USAF units overseas." Officially, the total number of countries represented is from fifty-five to sixty. And most students enroll in more than one course. By the late 1960s approximately 2,500 airmen were being processed annually.[77]

Albrook Air Force Base

Perhaps the oldest training program for foreigners is located near Fort Gulick in the Panama Canal Zone. After several years of orientation tours for high ranking Latin American officers, regular courses were initiated at Albrook in 1943. Named the Latin American School, it suspended operations temporarily in January 1964 due to the post-World War cutback in U.S. defense budgets. Thirteen months later, Albrook was reopened with a twenty-week curriculum as the U.S. Air Force School for Latin America.

Until the early 1960s its primary role was to administer low-cost training for noncommissioned officers in aircraft maintenance and similar areas. In 1961 eighteen courses were being given to about 250 men per year. High cost pilot and tactical trainees were sent to CONUS installations after completing English language instruction at Lackland Air Force Base in Texas.[78]

While Albrook continued to provide training in aeronautical occupational specialties, the U.S. shift in the early 1960s to counterinsurgency warfare was reflected by the establishment of a "Preventive Medicine Civic Action Training Program." The first class of forty trainees from nine countries enrolled in this course during July 1963. Its primary objective was to offset revolutionary "potential" by enhancing the frequently negative popular image of the armed forces. This was to be effectuated through the dispatch of airborne medical teams for short visits to remote areas. Aircraft, equipment, and supplies were to be provided by the U.S. taxpayer.[79]

A year later the commander of the USAF Southern Command directed that a study be made of the school's program. The subsequent report advocated a "Project Vision" which, among other things, proposed an integrated training program with newly standardized procedures and classifications.[80] In accord with the "image" emphasis in counterinsurgency tactics, the school was renamed during September 1966 as the Inter-American Air Force Academy, although its funding and direction remained exclusively North American. Other innovations included the opening of a USAF Tropical Survival School for both American and foreign trainees at Albrook. By July 1968, about 1,600 men had completed jungle and water survival courses.[81] The wedding of the Latin American military to their North American protectors was also reflected by a new Job Familiarization Training Program "which permits Latin American officers and air men to work directly with their US Air Force counterparts on the job in a variety of technical and nontechnical fields. Some 200 students were enrolled in this form of training during the past year."[82]

The post-Cuban Revolution expansion of Albrook's program is recorded by a sudden rise in enrollment, and an increase in course offerings from eighteen to twenty-two. During the 1943-65 period, the school graduated about 7,500 men or an average of 340 per annum. The corresponding average for 1966 through 1969 was 450. Fiscal year 1968 apparently marked the zenith with "more than 600 students from fifteen Latin American countries."[83] The containment or liquidation of guerrilla insurgencies in Venezuela, Bolivia, Colombia, and Guatemala during the 1966-68 period along with military fund needs elsewhere were responsible for a decline of enrollment toward the end of the decade.

Since unilateral orientation tours were first launched in 1939 in order "to counteract the residue of distrust and fear of the United States in the minds of many Latin Americans, particularly the Latin American military," great stress has continued to be placed upon "creating friendship and goodwill for the United States." In addition to Spanish or Portuguese language instruction and reinforcing anti-Communist propaganda on a limited scale, the airmen enjoy a very "active extracurricular program of cultural presentations, field trips, picnics and other social functions, and a vigorous sports program."[84] Because some Latin American armies are plagued by endemic factionalism, skilled deployment of bombers and fighters can be crucial to the success of attempted *golpes*.

Air University

Situated at Maxwell Air Force Base in Alabama, the Air University includes the Squadron Officer school, the Air Command and Staff College, and the Air War College. All of their foreign military trainees are officers.

The Air University Squadron Officer's School operates the largest program. During the early 1960s when about 2,800 officers were being processed annually, approximately twenty-four hours of instruction were devoted to "Public and International Affairs" which encompassed "world trouble spots, the theory and practice of Communism in the Soviet Union and elsewhere, United States foreign policies and United States organization for national defense."[85] Occasionally such films as "Communism and You" would be shown to all Air University students.[86]

Also located at Maxwell and established in 1946 is the Air Command and

Staff College. Until the mid-1950s when the first full-year program was begun, the courses were of short duration and subject to frequent modifications. With the 1954-55 school year, a forty-week, two-phase curriculum was introduced. Foreign officers were enrolled only in the fifteen-week first phase. Roughly 10 percent of their study time was devoted to "international relations . . . communism, fascism, and democracy. . . ." The ACSC which is intended "to prepare officers for the command and staff responsibilities to which they will be assigned as majors, lieutenant colonels, and colonels," used curricula that emphasized a "sharp distinction . . . between 'the enemy' and 'the free world.' "[87] During the early 1960s, approximately 600 U.S. and foreign officers were attending the ACSC annually.[88]

Less than one-third this number enrolled in the Air War College. These colonels and generals devoted a large portion of their time to geopolitical subjects. But as of the mid-sixties, the only foreign officers in residence were from such trustworthy nations as Britain and Canada.[89]

By that time organized tourist and social programs had been instituted for all foreign officers enrolled in any of the Air University schools. These assigned priority to meetings with selected business and civic leaders in surrounding communities and other cities. And Spanish/Portuguese editions of the *Air University Review* were being widely distributed in Latin America.[90]

U.S. Naval Programs

Naval War College

Located in Newport, Rhode Island, the U.S. Naval War College was the first American War College to admit foreign officer trainees. Its Command and General Staff Course for senior foreign naval officers differs from CGS courses offered by the other armed services in two ways. First, it is generally restricted to captains and other *senior* officers. Second, it is exclusively designed for foreigners, American naval officers have been excluded since this course was inaugurated in 1956.

On the other hand, there are a number of similarities to Army and Air Force Command and General Staff Schools. All utilize an indoctrinational approach which places considerable stress upon unquestioned acceptance of American strategic goals and rote memorization.[91] Similarly, the special NWC course devotes substantial attention to international affairs and anti-Communist indoctrination. Thus one graduate has written that it contributes to "a more integrated defense of the Free World against the cancerous infiltration of Communism."[92]

To encourage "open" discussion, only one student may represent a particular country. Throughout the year, they participate in a "research and presentation program" on international affairs problems. Course enrichment is provided by quite a few outside lecturers drawn from the Washington area and various American universities. Despite this academic stimulation and the limiting of classes to less than thirty students, the preoccupation with structuring pro-American attitudes clearly outweighed intellectual ends. Thus while hospitality, entertainment, and personal friendships were promoted with American officers, classroom instructors declined to criticize student presentations. In the humble opinion of one graduate, the classes functioned as "mutual admiration societies"![93]

Between 1957 and 1961, only five of 132 "selected" student "officers of great promise" who attended the course came from nonaligned nations such as Indonesia, Burma, and Sweden. Although the remainder were "allied," a majority were nationals of Asian or Latin American countries. In excess of half had previously studied in other American or British military schools, while virtually all boasted command or high level staff experience. By the early 1960s the course was in the process of being expanded so that many more officers from nonaligned African, Asian, and Middle Eastern nations could be incorporated. The success of the special course was reflected by the important positions which graduates were given—ones in which they might "diffuse" their new pro-American attitudes. More than 25 percent were assigned to instructional duties in their own naval war colleges or training institutions while about 33 percent received commands upon return to their own countries. Some were given high level civilian jobs in government. Of especial importance is the intensive "follow-up" program which includes free distribution of the *Naval War College Review* and other publications. Names and address are provided and updated by U.S. Naval Attachés in the students' countries of origin.[94] Concern with indoctrination in American foreign policy goals is reflected by the assignment since 1952 of a State Department political officer to the NWC. His activities include teaching and a "public relations role."[o]

Other Naval Installations

In common with quite a number of training institutions, the U.S. Marine Corps Command and Staff College (until 1964 called the Senior School) instituted major curricular reforms during the early 1960s. Since July 1963, its CGSC has

[o]"The assignment of a senior Foreign Service Officer to the staff of the Naval War College has numerous advantages: it makes available knowledge and background in foreign affairs to the students, faculty and President that may be obtained in no other way; it provides current knowledge on U.S. policy and policy formulation which are difficult to obtain in other ways; it provides valuable liaison with the Department of State; it provides the Department of State students at the College with a senior adviser from their own service; and it provides an experienced and knowledgeable adviser to the President of the Naval War College." U.S., *State Department Advisers*, Senate, pp. 4, 14, 20.

featured 64 hours of counterinsurgency (5 percent), 120 hours on the organization and functioning of national security procedures (10 percent), and 90 hours on geopolitics and the current world situation (8 percent).

> *Geopolitics and Current World Situations* is designed to give the student an understanding of the national objectives, foreign policy, and national military strategy of the United States. This is a very vital contribution to the students, since many will be receiving their final formal military education. The College must equip them with a well-founded knowledge of world, social, economic, and political factors to assure maximum usefulness to Corps and country. Time is allocated to each Allied officer student for a presentation on his country's geographical, cultural, political, economic and military situation.

In addition to flattering the foreign officer's self-esteem, such presentations are upon occasion useful sources of intelligence. "A close relationship between the instructors and the students is established through the use of conference groups. Each instructor acts as a faculty advisor to the same ten students during the year." These students are generally majors with twelve or more years of commissioned service, though in the case of foreign officers captains are also permitted to attend.[95] Fourteen officers from twelve countries matriculated in the Marine CESC course in FY 1970.

Significant numbers of foreign officers have enrolled at the U.S. Naval Training Center (Great Lakes, Illinois), and the U.S. Naval Intelligence School (Washington, D.C.). Both feature highly institutionalized social hospitality programs. A small number of Third World cadets are admitted with each entering class at the U.S. Naval Academy in Annapolis, Maryland. In 1969 these included a Costa Rican and a Philippine.[96]

The training programs surveyed in this chapter are among the major ones operated by the U.S. military for foreign officers. Although large numbers of student invitees are processed at other CONUS installations, there is probably little aside from differentiated technical specialties to distinguish them from the foregoing. The ideological and personal rapport goals are common to all. While geopolitical subjects represent only 10 to 20 percent of the curricula, it may be safely assumed that social and entertainment activities account for from 30 to 50 percent of the foreign officers' time. The higher the rank, the more such nontechnical diversions seem to be stressed!

6 American Military Representation Abroad

It has been said that in most of the non-Western societies 'there are spheres in which Western influences dominate, while there are others that have hardly been touched.'[1]

McKahin, Pauker, and Pye, 1955

Intended political consequences that derive from military assistance are probably harder to achieve than expected. . . . On the other hand, if military trainees are subjected to systematic political propaganda and organization, and return home with explicit theories as to how their countries' political structure should be changed, it is quite conceivable that such politically oriented training would improve the chances that the donor country would retain influence over their trainees.[2]

H. Wriggins, 1968

Long-term professional identification, the first function, bolsters the most widely acknowledged utility of MAAG's, their potential for influencing the host government through a trustful relationship. The U.S. Army has probably been the most successful service in developing long-term professional. . . .[3]

Paul Hammond, Rand Corp., 1970.

We have argued that the exertion of posttraining influence does not depend upon the transmission of "explicit theories as to how their countries' political structure should be changed." Intense fears of and antipathy for vaguely defined groups of "subversives" or "agitators" are more than sufficient for the structuring of a community of perceived interest with their American mentors. Other than the reiteration ad nauseam of a military "nation-building" role, the United States has yet to field a "theory" of socioeconomic or political development which is empirically applicable to the stagnating and exploited economies of the Third World.[4] Except for radical nationalist countries, the prevention of institutional restructuring more aptly denotes general U.S. objectives vis-à-vis the armed forces. Although this may be less difficult than inducing foreign armies to bring about major societal changes, even the manipulation of intentionally nurtured fears involves formidable problems. This is particularly true when the personal associations, new reference group identifications, and ideological orientations erode with the passage of time. Hence, any long-term "payoff" for the United States on the MAP training investment is in great measure determined by the extent and character of communications with foreign officers *after their departure* from CONUS or overseas training installations. Table 6-1 indicates

Table 6-1
Map Trainees by Country of Origin: FY 1950 – 1969

Country	CONUS	Abroad	Total
Congo (Kinhasa)	158	69	227
Ethiopia	2,561	111	2,672
Ghana	96		96
Guinea	4		4
Liberia	376		376
Libya	422	31	453
Mali	55	5	60
Morocco	641	686	1,327
Nigeria	321		321
Senegal	12		12
Sudan	118	8	126
Tunisia	164	138	302
Upper Volta	20		20
Africa Total	4,948	1,048	5,996
Argentina	2,208	395	2,603
Bolivia	383	2,065	2,448
Brazil	5,496	800	6,296
Chile	2,467	1,508	3,975
Colombia	2,087	2,266	4,353
Costa Rica	33	496	529
Cuba	307	214	521
Dominican Republic	580	1,868	2,448
Ecuador	1,482	2,562	4,044
El Salvador	167	786	953
Guatemala	603	1,555	2,158
Haiti	444	60	504
Honduras	157	1,311	1,468
Mexico	368	202	570
Nicaragua	522	3,231	3,753
Panama	35	2,985	3,020
Paraguay	238	664	902
Peru	2,784	1,939	4,723
Uruguay	884	707	1,591
Venezuela	1,249	2,473	3,722
Latin America Total	22,494	28,087	50,581
Cambodia	215	122	337
Taiwan	19,616	3,837	23,453
Indochina	408	26	434
Indonesia	2,599	396	2,995

Table 6-1 (cont.)

Country	CONUS	Abroad	Total
Japan	9,643	5,637	15,280
Korea	20,688	9,120	29,808
Malaysia	180	18	198
Philippines	8,588	4,315	12,903
Thailand	7,559	2,577	10,136
Vietnam	10,587	3,411	13,998
Classified	1,741	17,183	18,924
East Asia Total	81,824	46,642	128,466
Afghanistan	275		275
Ceylon	24		24
Greece	11,229	1,965	13,194
India[a]	24		24
Iran	8,173	1,712	9,885
Iraq	372	32	404
Jordan	407	12	419
Lebanon	169	1,188	1,357
Nepal[a]	11		11
Pakistan[a]	60		60
Saudi Arabia	984	216	1,200
Syria	23		23
Turkey	14,951	2,352	17,303
Yemen	5		5
Classified	4,026	563	4,589
NESA Total	40,733	8,040	48,773

[a]Fiscal Years 1967-69 only.

Source: U.S., Department of Defense, *Military Assistance and Foreign Sales Facts: March 1970* (Washington, D.C.: Office of the Assistant Secretary of Defense for International Security Affairs, 1970), p. 17.

that a serious effort to maintain contact with ex-trainees must necessarily be an enormous undertaking. As the foregoing table makes clear, less than a third of the FMTs have received training outside the continental United States. Also worth noting is the fact that it fails to list non-MAP trainees. Their inclusion would increase the total by at least 30 percent.[a]

While the data are not broken down according to rank, totals for the

---

[a]Between World War II and 1971, more than 320,000 foreign officers and enlistees were trained by the United States. Of these, about 100,000 received training at overseas installations. U.S., Congress, Joint Economic Committee, *Economic Issues in Military Assistance, Hearings*, before the subcommittee on Economy in Government of the Joint Economic Committee, 92d Cong., 1st. sess., January-February, 1971, p. 187.

FY 1964-FY 1969 period suggest that about 50 percent were commissioned officers. Many of the enlistees were accounted for by entire naval units and air force maintenance personnel. Of army personnel brought to CONUS installations, approximately two-thirds were officers, while an inverse ratio applied to naval personnel. The latter, incidentally, tend to play a less determinative role in military coups. The air force CONUS percentage is slightly in favor of officers, while that of the marine corps resembles the armies. Proportionately more senior officers (colonels and generals) are brought to the United States than those with field-grade status. If we assume that similar patterns applied to non-MAP trainee categories, a conservative estimate of ex-trainee *officers* for the twenty-year period would be 100,000.[b] Of these, *more than half* would come from countries that could be termed underdeveloped. While it is certain that ongoing relationships were nominal or nonexistent with many of them, the activities described in succeeding portions of this chapter suggest that they were maintained, deepened and personalized for a substantial though indeterminate percentage of the ex-trainee officers. (See Table 6-2.)

Media Communications with Ex-Trainees

The danger that cognitive increments and new reference group identifications will "wear off" after the FMT returns to his home country is a real one which has also plagued technical assistance programs. During the 1950s, a number of techniques were developed to cope with this problem. First, large numbers of officer trainees are invited to participate in a second training tour under CONUS or other unified area command auspices.[c] Second, many U.S. training installations began to send free copies of their newsletter to ex-trainees. Thus in 1960, Dunn recorded "foreign graduates receive the *Naval War College Review* and certain other College publications. A Peruvian officer writes of his appreciation of this service:

> Continuously, I have been receiving the Naval War College Review of Newport. Thank you . . . for sending it. I feel very pleasant each time I receive it, not only only because of the amusement and how instructive it is, but for the good news it always brings from Newport.

---

[b]This figure is arrived at by assuming that 50 percent of the 320,000 were officers and that one-third of these were repeaters on their second or third training or orientation tours. The first assumption is derived from Table 6-2's breakdown by rank. The second is based upon the estimate referred to in note c below.

[c]According to Brookings Institution senior staff member Ernest Lefever, an estimated 62,000 out of 312,000 FMTs had been programmed for two or more training tours. Hence, between two-fifths and one-half of all foreign officers had been recycled for intensive indoctrination. *Military Assistance Training, Hearings*, House, pp. 7, 34-35.

Table 6-2
MAP Trainees by Rank and Status: FY 1964 – FY 1969

	FY 1964	FY 1965	FY 1966	FY 1967	FY 1968	FY 1969
Senior Officer	890	655	583	712	760	388
	(264)	(71)	(156)	(119)	(112)	(81)
Officer	4,297	3,851	3,212	2,902	2,915	3,370
	(2,340)	(2,323)	(3,046)	(2,421)	(2,183)	(2,169)
Enlistees and NCO's	3,844	3,750	2,396	2,508	2,459	2,514
	(2,750)	(2,633)	(2,934)	(2,937)	(2,559)	(2,187)
Civilians	102	43	12	46	24	62
	(285)	(349)	(48)	(17)	(17)	(16)
Grand Total	9,133	8,299	6,203	6,168	6,158	6,334
	(5,639)	(5,376)	(6,184)	(5,494)	(4,871)	(4,453)

[a]Second line figures in parentheses indicate those who were trained or provided orientation at overseas bases.
Source: Organization and Training Div., ISA, Department of Defense.

The inspiration for the most interesting of the 'follow-up' activities of the staff is said to have originated with the students themselves:

> Upon completion of the course, many of the students requested that some form of maintaining contact with the staff and their fellow students be established. To fill this need a class letter is sent out each year.

The 'class letter' is in fact a compilation of individual letters written by former students and former or present staff members. The college maintains a current address for all graduates (or if one is lacking, uses the MAAG or attache in the country of the student in question as a contact agency) and in the fall of every year solicits a letter from each former student. Promotions or honors received, present and contemplated assignments, and other personal news of general interest are to be reported. The responses are published in a separate *Annual New Year's Letter* for each class, to which American officers who served as members of the staff for the period of class attendance at the War College also contribute messages.

The purpose of the *Letter* as stated in the foreword of those published in January 1960:

> The annual letter is designed to perpetuate the bonds of friendship, mutual respect and understanding generated among the many representatives of the navies of the Free World while they were here at the United States Naval War College. . . .

The response of most members of the various classes to this attempt 'to perpetuate the bonds of friendship' as evidenced by their letter, is in fact very enthusiastic. A Brazilian officer writes:

First, I would like to congratulate you on the excellent idea of exchanging New Year's letters. Initiative such as this has two main purposes: to join the great number of free countries through Naval Officers forming a FRIENDSHIP ALLIANCE, and consolidating therefore, even more, the basis of the DEMOCRATIC BLOCK led by the UNITED STATES OF AMERICA. Thus, congratulations!

A Korean officer tells of his appreciation of the institution of the *Annual New Year's Letter:*

I really appreciate this wonderful idea and service of the Naval War College to be able to exchange our New Year's greetings and news through this 'New Year's letter' with our dear friends of the Free World. If it were not for this opportunity, I am afraid that we might neglect to write one another once a year at least, impairing our warm association and friendships which have valuably been developed from the time at Newport.

Sixty-one of the seventy-six graduates of the first three classes of the Naval Command Course submitted letters for inclusion in the *Letters* published in January, 1960. Most messages emphasize the theme of enduring friendship, of continuing affection for their former classmates and their instructors at Newport.[5]

While similar programs have probably evolved for officers attending the command and general staff courses of the Army and Air Force as well as the Inter-American Defense College and the Canal Zone complex, follow-up efforts are likely to be more modest at such specialized schools such as Fort Bragg, Fort Belvoir, the Great Lakes Naval Training Center, etc. Probably exceptional are the previously noted "Leavenworth Clubs" in several countries.

Other types of follow-up activities have been referred to in earlier chapters. Thus the Navy administers a post-high school "correspondence course program" for foreign military officers. It encompasses both technical and political subjects, and was enrolling between 600 and 700 per year during the late 1960s.[6] We have already mentioned the Industrial College of the Armed Forces correspondence courses and have noted the free distribution of translated military journals such as Leavenworth's *Military Review* and the *Air University Review.*[d] Follow-up

[d]A recent U.S. Army Regulation stipulates that "School commandants are encouraged to maintain contact with graduates of career and similar top level courses after the student's return to their home country. Such programs may include—(1) forwarding letters from the commandant and annual school newsletters or similar school publications, encouraging students to enroll in nonresident extension courses, and encouraging informal correspondence between classmates; (2) providing professional publications for students enrolled in CONUS staff and career courses (each subscription must be appropriate to the course taken

activities also encompass the dispatch of high ranking U.S. officers to lecture at War Colleges in underdeveloped countries on such subjects as the danger of Soviet sponsored communism.[7] Similarly, in 1961 the U.S. Army organized a conference of Latin American army representatives on the subject of communism with particular attention to the Fidelista challenge to the hemispheric social status quo![8]

In Latin America there are also such activities as Pan American Rifle Matches. For underdeveloped countries worldwide, the Defense Department has since 1964 financed an ambitious sports development program designed to enhance the image of the U.S. and its military services. Closely coordinated with the State Department's Bureau of Cultural and Educational Affairs, these efforts have encompassed exhibitions, visits, and training advice. Thus "a former basketball champion from the University of Georgia was sent to Damascus and the result was the best basketball team ever fielded by Syria. Both coach and team received special praise from the U.S. Ambassador and the Syrian Minister of Defense." Other countries involved during the first two years of this program include the Ivory Coast, Iraq, Kenya, Libya, Uganda, Venezuela, and Pakistan.[9]

Diplomatic and Advisory Functions of MILGRP Officers

The primary responsibility for attaining American foreign policy goals which involve the use of armed forces in underdeveloped countries is borne by American officers stationed abroad as embassy attachès or members of Military Groups (MILGRPS). The latter classification includes the missions which have traditionally been assigned to Latin America as well as Military Assistance Advisory Groups (MAAGS) that have been utilized throughout the world. Also embraced by MILGRP are Defense Liaison Groups. They have been used in countries such as Indonesia where the United States has deliberately endeavored to maintain a "low profile" via partially concealed military presence.[10] During the latter part of the decade, there were approximately 10,000 American training or advisory personnel stationed in about eighty-eight underdeveloped countries who were "in daily contact with those eighty-eight military establishments."[11] In the following discussion of their functions, the terms MILGRP and MAAG will often be used interchangeably as they are in official circles. As will

---

by the student and will be initiated prior to the departure of the officer from CONUS; the subscription will be for a maximum of 1 year and will be funded under the Informational Program); (3) providing and maintaining current school publications to in-country counterpart military schools." U.S. Army, *Foreign Countries and Nationals: Training of Foreign Personnel by the U.S. Army* (Washington, D.C.: Headquarters, Dept. of the Army, Army Regulation AR 550-50, October 1970, Effective 1 December 1970), p. 2-19. It is believed that perhaps 50 percent of the articles appearing in the military journals of Latin American nations–including the larger ones–are translated from British and American counterpart publications.

be seen, there is no strict relationship between titles of positions and particular activities.

Historically, embassy military attachés have been intelligence specialists while military missions concentrated upon the training of dependent armed forces. In the course of recent testimony, a military expert reaffirmed this distinction:

Certainly they are weighed. Military officials themselves are more concerned, naturally, with the operation of the training program itself, but we have plenty of political and economic eyes watching in embassies, unified commands, Department of Defense, State, and AID, all of them alert to these factors.[12]

Actually, this traditional dichotomy is overdrawn since the supervision of training itself involves obtaining intelligence on the political attitudes of potential trainees.[e] In an earlier chapter we noted Lee's claim that "military intelligence for powers outside Africa has . . . become largely a matter of reporting upon the reliability of given individuals, or on the fragility of the political system."[13] According to Dunn, who was given access to classified and other generally unavailable sources, this type of intelligence is in fact performed by American officers who select foreign officers for training.[14] Some of the criteria employed by MILGRP members are revealingly set forth in the exchange below.

Mr. Fraser. I want to go to the question of training. What role or what part do we play in selecting the people who are going to be trained from some of these countries?

General Fuqua. In the country our military assistance personnel in conjunction with the local military select persons to be trained. They are screened as to where they stand in their own country, how well they can absorb instruction, how well they speak our language, or perhaps if they don't, we may have courses where we could teach the subjects in the native language of that person.

This is the process that has to be gone through, and after it is determined exactly who should go, what course he should take, a program is made up and the spaces are earmarked in the various schools of the United States, and the person comes over and takes the course.

Mr. Fraser. Do we make an effort to screen these people from the point of view of the particular ideology that they hold?

---

[e]This is also true of the CIA which uses the MILGRPs as a cover. According to former Deputy Director Richard Bissell, while "these two categories of activity can be separated in theory, intelligence collection and covert action interact and overlap. Efforts have been made historically to separate the two functions but the result has usually been regarded as 'a total disaster organizationally.' . . . Although there are many disagreements within CIA on matters of doctrine, the view is unanimous that the splitting of intelligence and covert action services would be disastrous, with resulting competition for recruitment of agents, multiple recruitment of the same agents, additional security risks, and dissipation of effort." Cited from the minutes of his address to a "special study group" of the Council on Foreign Relations in 1968. *Intelligence and Foreign Policy* (Cambridge, Mass.: Africa Research Group, 1971), p. 8. The use by Army intelligence of reports on FMT attitudes is clearly implied in U.S. Army, *Foreign Nationals*, pp. 2-5, 2-20.

General Fuqua. We try to. In some cases, such as in countries where there is a strong bloc influence, as I was discussing recently, Afghanistan, we don't expect to find purely Western ideals in some of these people. We hope, however, by giving them a course over here we can at least give them a military liking for the U.S. way of doing things.

We try to screen and get across the board the best person for the dollar that we expend in training him. A great many things have to go into the selection.

Mr. Fraser. Are we in a position of having the country recommend people and we turn them down?

General Fuqua. This has been done; yes, sir.

Mr. Fraser. Might we turn them down for ideological reasons?

General Fuqua. It is possible. I don't have any specific instances where this has been done. But a great deal of care goes into selecting them and perhaps that might be one of the considerations, although I can't give you an answer.

Mr. Fraser. I assume if that is the reason that we perhaps don't put it on that ground. We perhaps find another reason.

Brigadier General Stephen O. Fuqua held the title of director, Near East, South Asia and Africa Region (NESA) in the Office of the Secretary of Defense for International Security Affairs.[f] The colloquy is notable because it evidences the acquiesence of a leading House liberal with the manipulative objectives of MAP training. As we shall see, many if not most liberal congressmen and senators share a neocolonial consensus in varying degrees.

Particular care is exercised by the MAAGs in selecting prospective FMTs for the command and general staff courses at Leavenworth, Quantico (Marines), Newport (Navy) and Maxwell (Air Force). Either they must already hold a fairly high level position or be deemed to be *upwardly mobile* in the not too distant future. Although foreign governments must as a matter of course certify that candidates are non-Communist, the determinative factor is an estimate of probable amenability to American influence. Intellectual aptitudes are also considered, for poor performance has in the past engendered resentment by trainees.[15] And this in turn defeats the political goals of MAP training.[g]

After American intelligence agencies grant the prospective invitee one of several grades of security clearance—and this has been construed quite liberally

[f]The 1970 U.S. Army Regulation governing FMTs specifies that "the MAAG will furnish biographical data for each foreign officer student 30 days prior to his reporting date." Copies go to U.S. Army Intelligence, the DIA and to school or installation commanders. U.S. Army, *Foreign Nationals*, p. 2-7. The Fuqua-Fraser exchange appears in U.S., Congress, House, Committee on Foreign Affairs, *Foreign Assistance Act of 1963, Hearings*, before the Committee on Foreign Affairs, House of Representatives, 88th Cong., 1st sess., p. 725.

[g]Equally nugatory would be geopolitical instruction for officers with very low analytical capabilities. In a distinct but probably relevant context, "it has been suggested that more intelligent preadults will learn political information more easily than those who are less intelligent. Some evidence suggests that more intelligent preadults will learn radical or liberal attitudes more readily than conservative attitudes." If this has even attenuated validity for adults, then it is particularly important to preempt the ideological socialization of the more intellectually gifted officers. The quoted sentences appear in T.G. Harvey, "Political Socialization Research: Problems and Prospects," (paper presented at the annual meeting of the Canadian Political Science Association, McGill University, Montreal, June 1972), p. 28.

in the case of some nonaligned countries–the name is submitted to the "country team" at the embassy, to the unified commands and to the Defense and State Departments for possible revision.[16] When the foreign government is not bankrupt and is firmly pro-Western, it may be requested to finance part of the training expenses. Because some training spaces are budgeted under area designations, MAAGs may exercise considerable discretion in ascertaining how many trainees will be invited from a particular country.[h] These funds are "deliberately not allocated in advance to specific countries in order that the unified commander may have the flexibility required to permit maximum utilization of the limited funds available.[i] When training spaces are very limited as in some former Anglo-French colonies, the screening functions are entrusted to embassy military attachés.[j]

In most but not all cases the technical or military functions of MAAGs are specified in military assistance agreements, treaties, or protocols entered into between the United States and host governments. Thus several weeks after Fulgencio Batista overthrew the elected Cuban government in early 1952, the U.S. signed a bilateral military aid agreement with him. Little wonder that one of Fidel Castro's first acts was the expulsion of the American Military Mission!

As this study indicates, tacit understandings and secret agreements–sometimes made directly with the client state's armed forces–may be more consequential than provisions which are matters of public record. Thus informal arrangements with host countries sometimes provide that U.S. advisors can issue actual commands to local counterparts.[17] Similarly, although the written agreements stipulate that MAAG officers and their host country colleagues are to jointly establish the kind of forces and training a country needs, Deputy

[h]Prior to selecting the FMTs, "training survey teams" for each service are sent by the unified commands to all MAP countries. Upon arrival, they bring estimates of the number and type of training spaces which will be available at CONUS and Third Country installations. "The MAAG people then present their estimate of training needs for the coming year, the amount of in-country training which may be available, and the requests which have been made of them by the recipient country." When such "requests" are not generated spontaneously, "the MAAG may encourage the foreign country to submit requests for needed training which is known to be available from U.S. sources." The Joint Chiefs of Staff customarily accept the recommendations tendered by the unified area commands and the MAAGs. *Military Assistance Training in East Asia, Report*, House, p. 31. U.S. Army, *Foreign Nationals*, p. 2-1. *Economic Issues in Military Assistance, Hearings*, Joint Economic Committee, p. 157.

[i]*Foreign Assistance Appropriations for 1970, Hearings*, House, p. 653. In some countries, American ambassadors have exercised the authority to offer training spaces to eager foreign officers. According to a military specialist with NESA field experience, "I well remember how highly American Ambassadors prized the political tool which MAP training gave them." *Military Assistance Training, Hearings*, House, p. 102.

[j]Presumably they also administer very low profile programs in nonaligned radical countries which have officially terminated MAP affiliation. Thus, "we ceased military assistance, including training, to Yugoslavia some 11 years ago. We do have an arrangement with them in which they periodically send a few officers to our Command and General Staff School at Fort Leavenworth." *Military Assistance Training, Hearings*, House, p. 155.

Assistant Secretary of Defense for Regional Affairs Frank K. Sloan testified in 1964 that "our experience is that they pretty faithfully follow our recommendations as to what they do."[18] Apparently, U.S. control over foreign force levels, equipment allocations, and training plans has increased during the decade. Thus in 1970 the Peterson Report elucidated the extent of dependency by noting that "United States now makes the basic determination of the amount and kind of military equipment the receiving countries need, and U.S. military missions do most of the detailed logistical planning and costing for them."[19] The fact that foreign officers lacked the expertise for such analysis despite years of MAP training once more attests to the transcendence of ideological hegemony over technical goals as does the willingness of foreign commands to abdicate such important decisions to agents of a neocolonial power. Despite some failures then, there may be little exaggeration in mangement studies which generally conclude that MAP administration has been more efficient than that of civilian aid programs.[20] Indeed, the MAAGs can usually expect the active cooperation of embassy officers—for according to one former Foreign Service official, "the State Department expects its employees to work closely with military officials and employees of the CIA. . . ."[21] Former CIA Deputy Director Bissell has put it even more cogently:

> The problem of Agency operations overseas is frequently a problem for the State Department. It tends to be true that local allies find themselves dealing always with an American and an official American—since the cover is almost invariably as a U.S. government employee. There are powerful reasons for this practice, and it will always be desirable to have some CIA personnel housed in the Embassy compound, if only for local 'command post' and communications requirements.[22]

As we demonstrate in the next chapter, Bissell explicitly designates foreign officers who may conspire to overthrow existing governments as a CIA target elite. Since MAAG and attache personnel are *the only official Americans in regular contact* with such officers, it is almost certain that they constitute a primary if not exclusive source of cover.[k]

---

[k]According to a recent resignee from the staff of CIA Director Richard Helms, the far-flung agency "is exactly what those who govern the country intend it to be—the clandestine mechanism whereby the executive branch influences the internal affairs of other nations. . . . The notion that the CIA is primarily an espionage organization, preoccupied largely with technical and analytical matters, is a delusion fostered by the agency's leadership to deflect attention from the more questionable clandestine activities. . . . Helms and the CIA earn their keep not by collecting and analyzing secret information for the benefit of policy makers and planners but rather by carrying out paramilitary, political, propagandistic and other covert operations to advance U.S. policy." Marchetti goes on to state that approximately two-thirds of the agency's 18,000 career personnel "are engaged in or supporting clandestine (largely nonespionage) operations." While Bissell chooses not to denigrate the CIA's intelligence functions, he is quite lucid in delineating the scope of clandestine operations which encompass the "penetration" of such institutions as military establish-

The eclipse of the State Department's traditional primacy in manipulating the internal political alignments of weaker countries is a post-World War II phenomenon as is the parallel decline in its formerly exclusive preserve over American diplomacy. Particularly striking in this respect has been the maintenance of MAAGs in such countries as Mali, India, and Pakistan even after grant aid programs have been terminated. Their functions include the diplomatic one of "providing DOD representation to the host government"[23] and, one suspects, its armed forces. In such cases they do promote sales of American weapons when feasible. Traditionally and legally, these must be licensed by the State Department office of Munitions Control. In recent years, however, some of this authority has also been delegated to the Pentagon.[l] Even in such areas as the introduction of an "aggressive" civic action training program by the USAF, there is little evidence of any State Department control over Pentagon initiative.[24] We then tentatively agree with Ralph Dungan, Washington's former Ambassador to Chile (1964-67) that the State Department may not exercise operational supervision over the MAP.[25] At least since 1961 the MAAGs have not been under the exclusive authority of the ambassador—this despite statutory language vesting general direction and continuous supervision of military assistance with the secretary of state. The MAAGs report directly "up through the chain of

ments. "Covert intervention is usually designed to operate on the internal power balance, often with fairly short-term objectives in view. . . . The essence of such intervention in the internal power balance is the identification of allies who can be rendered more effective, more powerful, and perhaps wiser through covert assistance. Typically these local allies know the source of the assistance but neither they nor the United States could afford to admit to its existence. Agents for fairly minor and low sensitivity interventions, for instance some covert propaganda and certain economic activities, can be recruited simply with money. But for the larger and more sensitive interventions, the allies must have their own motivation." And it is precisely here that MAP indoctrination can play a vital role in creating or reinforcing anti-radical sentiments. Victor Marchetti, "CIA: The President's Loyal Tool," *Nation*, April 3, 1972, pp. 430-33. *Intelligence and Foreign Policy*, pp. 9-10.

[l]"Under current Department of Defense regulations, U.S. military advisors are also mandated to cooperate with private industry in promoting the sales of American arms abroad. Thus the Air Force manual on Military Assistance Sales indicates: 'When directed by appropriate authority, [MAAG personnel will] cooperate with representatives of specified U.S. firms in furthering sales of U.S.-produced military equipment to meet valid country requirements.' Not only are the MAAG's enjoined to provide private firms with information, but also to actively promote such commercial sales. . . . Concluding that significant host country purchases of U.S. arms will win Washington's approval and thus advance their careers, many MAAG officers develop a personal interest in the FMS program, and thus, as noted by the MIT Arms control Project, generate independent momentum for the sales effort." The author goes on to comment on MAP training for Third World officers: "although ostensibly this program is designed to improve the defense capabilities of underdeveloped countries, a very real—if unspoken—goal is to inculcate a familiarity with, and appetite for, American-produced weapons." Elsewhere, these sales have been viewed as "preserving and extending U.S. military and political influence." Michael Klare, "U.S. Arms Sales to the Third World: Arm Now, Pay Later," *NACLA's Latin America & Empire Report*, 6 (January 1972), 7. U.S., Congress, House, Committee on Foreign Affairs, *Foreign Assistance Act of 1969, Hearings*, before the Committee on Foreign Affairs, House of Representatives, 91st Cong., 1st sess., p. 703. Capt. Robert F. Farrington, USN, "Military Assistance at the Crossroads," *U.S. Naval Institute Proceedings*, 92 (June 1966), 76.

command to the Joint Chiefs of Staff, and orders to ambassadors involving military matters come via a State-Defense Department cable.[m] And in some South-East Asian protectorates local governments are reported to frequently take their "problems" to American military missions rather than to the ambassador. This is fully consonant with one of the central military aid goals in "less developed" countries: "to inject American officials into host governments internal political processes."[26]

Nevertheless, it would be erroneous to conclude as Senator Fulbright and former Ambassador Dungan do that Latin Americans are essentially correct in believing the U.S. government to be dominated by the Pentagon.[27] Charles Wolf and other defense experts have pointed out that "the important but often implicit political aspects of the programs" are in fact considered in the course of interagency reviews.[28] When unreconcilable differences of a serious nature arise, they are channeled up to interdepartmental committees and if necessary to the White House.[n] According to State's director of politico-military affairs, the role of his department "in the entire military assistance program, including the training elements of it, has been a major one."[o] Spiers has also testified to both

---

[m]It has been said that the MAAG's have "three bosses:" (1) the service headquarters; (2) the unified commands; and (3) the ambassadors. *Military Assistance Training, Hearings*, House, pp. 129, 152. U.S. Congress, Senate, Committee on Foreign Relations, *United States Security Agreements and Commitments Abroad, Hearings*, before the Subcommittee on United States Security Agreements and Commitments Abroad, Committee on Foreign Relations, 91st Cong., 1st sess., pp. 597-98.

High level functional expertise and considerable policy leverage is vested in Lt. General George M. Seignious, deputy assistant secretary of defense for security assistance, and in the Defense Security Assistance Agency (lodged in the International Security Bureau) of which he is also director. The corresponding positions during the 1960s were directorate of military assistance and the deputy assistant secretary of defense for military assistance and sales. U.S., Congress, House, Committee on Foreign Affairs, *Foreign Assistance Act of 1972, Hearings*, before the Committee on Foreign Affairs, House of Representatives, 92d Cong., 2d sess., March 1972, pp. 6-7.

[n]This pattern of collegial supervision is also said to operate with respect to CIA covert operations. According to Bissell, "the essential control of CIA resided in a Cabinet-level committee, comprising a representative of the White House staff, the Under Secretary of State, Deputy Secretary of Defense, and in recent years the personal participation of the Director of Central Intelligence. Over the years this committee has become a more powerful and effective device for enforcing control. It reviews all new projects, and periodically scrutinizes ongoing projects. . . . Projects are usually discussed in the relevant office of the Assistant Secretary of State, and, if at all related to Defense Department interests, at a similar level in DOD, frequently after consideration at lower levels in these departments. . . . Generally the Ambassador had a right to know of any covert operations in his jurisdiction, although in special cases (as a result of requests from the local Chief of State or the Secretary of State) the chief of station was instructed to withhold information from the Ambassador. . . . Of the 'blown' operations, frequently among the larger ones, most are known to have been approved by the President himself." *Intelligence and Foreign Policy*, p. 7.

[o]The militarization of previously civilian foreign policy branches has been further mirrored of late by the creation of a new coordinator for security assistance who will rank as one of the top three officials under the secretary of state. His status will be that of under-secretary of state. Other functions which this bureaucratic innovation may serve include lowering the

a *consensus* on most important issues and coordination between the DOD and State Department—and this is said to extend down to the level of the ambassador.[p] A Brookings Institution foreign affairs specialist after completing a recent study of MAP reported that "in all countries but one there appeared to be common understanding between the Ambassador and the military mission on the size and character of MAP aid needed."[29] And General Mather has recently noted that the ambassadors can use MAAG officers to communicate with the local military leaderships that play an important role in most Latin American political systems.[30] Generally, as in the case of the overthrow of civilian governments of Brazil (1964) and Indonesia (1966), there is complete accord between the State and Defense Departments. This may explain the implication in recent congressional hearings that MAAG and embassy officials have actively cooperated in subverting the stability of certain governments![31]

Although America's general counterrevolutionary foreign policy *goals* were for the most part elaborated by civilians, these do not necessarily suggest the most appropriate or desired means for coping with all situations. Hence, the problem of DOD autonomy arises when there are intrabureaucratic conflicts concerning the "best" way of controlling radical nationalist and/or Communist groups in particular countries. As one congressman has stressed, "without identifying any particular country, it was very obvious that there have been occasions, and fairly serious occasions, under which the Ambassador was not really in control of what was done by U.S. military representatives within the country."[32] In all likelihood, he was probably referring to coups against the governments of the Dominican Republic (1963) and Bolivia (1964) when there were sharp disagreements between the ambassadors and the MAAGs over the appropriate "action policy."

The involvement by MAAGs in domestic political conflicts of Third World countries is an inevitable corollary to relying upon military elites in the furtherance of American neocolonial aspirations. Indeed, it is fully congruent with the expanding role of a functionally specialized covert action agency like the CIA.[33] Thus, according to one experienced officer's candid role perception, "it would be a mistake to consider mission duty as a one-way street. . . . We

Pentagon's MAP profile and possibly preparing the groundwork for separating military supporting assistance from the provision of weapons systems and training in order to fragment critical factions within the United States Congress. In 1971, George S. Newman was appointed acting coordinator who is in daily contact with his DOD counterpart Lt. General George M. Seignious, deputy assistant secretary of defense for security assistance. *Foreign Assistance Act of 1972, Hearings*, House, pp. 6-7.

[p]And this seems to extend to CIA covert interventions. According to Bissell, "it was rare to take an issue before the Special Group prior to discussion at lower levels, and if there was objection at lower levels, most issues were not proposed to the Special Group—excepting large projects or key issues, which would be appealed at every level, including the Special Group." Spiers remarks appear in *Military Assistance Training, Hearings*, House, pp. 152-55. Bissell's remarks are published in *Intelligence and Foreign Policy*, p. 7. The CIA role is also mentioned in: Dunn, "Military Aid," pp. 138-42; and *Foreign Assistance Appropriations for 1970, Hearings*, House, pp. 830-31.

offer them U.S. experience and they offer us understanding of U.S. aims and policies that makes for enduring friendship."[34] Or put somewhat differently, the task of MILGRP officers is to "influence the thinking at the highest echelons of the host government's Army," particularly when the advice they have to offer is unsolicited.[35] Although in some countries, foreign counterparts are in the habit of asking "what do we do next?" control is more problematic in others–particularly those with large military establishments.[q]

Contrary to what might be generally expected, foreign military personnel are not necessarily eager to accept and implement the advice of their U.S. advisors. As a result, the advisor resorts to persuasion, particularly by developing a close personal relationship with his foreign counterpart. This relationship, termed 'rapport,' is heavily emphasized both in the advisor's preparatory training and in published instructions on advisory techniques.[36]

This political objective was evidenced by the assignment of military missions to Latin American countries years before the shipment of significant quantities of equipment, and by the comment that the MAP advisor must "present the best possible image for the United States." There is also ubiquitous thematic stress in the literature on the "diplomatic" and "personal influence" aspects of the role.[37]

Access to key foreign officers is furthered by the MAAG function–shared with U.S. intelligence agencies–of supervising the follow-up for returned officers back from tours in the United States.[r] Under MAP regulations, for at least two

[q]*Military Assistance Training, Hearings*, House, p. 45. Jose Nun, *Latin America: The Hegemonic Crisis and the Military Coup* (Berkeley: University of California, Institute of International Studies, Politics of Modernization Series, no. 7, 1969), p. 54.

Yet even in so large a country as Brazil, it has been reported that mission officers' "advice is often solicited on military problems and, at times, through personal contacts, individual U.S. representatives have been able to exert beneficial influence on authorities." U.S., Congress, House, Committee on Foreign Affairs, *Reports of the Special Study Mission to Latin America on Military Assistance Training and Developmental Television*, submitted by Hon. Clement J. Zablocki, Hon. James G. Fulton, and Hon. Paul Findley to the Subcommittee on National Security and Scientific Developments of the Committee on Foreign Affairs, House of Representatives, 91st Cong., May 7, 1970, p. 8.

[r]"The MAAG interviews the returnee, noting his evaluation of the program: its strengths and weaknesses, the student's personal problems, the effectiveness of the advisory officers and orientation tours. This information is helpful in determining future training needs. MAAG personnel are urged to maintain informal social contact with returned students and to visit them as frequently as possible." Institute of International Education, *Military Assistance*, pp. 20-21. As noted in Chapter Three, academic reports which include the FMTs extracurricular involvement and general attitudes are filed to appropriate intelligence offices as well as with the MAAG. U.S. Army, *Foreign Nationals*, pp. 2-5, 2-20. Cf.: notes d and f to this chapter; Dunn, "Military Aid," pp. 206-8; L. Fletcher Prouty, "The Secret Team and the Games They Play," *Washington Monthly*, May 1970, pp. 11-19; U.S. Army Command and General Staff College, *Strategic Subjects Handbook* (Ft. Leavenworth, Ka.: RB 100-1, August 1969), p. A-20, Lecture R250; *Foreign Assistance Appropriations for 1970, Hearings*, House, p. 681; Fox, "Military Representation," in *The Representation*, pp. 132-33.

years following his return, the ex-trainee is expected insofar as possible to ensure the further transfer of his newly acquired skills. This is often accomplished by their assignment as instructors in local military schools.[38] At higher levels, MILGRP chiefs and their assistants dispose of *entertainment* allowances in excess of those available to AID or the Embassy.[39] In some cases, the consequential rapport has verged upon affection. Senator Church, for example, has reported that on his trips to Latin America the identity of interest was so great that Americans "espouse and . . . parrot the viewpoints of the local military people."[40] A group of congressmen who recently visited Brazil described the MILGRPs attitude in a similar vein: "rather than dwell on the authoritarian aspects of the regime, they emphasize. . . ."[41] While an indifference to a terroristic dictatorship may well connote reciprocity of influence, it may also be construed as manifesting a prevalent approving disposition toward military rule in the Third World. The tacit yet consistent denigration of civilian supremacy in the course materials surveyed in Chapter Four and statements cited in all of the previous chapters save the first suggest the utility of an interpretation that incorporates both variables.

Brazil symbolizes a MAAG function which is central to the thesis of this study. Phrased euphemistically, it is to maintain stability in "countries festering with political unrest."[42] The stability of an open door to American corporations requires the overthrow of an elected radical government. Thus according to one arms control specialist, the United States began to systematically support all Brazilian rightist opposition groups in June 1963. This included the military factions which identified most closely with–and had trained in–the United States.[43] After they deposed Goulart in April 1964, stability came to mean ensuring that a pro-American government remained in power. This is the operational meaning of the term in underdeveloping areas!

Regimes are strengthened by eliminating radical military factions through dismissal, retirements, and by means of ideological indoctrination. Thus about a decade ago U.S. military missions used their not inexhaustible reserve of responsiveness to induce Latin American generals to request Washington's "aid" for antiradical ("Communist") indoctrination of their officer corps and troops. These diplomats in uniform prudently adopted an indirect "soft-sell" approach. Symptomatic of the messages contained in the subsequent propagandization was

---

"In addition to the [MAP] programs described above, the Army, Navy, Air Force and Marine Corps have authority to train certain foreign personnel with their own funds. The Draper Committee estimated that some 4,400 men received such training from 1950-1958. The programs are normally administered by Embassy Attaches in the field and by the United States military departments in Washington." Harold A. Hovey, *United States Military Assistance: A Study of Policies and Practices* (New York: Praeger, 1965), p. 175. With regard to attachè functions relating to "politico-military" as well as "military" intelligence, see: U.S., Congress, House, Committee on Appropriations, *Department of Defense Appropriations for 1970, Hearings*, before a subcommittee of the Committee on Appropriations, House of Representatives, 91st Cong., 1st sess., p. 186. Cf., *U.S. Military Policies in Latin America, Hearings*, Senate, pp. 75, 78.

Capt. Cunha's 1962 testimony that *no military regime* had objected to the "democratic content of the course on the grounds it might engender discontent within the ranks." According to the same MAP administrator, it was quite "within our bounds . . . to influence the military of another country"–but not their civilian officials–on political issues.[44]

Insofar as the challenge of radical nationalists is concerned, it is commonly unnecessary to *overtly* intervene in any way. As one specialist has testified:

I think that American involvement in politics in Ethiopia is something which is not very often talked about, but is obviously very real. The presence of American military personnel in Ethiopia is a very clear sign to anyone who would challenge the right of the Emperor that obviously the United States has a presence there. Whether or not Americans would become actively involved or not is less important than the fact that our program does manifest a great American interest in the existing regime.[45]

Unfortunately for the MAAGs, this is not always a sufficient deterrant to opposition elements. Thus even in Ethiopia–which along with Morocco has received a preponderance of MAP aid in Africa–there was an attempted Nasser type coup by the Palace Guard in 1960. Although this unit and the Air Force had been trained by the United States, rapport was strong only with the latter service. MILGRP officers actively advised Air Force commanders in their successful campagin against the neutralist rebels. MAAG personnel did the same thing in Indonesia when the "political situation was still unsettled" some years ago. There the Defense Liaison Group of ten officers "was able to provide the anti-Communist forces with moral and token material support" in their defeat and massacre of left-wing nationalist and Communist groups.[46]

Elsewhere, challenges by insurgents to autocratic regimes may be defeated with U.S. advice before the revolutionaries make much headway. The Guevara disaster in Bolivia is a case in point. Less widely known is the successful effort to ensure the safety of "Papa Doc" in Haiti between 1959 and 1963. There the American

mission engineered the Haitian Army's destruction of an invading force in April 1959; it trained the Presidential guard; it served to immobilize internal resistance to the regime during the critical period of its consolidation.[47]

In Laos, U.S. Air Force attachés assist the king's U.S. trained pilots in targeting against radical nationalist insurgents.[48] Often the MAAGs call for "mobile teams" to provide training or tactical direction in the field. Thus they assisted the Colombian army in crushing "insurgents" during the late 1960s, and in so doing preserved the stability of that country's oligarchy.[49]

An extremely important function of the MAAGs is the maintenance and consolidation of pro-American influence at command levels of the armed forces in Third World countries. Only a committed and ideologically cohesive officer corps is safe from: (1) attempted coups by nationalistic or radical factions; and (2) possible defections to radical insurgent forces. This key political role of American military representatives was explicitly acknowledged by a presidential committee a decade and a half ago.

> Military assistance to Lebanon and Iraq did not prevent the coming to power of regimes less friendly toward the West than those they displaced. In Lebanon, the new Chehab regime is markedly cooler toward us than the government of Camille Chamoun. Even more serious is the situation in Iraq, where the Kassem regime seems to have given its tacit approval to the resurgence of domestic communism. It should be borne in mind, however, that U.S. military assistance to Iraq was limited to the supply of equipment; a full-fledged training mission might have produced different results.
>
> . . . .
>
> The task of the MAAG officer often proves more difficult than that of his diplomatic counterpart and his responsibility for furthering the national interest of the United States may actually be greater.

Hence, the Draper Report went on to warn in characteristically euphemistic terms that "to abandon the [MAP] program, for errors of execution or for any other reasons would be to abandon the Free World and to lose the Cold War."[50]

The contemporary neocolonial *function* of military aid in the under-developing areas can be readily inferred from the distribution of MILGRP strength in Tables 6-3 and 6-4. While this global distribution in strength seems to have been maintained through 1970, the Foreign Assistance Act of 1971 prescribed a mandatory reduction of 15 percent in MAAG personnel.[51] Within several years of the Cuban Revolution, mission strength in Latin America had swelled to more than a thousand. The defeat or containment of most insurgencies, liberal criticism, and reallocations forced by Vietnam resulted in a sharp reduction between FY 1968 and FY 1970. In the latter year there were only 458 Americans assigned to forty-four MAAGs, Missions, and MILGRPs in the region.[52]

Resurgent nationalism in some countries was also a factor. Thus Peru's military seized government office in 1968 and promptly expelled the entire mission of forty-seven officers and enlistees. In order to continue to receive replacements and spare parts, the regime was induced to receive a new seven man mission. But none of the personnel associated with the previous mission were allowed to return! The largely interventionist function of the earlier mission was also reflected by the fact that the drastically reduced mission informed a visiting congressional delegation that it had enough manpower to fulfill most of their public (statutory) duties![53]

The political roles of missions are also displayed—albeit less nakedly—by considering their budgetary allocations. In 1969 expenditures were about $2,000,000 a mission with an average of $20,000 per MAAG member.[54] For

Table 6-3
Average MAP Personnel Strengths[a]

	FY 1965			FY 1967		
	Mil.	U.S. Civ.	Local	Mil.	U.S. Civ.	Local
Departmental & Field (U.S.)	147	693		135	605	
Overseas & Unified Commands						
Australia	–	–	–	–	–	–
Belgium/Luxemburg	25	7	18	22	7	14
China (Taiwan)	779	4	120	594	23	109
Congo (K)	21	–	3	42	–	5
Denmark	34	3	9	28	3	8
Dominican Republic	26	–	–	49	–	2
Ethiopia	105	3	28	168	3	34
France	9	1	9	9	1	9
West Germany	102	21	18	82	19	15
Greece	108	11	66	108	12	66
Guinea	–	–	–	7	–	–
India	108	–	19	no information[a]		
Indonesia	32	–	15	5	–	–
Iran	360	14	176	382	16	191
Italy	40	7	10	35	7	7
Japan	100	29	–	62	20	0
Jordan	4	–	1	8	–	3
S. Korea	1,267	73	471	1,187	84	459
Liberia	17	–	5	18	–	9
Libya	26	1	11	36	1	9
Mali	24	–	3	1	–	2
Morocco	26	–	6	84	1	6
Nepal	4	–	–	no information[a]		
Netherlands	26	2	7	23	2	7
Nigeria	–	–	–	2	–	1
Norway	41	6	19	34	4	14
Pakistan	112	3	72	no information[a]		
Philippines	79	8	15	76	9	16
Portugal	25	5	17	19	5	16
Saudi Arabia	222	–	–	218	–	4
Senegal	4	–	8	2	–	3
Spain	77	14	30	80	13	26
Thailand	488	2	95	619	5	36
Tunisia	4	–	–	4	–	–
Turkey	355	52	195	315	53	177

Table 6-3 (cont.)

	FY 1965 Mil.	FY 1965 U.S. Civ.	FY 1965 Local	FY 1967 Mil.	FY 1967 U.S. Civ.	FY 1967 Local
United Kingdom	4	1	–	–	1	–
Classified Countries	4,641	–	24	86	2	103
Cmdr-in-Chief, Europe	50	27	–	48	20	–
Cmdr-in-Chief, Pacific	78	2	–	89	2	–
Cmdr-in-Chief, MEAFSA	41	13	–	40	14	–
Cmdr-in-Chief South	30	7	–	30	7	–
U.S. Regional Org.	31	38	–	24	43	–
USAF, Europe	5	3	–	3	1	–
USA, Europe	10	–	–		no information[a]	
USA, Pacific	–	4	–	–	10	–
USAF, Pacific	12	2	–	12	2	–
US element–Cento	26	3	6	27	3	6
SEATO Perm. Mil. Plng. Staff	9	–	5	3	–	3
Total Overseas	9,587	355	1,481	4,681	383	1,360
Grand Total	9,734	1,059	1,481	4,816	988	1,360

[a]Administrative and Mission Training. As the classified countries item suggests, much of the decline may be explained by the removal of Vietnam and Laos from the MAP designation in FY 1967.

Source: U.S. Department of Defense, *Military Assistance Facts: 1 May 1966* (Washington, D.C.: Office of the Assistant Secretary of Defense for International Security Affairs, 1966), pp. 24-25; ibid., March 1968, pp. 20-21.

those assigned to Latin America, it has ranged from $25,000 to $50,000 per man.[55] In some countries such as Costa Rica recently, MILGRP administrative costs were substantially in excess of the value of all equipment deliveries. The high cost per member in Latin America may have been occasioned by the endemic nationalism inherent in all hegemonic interstate relationships–a reaction given some impetus by the Cuban Revolution. Ironically, the official entertainment allowances listed in Table 6-5 appear quite modest. Their allocation, however, is much more closely associated with nationalism or leftism in particular countries than with the size of their armed forces.[s]

The worldwide total for FY 1970 entertainment allowances for MAAGs was

[s]Such firmly rightist regimes as Brazil, the Dominican Republic, and Argentina were in contrast to those in Bolivia, Ecuador, and Peru where nationalistic and moderately leftist military factions had appeared.

Table 6-4
Military Missions in Latin America: U.S. Personnel (FY 1959)

Countries	Officers	Warrant Offs.	Enlistees
Argentina	16	–	7
Bolivia	10	2	11
Brazil	41	1	57
Chile	14	–	18
Colombia	28	1	23
Costa Rica	2	–	4
Dominican Republic	–	–	–
Ecuador	17	1	18
El Salvador	10	–	7
Guatemala	8	–	7
Haiti	8	–	9
Honduras	4	1	11
Nicaragua	7	–	8
Panama	2	–	–
Paraguay	10	–	9
Peru	27	1	26
Uruguay	10	–	10
Venezuela	35	1	46
Total Latin America	249	8	271

Source: The President's Committee to Study the United States Military Assistance Program, Composite Report, II (Washington, D.C., 1959), p. 347. cited in Dunn, "Military Aid," (dissertation), p. 167.

$87,825, while Germany boasted the highest per country allocation at $5,775![56] Allowances for Latin America in FY 1970 actually exceed the total for the remainder of the world.[57] It is also likely that in some underdeveloped countries PL480 counterpart local currency is made available by the embassy for MAAG and attaché entertainment activities.[58] As mentioned earlier, embassy political and intelligence officers cooperate closely with the MAAG chiefs of mission, as does the local USIA when mutual interests are paramount.[59] Thus one careful student of American foreign propaganda reports that "in the field, the CPAO (embassy country public affairs officer) and the USIS staff are also responsible for insuring an adequate local understanding of the work of any United States Military Assistance Group in the host country, working in cooperation with the military, naval, and air attaches of the American embassy."[60]

The MAAG chief also sits with representatives of AID, the CIA, and other U.S. agencies on the "country team," chaired by the ambassador. Because "during its 17-year history, and an expenditure of over 32 billion dollars, MAP

Table 6-5
U.S. MILGRP Representation Allowances and Armed Forces Strength

Country	Authorized Ceiling FY 1969	Manpower 1968
Argentina	$ 9,050	118,000
Bolivia	5,950	15,000
Brazil	10,800	194,000
Chile	7,150	60,000
Colombia	7,550	63,000
Costa Rica	1,300	3,000
Dominican Republic	1,685	19,300
Ecuador	7,150	19,000
El Salvador	4,100	6,600
Guatemala	6,150	9,000
Honduras	3,900	4,725
Nicaragua	3,850	7,100
Panama	1,400	3,425
Paraguay	4,550	13,000
Peru	9,000	54,650
Uruguay	5,500	17,000
Venezuela	9,100	30,500
Total	98,185	

Source: U.S., Congress, House, Committee on Foreign Affairs, *Foreign Assistance Act of 1969, Hearings*, before the Committee on Foreign Affairs, House of Representatives, 91st Cong., 1st sess., p. 653. Robert C. Sellars, ed., *Armed Forces of the World* (Garden City, N.Y.: Robert C. Sellars Associated, 2d. ed., 1968).

has been viewed principally as an element of U.S. diplomacy,"[61] senior U.S. advisors are known to go to the U.S. Embassy regularly, together with the chiefs of other U.S. mission agencies, to meet with the ambassador to whom they are clearly subordinate."[62] How this interactive process works out in practice is superbly demonstrated by the 1963 overthrow of the Diem regime in Saigon:

The Vietnam commitment limped upward, although under great official scrutiny, until the summer of 1963—when Diem's repression of demonstrations by the Buddhist majority sent his government tottering again. Diem's brother, Nhu, raided pagodas with his secret police apparatus, while Madame Nhu enjoyed self-immolations by Buddhist monks. Her description of them as "barbecues" was taken to reflect the sentiments of the ruling family. This barbarity and the crisis of the government got the computers running and the memos flying back in Washington, and in the midst of general indecision, a cable authorizing the instigation of some kind of coup went out to Ambassador Lodge. Taylor, by then elevated to Chairman of the Joint Chiefs of Staff, was so furious at this move that he pinpoints the authors of the cable as the "anti-Diem clique" in the State Department, led by Averell Harriman and Roger Hilsman. (The cable appears in the Pentagon Papers, but the authors are not identified.) The general brands the maneuver as a bureaucratic "end run" by the perpetrators, who, on a Saturday when top officials were home in the suburbs, "drew up this cable, cleared it with Undersecretary George Ball on the golf course, and obtained a telephone clearance from President Kennedy in Hyannisport." Without any wide circulation says Taylor, the message went off, and by Monday morning Lodge had informed various Vietnamese generals through the CIA that the U.S. would look favorably on a new team.

Although the expected August coup did not come off, the American policymakers deemed themselves beyond the point of turning back, so Taylor and McNamara were dispatched to Saigon to scout out possible Vietnamese kingmakers. In a carefully arranged diplomatic meeting at the Saigon Officer's Club, Taylor records that he played tennis with the most eligible plotter, General "Big" Minh, while McNamara watched and calculated nervously on the sidelines. The two eminent Americans dropped decreasingly casual hints of their interest in a coup on Minh–feelers between games, seditious innuendos while retriving stray shots–but Minh only concentrated on his forehand. The twosome returned to Washington convinced that nothing was afoot, but soon learned that the CIA messages had taken hold after all (Minh was just not ready to talk). Lodge was instructed to review Minh's arrangements ("except for assassination plans"), and Diem was soon dead.[63]

Here then is a case of American promoted instability which was necessitated by counterrevolutionary goals that were not being effectively promoted by a pro-American dictator who himself had been previously installed in power by the United States! Diem's political skills were notably deficient and consequently it became necessary to dispose of him "with prejudice."

Preparing MILGRP Officers

While there are historical grounds for questioning whether mission officers have always acted in a "clearly subordinate" fashion, there is no doubt as to the diplomatic talents which their roles mandate.[64] In the past one major problem encountered in selecting trainees and cultivating posttour friendships or rapport has been the very limited linguistic aptitudes and cultural ethnocentrism of American officers assigned to the MAAGs.[65] For well over a decade there has been a growing recognition that "the ability to exchange amenities at receptions . . . will not carry the individual when he is discussing military-political relationships."[66] Yet the incessant flow of requests for more "psycho-cultural" and political preparation has not abated! As late as January 1971, a Rand Corporation expert informed a congressional committee

that the solution lay in special training for military assistance teams so that their 'inevitable involvement' in the political life of the countries where they are stationed would be 'a conscious, deliberate process in the hands of competent, politically informed military-political analysts'![67]

The apparent inertia manifested by executive decision-makers could well have been founded upon an experientially–based conception that social skills and a field briefing were in most cases adequate for the particular tasks in question. And when they weren't, the functionally specialized CIA would contribute the requisite expertise. Furthermore, while not comparable to the four or more months of intelligence and linguistic preparation that attaches receive, MAAG assignees had in fact been exposed to increasing amounts of specialized training.

Before looking at this in some detail, mention should be made of a recruitment problem which represented a more serious hindrance to effective mission performance. During the fifties, the most ambitious officers avoided mission duty because command assignments weighed more heavily with promotion review boards. The result was that both language proficiency and educational levels of MAAG applicants were below average for the services. The military tradition that command experience must be given preference for promotions was so ingrained that it was not until the late sixties that new directives were issued. These required that promotion boards give greater consideration to MAAG assignments.[68]

The Pentagon's response to the Draper Report's recommendation that *specialized MAAG training* be instituted was almost instantaneous since here no conflict with martial values was involved.[t] In September 1958 a Military Assistance Institute was opened at Arlington Towers, Virginia. It provided a formal month-long training course for officers assigned to prospective MAAG duties. Operated under contract with the Defense Department, the MAI was staffed by twelve retired officers–six of whom lectured on a part-time basis. More than a hundred MAAG appointees were instructed during each four-week course so that the annual output soon exceeded a thousand.[69] Shortly after the institute began operating, its curriculum was faulted for being insufficiently geopolitical–despite the fact that many officer students had already received such exposure at command and general staff schools or in service war colleges. According to one scholarly critic:

> In many countries MAAG's are closer to the center of political influence than any other American agency and in every country they have close contact with their native opposite numbers in the military command. Yet more than half of the new Military Assistance Institute's program . . . deals with the structure and procedures of the military-aid program narrowly conceived.[70]

The remainder of the curriculum consisted of "a detailed study of the country to which the specialist is to be assigned–its history, geography, economics,

[t]The Draper Committee had not only proposed broadened training but also warned that those who volunteered for MAAG assignment "should, as well as being technically qualified, understand, accept and have the will and ability to promote the basic purposes of the Mutual Security Program." A euphemistic statement if there ever was one! Cited from: U.S., President's Committee to Study the United States Military Assistance Program, *Composite Report*, vol. 1 (Washington, D.C.: GPO, 1959), pp. 88-89, 168.

politics, military forces, and the status of U.S. and Soviet aid programs.[71] The instruction was garnished with top secret briefings by State Department and CIA officials on the actual (i.e., nonpublic) objectives of the United States in particular countries and the roles of their agencies abroad. Officer students at the MAI whose average rank was Lt. Colonel also were provided with lectures on the virtues of American capitalism and on Communist goals. The latter were delineated by experts from the FBI, CIA, and military intelligence agencies. During the early 1960s, a pamphlet *Facts about the United States* was distributed, as was an "Outline Country Criticism" by a USIA official.[72] This contained currently fashionable critiques of the United States which the MAAG officer might encounter in his country of assignment. Techniques for manipulating foreign counterparts were also presented and evaluated.[73]

The fact that the large majority of MAAG officers and MAI students are army men reflects the key political role of Third World armies as it does the secondary role which the U.S. assigns to naval and air forces equipped in most cases with obsolete military equipment.[u] Among the services, the U.S. Army has taken the greatest pains to prepare officers for their technical and quasi-diplomatic roles in other countries. As early as 1960, the U.S. Army Language School at the Presidio de Monterey in California was processing 2,000 officers and enlistees annually in twenty-eight foreign languages.[v] Indicative of the progress made with easier foreign idioms is that by 1966 over 85 percent of those assigned to the 8th Special Forces at Fort Gulick were bilingual.[w] Research was also being commissioned to develop new techniques and guidelines for psychocultural" manipulation of foreigners of non-Western cultural backgrounds.[74] By 1966 the Army had instituted a very rigorous Foreign Area Specialist Program. Admission required a high aptitude for foreign languages as well as a university degree. The period of study ranged from two and a half to four years and offered an M.A.

---

[u]This is not to say that the Pentagon is wholly indifferent to professional capabilities for assisting U.S. forces in such countries as the Dominican Republic, Laos or Vietnam. And a genuine military role of sorts is envisaged for "Forward Defense" countries like Turkey, Greece, and South Korea.

[v]And by 1970, the output had been increased to 5000 per year. A second language training facility was opened during the mid-1960s at the Anacostia Naval Annex near Washington, D.C. U.S., Department of the Army, *Army Information Program: Supporting Materials* (Washington, D.C.: GPO, DA Pamphlet 355-201-1, July 1960), pp. 49-50. U.S., Congress, House, Committee on Appropriations, *Department of Defense Appropriations for 1970, Hearings*, before a subcommittee of the Committee on Appropriations, 91st Cong., 1st sess., pp. 135-36. According to another source, "enrollment in language training under the Defense Language Institute, for instance, has increased from 5,200 students in 1964 to 9,500 in 1967." Col. Wallace W. Wilkins, Jr., USA, "You are What You Say," *Military Review* 48 (June 1968), 95.

[w]"It has been said that one can buy anything anywhere in English, but that to sell things—especially ideas—one must speak the other man's language.

This statement directly applies to the modern US military officer whose future will likely find him increasingly involved in the developing areas of the world. If the Army is to accomplish its mission of selling ideas, the Army officer must speak the language native to his assigned area." Richard D. Clarke, "U.S. Military Assistance in Latin America," *Army Digest*, 21 (September 1966), 19, Wilkins, "You Are," p. 94.

from an accredited university level institution. During the late sixties, emphasis was placed on Africa and South East Asia. Although 500 officers had completed the program by 1970, the long-range target was 1,300.[x]

Defense expenditures reallocations occasioned by congressional resistance to the rising costs of the war in Indochina forced the closing of the Military Assistance Institute in July 1968. Between that date and 1970, the unified commands provided several weeks of preassignment orientation to MAAG chiefs and some senior officers. Others received little more than briefings.[75] Despite limited appropriations, the inadequacy of this approach was quickly perceived and a new Institute of Military Assistance was opened at Fort Bragg. It currently provides a twenty-week command and staff course for officers from all services who have been assigned to MAAG duty. The curricula includes techniques of psychological warfare and administration of civil affairs in foreign countries.[76] By early 1972 attention was being directed at the need to upgrade managerial and programming instruction for future MAAG officers. And applicants with appropriate experience were being actively sought.

If the political objectives of the MAP and the administration role of the MILGRP officers assigned to Third World nations are now fairly evident, there remains the question of just how efficacious the United States has been in attaining them. In the following chapter, we shall examine evidence which indicates that while Washington has by no means been omnipotent, it has been more successful than generally believed in manipulating domestic political conflicts in many Third World nations.

---

[x]This included 6-12 months of language instruction, 9-12 months of academic foreign area studies at a U.S. or foreign university, and 12-36 months of travel, research or study in a particular region or country. According to Rosser, area trainees "will work in intelligence, in plans and operations, in military assistance, as advisers to allied forces, as attaches." U.S., Department of Defense, Office of the Assistant Secretary of Defense for Manpower, *Officer Education Study* 2 (Washington, D.C.: July 1966), pp. 193-95, referred to in Col. Richard F. Rosser, "The Air Force Academy and the Development of Area Experts," *Air University Review*, 20 (November-December 1968), 28-29. "Army Seeks FAS Officers, *Armed Forces Journal*, 16 May 1970, p. 14.

7 Effectiveness of the Program

> Young officers trained abroad are beginning to show evidence of discontent with the abilities of the ruling group.[1]
>
> Peter B. Riddleberger, March 1965

> We have recently completed a worldwide training conference and have placed a deal of increased emphasis on training during the coming year. We are going to make a determined effort to single out potential future leaders for our training.[2]
>
> Vice Admiral L.C. Heinz, Director
> of Military Assistance, March 1966

> But there is a tendency to classify everything that might be embarrassing.[3]
>
> Sen. J.W. Fulbright, January 1971

In our first chapters the instrumental relationship of military aid to the attainment of counterrevolutionary foreign policy goals was delineated. Our hypothesis was that the attitudinal modifications and personal friendships structured by one or more MAP training/orientation tours were not only intended to but in fact significantly reinforced and diffused conservative tendencies within foreign military establishments. By creating new external reference groups and definitions of national interest, the MAP has transformed dependent officer corps in many nominally independent Third World countries into servants of America's informal empire.

Since World War II, one of the major threats to control over the raw materials and markets of underdeveloped countries by American dominated corporations has been the occasional rise to power of "nationalistic" regimes. In an attempt to retain greater economic surplus for development, some of these governments have in varying degrees limited or closed the open door. Their ability to embark upon–though not necessarily to succeed with–such policies has required that they adopt independent or nonaligned foreign policies so that the West could be "played off" against the East.[a]

The problem at hand is to assess the effectiveness of the training-induced relationships in preventing the emergence of such governments on one hand, and in deposing them or forcing "moderation" of their policies on the other. Because

[a]This is not to imply a pattern of indiscriminate support by Soviet Bloc nations or China for either radically-oriented regimes or revolutionary movements in the Third World. See the insightful analysis by David Horowitz, *Empire and Revolution* (New York: Random House, 1969).

of the pervasive censorship of relevant data and the multiplicity of variables which are often operative, it is not possible to isolate and measure the effectiveness of this foreign policy instrumentality with great precision. Even if the vital interpersonal relationships were not systematically concealed, the immensity of the research effort would mandate nothing short of a Defense Department financed task force!

We shall therefore be compelled to rely upon a number of partial indicators which can support only the most tentative conclusions. Before surveying official testimony and the views of military aid experts, consideration will be given to available quantitative data which are relevant to the thesis. Optimally we should have a substantial number of countries which did not receive American military aid. These could then be compared to those which did in order to assess the impact of MAP upon military political behavior. Unfortunately, because most nations in the Third World have received military aid from the United States or other Western great powers, it is not possible to establish such a "control group."

Secular Trend Data

Several years ago, a noted specialist on Latin American affairs compiled a time-series chart based upon the changing character of *golpe de estados* in the Western Hemisphere. Scrutinizing an area where a higher ratio of officers are U.S. trained than in most other underdeveloped countries, Needler found a very pronounced increase in the percentage of antipopulist and anti-Constitutional *golpes*. Similarly there was a marked decline in the incidence of reformist military takeovers. While the definition of reformism used in Table 7-1 may include some moderates who did not intend to expropriate or otherwise restrict American investor prerogatives, the overall trend is consistent with our hypothesis. And the data support the impressionistic conclusion of John J. Johnson "that without the military, every government in the Latin American orbit would be further to the left than it is now."[4] The fate of Chile's elected left-wing nationalist government is the acid test of General Porter's "insurance policy." For there is little reason to doubt that Washington's "action policy" is to encourage a military seizure of power in that *newly independent* country.[b]

---

[b]So bold an assertion is premised upon: (1) the historic pattern of U.S. hostility and often successful subversion of radical regimes less nationalistic than Chile's; (2) Bissell's and Marchetti's experience as related in Chapter Six; (3) the texts of the ITT Papers; (4) due inferences from the sources cited in Chapter One, footnote m; (5) the deterioration of American-Chilean relations since 1969. Before a second golpe is attempted, three conditions must be met: (1) an imminent Marxist victory in the 1973 congressional elections or a plebiscite to reorder the national government along unicameral lines; (2) leftist unwillingness or incapability of mobilizing several hundred thousand demonstrators who would not shrink from physical confrontation with the Armed Forces and Carabineros; (3) widespread CIA supported attacks against leftist political campaigners which finally induced retaliation.

Table 7-1
Successful Latin American Golpes: 1935-64

Characteristics	1935-1944		1945-1954		1955-1964	
Reformist	8	50%	5	23%	3	17%
Low in Violence	13	81	15	68	6	33
Overthrew Const. Govts.	2	12	7	32	9	50
Around Elections	2	12	7	32	10	56

Source: Martin C. Needler, "Political Development and Military Intervention in Latin America," *American Political Science Review* 60 (September 1966), 616-26.

Because of limited financial resources and the availability of Needler's study, the data presented omit Latin America. They are based upon the classification of coups or quasi-coups in the Middle East, Africa, and Asia during the post-World War II period.[c] Between January 1, 1946 and December 31, 1970 there were fourteen radical interventions and fifteen rightist ones. Rightist is here defined as involving a change in foreign and/or domestic policy. The former may encompass a shift from militant neutrality to moderate nonalignment, or from the latter to a *de facto* pro-Western orientation. Sometimes, as in the case of Indonesia or Ghana in 1966, the shift is from militant neutrality to the pro-Western category. As for domestic policy, it encompasses a shift from national–usually government–ownership of basic économic resources and activities to a mixed economy where there is limited though substantial opportunity for control by foreign corporations. It may also refer to a further shift along this continuum to a commitment by the regime to maximize incentives for foreign investors, i.e., to create a "healthy" investment climate. Again, the rightward tendency may omit the intermediate stage. Such policies, especially those in the domestic sphere, usually are supported by curtailment of lower class sector real income (i.e., enforced wage restraint), suppression of unions and radical leftist organizations, outlawry of Communists, tax concessions to foreign investors, currency exchange guarantees and demagogic promises of reform. Radical coups and semicoups occasion leftward policy changes on these often related continua.

In addition to the twenty-nine interventions which we have classified as ideological, forty-four have been excluded from the following tables.[d] Designated "other," they were judged to have been motivated by objectives of an

[c]A "quasi-coup" is a change in government policy prompted by the threat of a coup (an overt military seizure of civil office). Both are referred to as interventions in the text.

[d]Algeria (Radical–7/62); Burma (Rightist–9/58); Cambodia (Rightist–3/70); Congo–B (Radical–8/63); Congo–K (Rightist–9/60); Egypt (Radical–3/54); Ghana (Rightist–2/66); Greece (Rightist–4/66); Indonesia (Rightist–3/66); Iran (Rightist–8/53); Iraq (Radical–7/58, Rightist–2/63); Jordan (Rightist–4/57); Laos (Rightist–1/60, Radical–8/60, Rightist–4/64); Libya (Radical–9/69); Mali (Rightist–11/68); Pakistan (Rightist–10/58); Somalia (Radical–10/69); South Yemen (Radical–5/68); Sudan (Radical–10/64, Radical–5/69); Syria (Rightist–3/49, Radical–2/54, Rightist–9/61. Radical–2/66); Thailand (Rightist–11/51); Zanzibar (Radical–1/64).

essentially nonideological nature: discontent with civilian corruption or incompetence, personal antipathy or ambition, service rivalry or factionalism, a desire for a larger budgetary allocation, ethnic rivalries, prestige or dominance aspirations, etc.[5] While some of these motives and opportunism may also play a part in motivating officers who participate in ideological coups, most of them are compatible with *either* radical or rightist interventions. Hence, in contradistinction to MAP indoctrination they cannot account for or explain the clearly ideological character of such coups.[e] Furthermore, we agree with the Defense Department that such interventions are normal rather than deviant attributes of "political cultures" in underdeveloped areas. And as we have seen, MAP indoctrination has never laid great stress upon transforming such institutional patterns. Western neocolonialism, like imperialism, is oriented towards securing desired policy outcomes in *only limited and specific areas*.[f] Thus consideration of such coups would not bear upon assessing the efficacy of MAP training.

In Table 7-2, we have separated long-term military aid recipients from those whose intermilitary ties were more recently established. Over the long run both MAP countries and those aided by other Western powers exhibit a pronounced trend toward rightism. Hence, in the absence of nonaligned or Eastern donors, the MAP may be considered moderately effective in orienting the political behavior of officer corps in these areas. These data are consistent with Needler's for countries which likewise did not receive Eastern aid, and they would tend to support our hypothesis.[g] On the otherhand, the impact is somewhat weaker in those nations which have sought to diversify their ideological donors. But even if the small number of cases were regarded as insufficient to warrant great confidence in the pattern of association–given the magnitude of change–the trend is nevertheless in the hypothesized direction. While one is not justified in postulating monocausal relationship, the testimony examined in Chapter Two and evidence considered elsewhere in this chapter enable us to conclude that the relationship is probably not spurious.

[e]It is conceivable that a systematic analysis of these variables would reveal an association of some strength between radical interventions and discontent with corruption and similarly between budgetary discontents and conservative coups. As in our own case, public *declarations* must whenever possible be evaluated in terms of not necessarily successful *behavior*.

[f]In this respect it bears marked resemblance to Soviet neocolonialism within that country's East European security zone. While policy in economic and other system transforming areas was crucial during the forties and early fifties as was Soviet hostility to all manifestations of nationalism, the preservation of Party hegemony and Warsaw Pact commitments seem to have emerged as the key determinants of intervention since 1955. Hence, considerable if not absolute policy autonomy in socioeconomic and diplomatic areas has been exhibited in recent years. In ending its brief flirtation with economic imperialism, the USSR has begun the long road of adjustment to the aspirations for self-determination within her own sphere of influence. Those who shape American policy have yet to even begin this process.

[g]They also are congruent with Price's reference-group hypothesis that training experiences are more determinative than social origins or prior education in conditioning military political roles. See: Robert M. Price, "A Theoretical Approach to Military Rule in New States: Reference-Group Theory and the Ghanaian Case," *World Politics*, 23 (April 1971), 399-430.

Table 7-2
Interventions in Third World Countries

Source of Military Aid	Interventions Rightist	Radical
United States		
0-8 years	50% (N = 4)	50% (N = 4)
9-20	75% (N = 3)	25% (N = 1)
United States[a]		
0-7 years	58% (N = 7)	42% (N = 5)
8-22	66% (N = 6)	34% (N = 3)
Western[b]		
0-8 years	36% (N = 4)	64% (N = 7)
9-22	71% (N = 5)	29% (N = 2)
Western[a]		
0-8 years	47% (N = 7)	53% (N = 8)
9-22	63% (N = 5)	37% (N = 3)

[a]Also received military aid from Eastern or nonaligned countries.
[b]Includes a few countries which did not receive U.S. military assistance.
Source: Appendix 10.

It is not wholly clear why the number of rightist coups declines as the aid relationships is extended. Since the political systems of these regimes were in many if not most instances open to nonmilitary U.S. foreign policy inputs, it may be that over time even more systemic bias was mobilized against the resources and opportunities available to radical elements.[h] The fact that there were fewer ideological interventions *and a sharp decline in radical coups* in countries which did not receive non-Western aid probably reflects the neutralization and elimination of radical factions within the officer corps. On a long-term basis, this as well as the decline in *relative* societal strength and provocative potential of radical civilian sectors would account for the reduction in ideological coups. The high percentage of radical coups for the initial period during which some countries received aid only from Western sources might be accounted for by the legacy of colonial bitterness and the commitment to demonstrate genuine independence.

---

[h]To the extent that civilian institutions and political organizations can be strengthened and induced to both suppress radicals and further "responsible" or "mature" economic policies, an "unstable" situation likely to provoke rightist military intervention is unlikely to develop. This approach was first used to reestablish capitalist hegemony in Western Europe after World War II and then less successfully extended to the underdeveloped areas. A recent statement of the policy appears in footnote a to Chapter Two. For a case study, see my *Cuban Foreign Policy and Chilean Politics* (Lexington, Mass.: D.C. Heath, 1972), Chapters Five, Six, and Eight.

Aggregate Data

The trend data presented in the preceding section are based upon the length rather than the magnitude of the MAP relationship. As indicated in the preceding chapter, some countries sent a relatively small number of officers for U.S. training/orientation tours. Table 7-3 presents a more sensitive or direct indicator of the MAP impact upon Third World officer corps. While approximately 55 percent of the interventions were rightist in those countries with less than 1 percent of their Armed Forces trained in the continental United States, the percentage rises to eighty-six for the category with more than 1 percent. Again as in the previous table, the number of radical coups declines sharply as the proportion of officers trained increases, while the number of conservative coups remains fairly stable. The reasons for this are probably similar to those suggested earlier.[i]

Finally, brief mention should be made of Nordlinger's study of seventy-four MAP-aided Third World countries. He correlated degrees of military dominance between 1957 and 1962 with various quantitative indices of socioeconomic modernization. Except in the most backward category of countries where radical led mass organizations were virtually nonexistent as was any real middle class, the military contributed little (0.05 correlation) to modernization and played a "conservative" socioeconomic role.[j] And even reformist coup leaders frequently move towards greater moderation once firmly entrenched.[6] If this finding goes

Table 7-3
Interventions and CONUS-Trained Proportions

CONUS Trainees (1950-69) as a % of Armed Forces Manpower (1968)	Interventions	
	Rightist	Radical
0.00 – 0.49	3	2
0.49 – 0.99	3	3
1.00 – 1.99	3	0
2.00 +	3	1

Sources: U.S., Department of Defense, *Military Assistance and Foreign Sales Facts: March 1970* (Washington: Office of the Assistant Secretary of Defense for International Security Affairs, 1970), p. 17; Robert C. Sellars, ed., *Reference Handbook of the Armed Forces of the World* (Garden City, N.Y.: Robert C. Sellars Associates, 2d. ed., 1968.)

[i] If the decline in successful radical coups can be attributed to the gradual erosion of such factions, and the failure of conservative coups to rise can be explained by the relative weakening of radical civilian forces which are perceived by rightist officers as threatening, how then can we account for the fact that conservative coups don't also decline? We shall attempt to come to grips with this seeming anomaly in the next chapter since it highlights a dynamic tension which is ubiquitous in the American neocolonial subsystem.

[j] We are not implying higher "modernization" capabilities for undifferentiated civilian regimes in these areas. Quantitative analysis of this problem by Philippe Schmitter at the University of Chicago reveals only minor outcome differences on indicators similar to those used by Nordlinger.

part of the way in discrediting the American imputed "nation-building" role for Third World military establishments, it goes even further in suggesting (1) the necessity for a truly radical break with existing relationships of dependency upon transnational capitalism; and (2) American wisdom in selecting so socio-economically conservative an elite for its hospitality and indoctrination.

Military-Executive Testimony

Defense and State Department officials with access to classified information almost invariably contend that MAP training is an effective source of influence. Thus,

> Lt. Gen. Robert H. Warren, Deputy Assistant Secretary of Defense for Military Assistance and Sales, recently stated that, although he places a high priority on contacting and influencing leaders in foreign societies, 'long experience has indicated that *training* is one of the most productive forms of military assistance investment, in that it fosters attitudes on the part of the trainee which lead to better mutual understanding and greater cooperation in collective security undertakings.[7]

Since Warren's office administers the MAP, a possibly self-serving statement of this nature cannot be regarded as conclusive in and of itself. Almost identical views, however, have been proffered by the State Department's director of politico-military affairs–clearly the junior partner insofar as bureaucratic interests are concerned. According to Spiers,

> our own investigations show that in almost all cases those trained in the United States leave here with at least a very warm regard and respect for us and our institutions. Although we are by no means assured of their support, these trainees are much more likely to understand our objectives and motives than someone who has had little or no contact with us.
>
> I believe that the development of such attitudes on the part of potential leaders is very significant to our foreign policy objectives as these people rise to positions. . . .[8]

As noted in an earlier chapter, Spiers does not regard the promotion of political democracy abroad as a purpose or necessary consequence of controlled exposure to America's "free enterprise" system.[9]

Although there is a willingness to acknowledge that occasionally "hand-picked" future leaders such as Camaano in the Dominican Republic and Yon Sosa or Tureios in Guatemala turn out as undesired revolutionary nationalists, they are classified as *exceptions*: "Unfortunately we have one or two like that, but we think the record is very good."[10] In 1965 General O'Meara attributed "predominant" U.S. control over Latin American military establishments to the MAP.[k]

With respect to the character of coups in underdeveloped countries, American officers are well aware of and unhappy at the fact that some of these have been "radical." It is stressed that only "a few" fall into this category, while "most coups are from the 'right' or from conservative elements." The latter are viewed favorably because "the regime they displace may be extremist leftist."[11] Concern over the ideological direction of coups has been expressed at the highest levels.[12]

Particular rightist coups are frequently cited as evidence of MAP's effectiveness. Thus the overthrow of Sukarno in Indonesia has been directly linked to the training of 1,200 officers and the maintenance of communications with them during the country's period of militant neutralism. Military training paid "enormous dividends" and was the "most productive form of aid."[13] Similarly, according to Pentagon sources, "United States military influence on local commanders was widely considered as an element in the coup d'etat that deposed Brazil's leftist President Joao Goulart in 1964."[14]

Another index of MAP efficacy maybe is the failure of radical military factions to successfully bring off their own attempted coups. Of thirty-seven political crises in underdeveloped countries between 1958 and 1967, twenty-one were abortive coups and military rebellions. Of these, according to Admiral Heinz, fourteen were "put down or forestalled by military establishments receiving some U.S. Military Assistance and which remained loyal to the incumbent regime."[15] In Chapter Six we discussed the importance of rapport and the role of MAAG officers in putting down such coup attempts as occurred in Ethiopia during 1960. Increasing American penetration of and control over Third World military establishments may well account for the secular decline in the number of radical coups which was noted earlier in this chapter.

A third index of effectiveness which was examined in an earlier chapter wherein the discussion focused upon Latin America, pertains to the Americanization of military education in recipient nations. Hence, after more than 1500 Thai military men had received overseas training, the schools of that country

---

[k]Insofar as policy scopes were concerned, a security specialist in the State Department stated in May 1972 that the in-house phrase "broad but shallow" most aptly portrayed this influence over Latin American officers. Over a broad range of issues, then, American MAAGS can utilize good will to encourage or hint at the utility of particular decisions. But a hard sell or too much pressure can easily transform responsiveness into dysfunctional resentment. Our specialist who asked that his identity be concealed was less categorical than this superior Secretary of State Rogers who some weeks earlier informed the House Foreign Affairs Committee that "the record of the last 3 years shows conclusively, I believe, that security assistance works." Taking a global perspective, he particularized the claim by noting the "rapport we have had with the present Government of Cambodia" which was attributed to military assistance. *Foreign Assistance Act of 1965, Hearings*, House, p. 346. U.S., Congress, House, Committee on Foreign Affairs, *Foreign Assistance Act of 1972, Hearings*, before the Committee on Foreign Affairs, 92nd Cong., 2nd sess., March 1972, pp. 2, 8.

were "staffed largely by graduates of USMAP training courses." In resolving to completely reorganize the Naval Academy of Thailand during 1962, Thai officials also committed themselves "to erect modern buildings and overhaul the curriculum . . . to reflect that of the U.S. Naval Academy. . . ."[16] Not coincidentally, a recent congressional report urged that "consideration should be given to establishing in Thailand, with U.S. support, a common facility for training military personnel from allied and friendly nations in Asia."[17] The degree of American influence over the Thai military regime has been so great that "Thai ground forces . . . have been deployed in Vietnam since August 1967."[18]

The view that MAP training as a "key instrument of United States foreign policy . . . has paid good dividends,"[19] is well nigh universal among officers and civilian officials associated with the program. Even when the wrong types of coups occur–those in which the leaders oppose some U.S. policies–it is argued that MAP ex-trainees tend to be more friendly and "favorable to American interests" than officers trained in Eastern countries. The general position is that in the absence of exposure to selected aspects of American society, such military officers would tend to be more nationalistic and/or radical.[1]

Advisory Commission and Congressional Assessments

Like the 1957 Fairless Committee Report, the Draper Report released two years later emphasized the long-run value of military assistance. In vigorously advocating an expansion of MAP political indoctrination and training, the Draper Committee argued that this would increase the likelihood of "stability" which was the "only alternative to domestic chaos and leftist coups" in the underdeveloped areas. The confidence of the Draper Committee in MAP's efficacy was reflected by its disinclination to recommend that primacy be given to economic aid in the struggle against insurgency. In this it differed from the Fairless Committee.[20]

---

[1]The Sudan and Indonesia were considered to be illustrative. Drew Middleton, "Thousands of Foreign Military Men Studying in U.S.," *New York Times*, November 1, 1970, p. 18.

Yet other officials do not share so flexible a perspective. Thus General Warren has viewed the nationalist military regime in Peru as evidence that MAP had been "ineffective" there–perhaps because a large majority of the coup leaders were American trained. In the same context, he referred to the existence of certain neutral regimes elsewhere in the underdeveloped areas. *Foreign Assistance Appropriations for 1970, Hearings*, House, p. 796. Cf., U.S., Department of the Army, *U.S. Army Area Handbook for Venezuela* (Washington, D.C.: GPO, 1964), p. 523.

Another pessimist of sorts is the former deputy director of the CIA, Richard Bissell. While addressing a Council on Foreign Relations "confidential discussion group" on January 8, 1968, he complained that "again and again we have been surprised at *coups* within the military; often we have failed to talk to the junior officers or non-coms who are involved in the coups." *Intelligence and Foreign Policy* (Cambridge, Mass.: Africa Research Group, 1971), p. 12.

Other official bodies have also stressed the utility of the MAP. In the course of reviewing the use of cultural exchange as an instrument of foreign policy, the Committee on Educational Interchange Policy cited the training program's "indirect" and "long period" dividends. Defining military assistance training as a form of cultural exchange, the 1959 Report optimistically noted that the Pentagon was bringing FMTs to the United States at an annual rate of 20,000 with an important "collateral attitudinal impact."[21]

The Clay Committee Report (1963) obliquely acknowledged the long-range impact of military aid on political roles of officers in Third World countries by recommending that–unlike economic aid–it be continued even if recipients failed to establish a hospitable climate for foreign investors.[22] A year later the Committee on Educational Interchange Policy released a study of American military assistance programs which unreservedly endorsed the Draper Committee's finding that " 'there is no single aspect of the military-assistance program which produces more useful returns for the dollars expended than these training programs.' "[23]

Despite the appearance of widespread congressional criticism directed at military aid, this has not focused upon the counterrevolutionary and neocolonial aspects of the program. While the objectives listed in Chapter Two have reflected a firm consensus between conservatives and all save a handful of liberals, they have indeed been occasional manifestations of concern over the global *impact* of the program. Some MAP enthusiasts have demonstrated unusual alacrity by requesting quantitative data which would hopefully disarm their more skeptical colleagues.

Senator Pell. Mr. Secretary, . . . I was struck with your point about the number of coups that had occurred.

Secretary McNamara. I was speaking from memory. There were 31 military coups in the years 1961 through April of 1966.

Senator Pell. Right. Would you be kind enough to submit for the record–it would be too much to ask you to know the answer here–how many of those coups were coups to the right and how many were coups to the left?

Secretary MacNamara. Yes. I tried to examine it. Let me submit a brief analysis of each coup. I could not fully analyze all of them, but 12 occurred in countries for which we had military assistance programs, and 19 did not.

[The information referred to is classified and in the committee files.]

Senator Pell. By coincidence I went through the same exercise with Latin America. My recollection is there were about 10 or 11 in the same period, but there was only one or two to the left, and the majority were to the right in that particular area.[24]

Nor have congressmen generally disagreed with military evaluations of effectiveness.[25] Thus a recent investigation of military assistance training by a Foreign Affairs Committee subcommittee concluded:

The subcommittee study has found two principal effects of the exposure of foreign military trainees to the United States: (1) the formation of some lasting friendships and associations, and (2) the development of more realistic views (positive and negative) of American life and society.

Lasting friendships and associations are a common result of the MAP training program. . . .

. . . .

When the average MAP trainee returns home he may well have some American acquaintances and a fair appreciation of the United States. This impact may provide some valuable future opportunity for communication, as occurred in Indonesia during and immediately after the attempted Communist-backed coup of October 1965.

. . . .

To cite Peru and Libya as examples of the failure of MAP training is to offer an oversimplified and distorted analysis of the complex political considerations involved in both situations.[26]

We see then not only a recognition of the program's effectivenss, but also cognizance of the problem posed by nationalist sentiment which at times escapes the best efforts of the United States to control or neutralize. A study mission from the same House committee which visited Peru similarly concluded that the loss of neocolonial control was partial and not a cause for undue pessimism.

The junta group has related its reform program to the goals of the Alliance for Progress.

. . . .

Despite the attendance of top Peruvian officers at American military schools, despite U.S. influence on Peruvian military thinking and despite the close friendships between members of the general staff and U.S. military personnel, the ascendancy of the military government to power in Peru has resulted in very difficult problems for United States-Peruvian relations. This situation raises a serious question of how much leverage our military assistance training programs really provide when strong national feelings are involved.

. . . .

It is noteworthy that despite all frictions between the United States and Peru, the junta government is still favorable to sending its officers and enlisted men to U.S.-run schools in CONUS and the Canal Zone schools.

Also on an optimistic note, the mission reported that "in more recent months the junta has slowed the pace of reforms and focused on other issues, particularly the need [*sic*] to restore confidence and stimulate domestic and foreign investment, particularly in the minerals field. . . .[27] Needless to say, there is an extreme paucity of congressional criticism of MAP's tendency to encourage *rightist* coups in Latin America.[28] And in the House of Representatives, objections to the use of military aid to insure oligarchic social orders have been voiced by proportionately fewer foreign policy specialists than in the Senate.[29] Most liberal critics–and there haven't been many in either the House or Senate–have endeavored to link the MAP to the fact of military intervention in the Third World. The Defense Department has been more successful in refuting this allegation than it has the equally prominent claim that its aid strengthens repressive dictatorships.[30] For some reason, American officials–as we have seen—appear incapable of perceiving any real meaning in moral arguments. One wonders who are the "true" historical materialists![31]

Expert Opinion

Similar concerns have consumed the energy of many military aid specialists. While systematic analysis has failed to indicate a significant relationship between military assistance and electoral intervention *or economic development,*[32] surprisingly little attention has been directed at the occurrence of particular types of ideological coups. And no quantitative tabulations have been compiled with the exception of Needler's and another by the military historian, Edwin Lieuwen.

According to Lieuwen, since 1930 the "armed forces have represented a static or reactionary social force" in Latin America. Only eleven of fifty-six coups were "reformist." This is traced to a number of factors. The officer corps of these countries have been functionally allied with local oligarchies which are both incapable and ill-disposed toward the introduction of significant social reforms. It is argued that military intervention is often catalyzed by an intensely held simplistic image of communism which has been reinforced by MAP training. A consequential inability to distinguish between "democratic reformers" and Fidelistas triggers the overthrow of the elected government. Thus the MAP actually contributes to the suppression of populists who are the "principal threat to internal security." And "internal security" is equated with the perpetuation of military institutional privileges and career aspirations.[33]

In the Latin American context, both Jose Nun and Irving Louis Horowitz have presented views rather similar to those of Lieuwen. Horowitz's analysis emphasizes the external factor and seems to imply it is highly effective:

> The United States has increasingly come to regard support for the military as insurance for its general political lines, and for the execution of its economic wishes in the area. In nations such as Brazil and Argentina, when 'showdowns' with civilian oriented, public sector dominated regimes did occur, the United States' faith in military conservativism did pay off. . . .
>
> . . . .
>
> While each of the Latin American military elites might employ such themes to justify its own behavior . . . basically they represent supposed United States needs in the area. This supposition is in itself the most decisive aspect of the present situation—namely, the breakdown of neocolonialism and its replacement with imperial politics of a more classic vintage. The present turn to counter-insurgency as a style of politics marks a return to military solutions of economic problems, rather than economic solutions to military problems. While the form of colonialism may be classical, the content is quite new. . . .
>
> . . . .
>
> What has taken place in increasing degrees is the external or foreign management of internal conflicts in Latin America. . . .[34]

Nun's assessment is closer to Lieuwen's which emphasizes the historical institutional roles of the military and their identification with a non-"hegemonic" middle class. Thus the MAP has been effective primarily because of a favorable and receptive domestic environment.[m] Nun goes on to suggest that military missions are particularly influential in the smaller countries while training experiences are more crucial for large ones "where direct institutional manipulation is more difficult to achieve."[35]

Other scholars who have not confined their attention to a particular region also isolate the MAP as one factor among many and conclude that its effectiveness has been no more than moderate. Thus one expert with access to classified data states that because of the MAP and despite ubiquitous nationalism in the underdeveloped areas, U.S. influence "is very strong in many nations." According to Dunn, it varies from time to time and from issue to issue yet there can be no doubt concerning its existence. He acknowledges that anti-Communist officers may be incapable of distinguishing non-Communist progressives from Communists.[36] Another expert on military aid, Edgar Furniss, concluded in the late fifties that because of the MAP, a "slightly more favorable" investment climate had been structured, while in the Middle East there had been a failure to induce "unwilling recipients" to accept U.S. military aid. Only Turkey, Iran, and Iraq had accepted the program, and in these countries it had paved the way for economic aid and consequential U.S. influence over economic policy-making.[37]

More recent studies of MAP's political efficacy *have ignored* its relationship to nationalistic economic or foreign policies. Appraisals of "success" with regard to undifferentiated influence over foreign officers are always qualified and

---

[m]A similar emphasis is central to a study of militarism in Brazil and the Dominican Republic by a former Rand researcher. After stressing intrasystemic factors, Kaplan notes that externally induced anti-Communist socialization resulted in unduly expanded definitions of Communism which heightened military antipathy toward the reformist and nonsuppressive Bosch and Goulart regimes. In the latter case, counterrevolutionary fanaticism was directly linked to the U.S. established ESG (Brazilian War College). Given the cumulative and intensity laden nature of the value conflicts, the author argues that MAAG/Attaché encouragement was unnecessary to catalyze these interventions–though he does not deny that such advice was given.

"Had there been no U.S. mission at the ESG and no military aid over the years to keep up the status of U.S. military men and the sense of identity with the United States, the FEB linkage may not have been built upon and the Brazilian ESG group may have been somewhat more relaxed regarding signs of Communism and not had such a cold war orientation. As a result, then Col. Vernon Walters–who was an Army attaché rather than a member of the U.S. training mission–may not have been able to retain a close relationship with Brazilian officers with whom he had served as a liaison to the FEB in World War II. Nevertheless, even if Walters explicitly expressed Embassy support for a coup, as the situation had developed by 1964 the military did not need nor would they have waited for Washington's approval." The author is apparently unaware of the fact that since mid-1963, U.S. agencies had been deliberately fomenting "the situation [which] had developed by 1964." Stephen S. Kaplan, "United States Military Aid to Brazil and the Dominican Republic: Its Nature, Objectives and Impact," (paper prepared for the Arms Control and Foreign Policy Seminar, Center for Policy Study, University of Chicago, May 26-27, 1972), p. 45.

generally regarded as quite modest. There may be some understatement since none of the establishment-subsidized experts ever advocate abandonment or even curtailment of the training program or its concomitant political indoctrination. Quantitative analyses by Amelia Leiss and Geoffrey Kemp have revealed a strong correlation between the proportion of a country's officers trained and the ability of the United States to control weapons acquisitions, force levels, and success in the Americanization of foreign military doctrine and tactics.[38] This is consistent with our own comparison of proportions trained and the ratio of rightist to leftist coups.

One expert who is careful to avoid any exaggeration of MAP's impact has concluded that a new "kind of military culture that cuts across national lines" is probably structured for only two categories of trainees: (1) the 3,200 foreign officers who took command and general staff courses with U.S. officers; and (2) the estimated 62,000 (out of 312,000) FMTs who experienced more than one tour of U.S. MAP training.[n] With the exception of these 65,200 officers, Lefever contends that "the kinship created by the large number of men trained by America is not much deeper than vocational identity." Nevertheless, he also notes:

According to interviews with returnees and others who have observed the program, it appears clear that the great majority come away with greater understanding of the United States, and many times the understanding borders on some degree of affection, especially among those who have been here for periods of 6 to 12 months.

Yet Lefever goes on to deny the existence of any proof one way or another that MAP has contributed to any kind of coups in Third World countries.[39]

Another skeptic whose approach appears quite impressionistic has argued that the MAP is simply incapable of subverting nationalist aspirations and in some cases has even contributed to them. Thus Ross K. Baker has testified:

I would submit that the cause of sustaining friendly governments through training assistance was not demonstrated by the events. Nor, I might add, was the objective of creating a military elite favorably disposed toward the United States.

The Libyan military, if anything, has become one of the most avowedly anti-Western in the region. Against such powerful intangible forces as Arab nationalism, U.S. military training assistance seems to possess little cogency.[40]

After interviewing two hundred Latin American officers, Alfred C. Stephan concluded that the American decision to force civic action roles on Latin American forces had in fact engendered considerable nationalist resentment

---

[n]Similarly, "evidence suggests that these [VIP orientation] tours, especially if they follow or precede one or two actual U.S. training periods, have been very valuable." *Military Assistance Training, Hearings*, House, pp. 7, 34-5.

because of its implicit denigration of their traditional role as defenders of the nation.[41] He is not alone in viewing counterinsurgency training as a source of politicization of the military in Third World countries.[42]

Particular Interventions

The problem of direct proof to which Lefever alluded is in most cases an insoluble one for the independent investigator. For obvious reasons *the roles of MAAG and/or CIA operatives* are concealed whenever possible. Hence, one must usually settle for such hearsay as that of Eliezer Be'eri:

> These interventions, of every kind and from many quarters have been numerous. However, they are intrinsically difficult to reveal and even more difficult to prove. Not everything that has happened is known, and not everything told is necessarily true.[43]

Occasionally, experts or government officials do refer to such complicity or encouragement by MAAG officers. This has occurred for overthrows of radical nationalist regimes in such countries as the Dominican Republic (1963), Iran (1953), Indonesia (1966), and Brazil (1964).[44] In the following paragraphs we shall summarize the policy changes which resulted from these four coups against elected governments in these four nations. Terrorism and/or military dictatorship are today characteristics of the political regimes of these countries.

Brazil, it will be recalled, was the only Latin American country that furnished army units to aid the United States in its invasion of the Dominican Republic during 1965. A year earlier, prior to the coup,

> certain U.S. business objectives—a deal for the Hanna interests, a deal for the American & Foreign Power Company, the course of profit-remittance legislation, and the imposition of an investment-guarantee treaty—were being effectively balked by the constitutional government of Brazil. Accordingly, the State Department leaked to the press its decision to modify the policy of rigid hostility toward military juntas, and alerted thereby the military everywhere that military seizures of political power had become an acceptable substitute for constitutional procedures. . . .
>
> Contrary to the popular impression, the Defense Department wanted it known, the change-over in Brazil as a result of the military coup actually should be referred to as an 'election for the new president!' . . .
>
> The 'Program of Economic Action of the Government,' outlined by Roberto Campos, predicted that, in response to this (post-coup) policy of stimulation and protection of foreign investment, capital would flow in from abroad to give an impulse to the development of Brazil and contribute to its economic and financial stabilization. One hundred million dollars of new direct investment of foreign origin was predicted for 1965. Seventy million came. One hundred twenty million was predicted for 1966. Seventy-four million appeared. And in

1967, a new investment totalled $70 million. In this same period, the firms remitted much larger amounts to their home offices in earnings and dividends: in 1965, $102 million; in 1966, $127 million. And in 1967 the outflow of earnings and dividends was almost double the new investment: $130 million fled. To this total must be added the huge sums paid out for management, technical assistance, patents, royalties, use of trade-marks, and import commissions; in 1967, the firms extracted $170 million under these headings.

Meanwhile, the external debt grew until it reached $4 billion; interest and amortization payments also grew, amounting to an overflow of $500 million in 1968. In Brazil, as everywhere in Latin America, the loans received under the terms of the Alliance for Progress required the purchase of goods in the United States (where the most expensive machines in the world are sold) and their transport in ships with North American flags (which charge an extra $10 per ton of freight unless insurance is taken out with North American firms). Brazil paid $100 million in 1967 for freight and insurance. And to all these drains the illegal remittances must still be added. In its 1968 report the Central Bank admits that $180 million left Brazil extralegally in 1966 and $120 million in 1967.[45]

The military regime had of course renounced the "independent" foreign policy which had been followed by the elected governments headed by Quadros and Goulart. Among the core groups of generals active in the 1964 military government *80 percent were U.S. trained*, while only 21.7 percent of the nonparticipant generals had received this indoctrination.[46]

The case of the Dominican Republic did not involve more than an incipient independent or nonaligned foreign policy. Although Juan Bosch was a firm anti-Castroite and a prominent supporter of the U.S.-sponsored Alliance for Progress, his domestic policies like those of Goulart fell afoul of the investing corporations.

As a skeptical business community looked on, Bosch negotiated a $150 million line of credit with a Zurich-based consortium to finance his larger development projects, a departure from the usual U.S. sources. Next he cancelled an oil refinery contract, which Esso, Texaco, and Shell had negotiated with Trujillo and the Consejo, because of the large profits which would leave the country.

Despite these initiatives toward financial independence, Bosch found it necessary to seek United States aid. The Kennedy Administration hoped to make the Dominican Republic a "showcase of democracy" as a counterweight to Cuba. Nevertheless, Bosch found U.S. financial backing hard to obtain. Former businessman Newell Williams, AID director for the Republic, commented: "Ever since Bosch has been in, we've been turned down. Later, toward the end of Bosch's first one hundred days in office, Ambassador Martin noted, "We had committed something over $50 million to last year's Consejo but not a cent for Bosch."

Several aspects of his administration alarmed native and United States investors. An effort was made to rescind several sugar contracts negotiated by the Consejo. While advantageous for U.S. sugar purchasers, these contracts meant a loss of several million dollars in foreign exchange for the Dominican Republic. Former Ambassador Martin wrote:

The Department [State Department] instructed me to inform the Dominican

Government that its failure to honor legitimate contracts with U.S. sugar firms would certainly have more serious repercussions for the Bosch government and might even lead to an invocation of the Hickenlooper amendment which would end AID to the Republic.

Within a week of Bosch's inauguration, *Business Week* attacked him for proposing a "revolutionary constitution" and land reform "which would prohibit operations of U.S. owned sugar companies." An official of La Romana expressed the fear that Bosch's government might make the Puerto Rican "mistake" of subdividing the cane lands. "To break up the Company's lands would wreck Romana." Concern over constitutional provisions about business operations was a constant theme in discussions between U.S. representatives and all Dominican governments. Much later during the U.S. invasion and occupation of the Republic in 1965, U.S. negotiators demanded that the "rebels," who were fighting for a reinstatement of the 1963 Constitution, modify several of its articles. They were especially troubled by Article 19, giving workers the right to profit sharing in both agricultural and industrial enterprises; Article 23, prohibiting large landholdings; Article 25, restricting the right of foreigners to acquire Dominican land; Article 28, requiring landholders to sell that portion of their lands above a maximum fixed by law, excess holdings to be distributed to the landless peasantry; and Article 66, prohibiting expulsion of Dominicans from their own country. American negotiators in 1965 proposed an amendment to exempt owners of sugar plantations and cattle ranches, the largest of these being South Puerto Rico Sugar's La Romana.

• •

In short time, inflated anti-Communism flourished and began to undermine Bosch's legitimacy. Even civil-liberatarian Ambassador Martin kept constant pressure on Bosch with advice like the following:

I recommend Bosch couple any changes with other measures–repeal of the old de-Trujilloization law and enactment of a law providing for the trial of military personnel by military tribunals. . . . I recommended that Congress adopt a resolution declaring it the sense of the Congress that communism was incompatible with the Inter-American system . . . and that it enact a Dominican version of the Smith act. . . . I recommended he hold back on agrarian reform if it entailed confiscation laws.[47]

Apparently, Bosch took the goals of the Alliance for Progress too literally. Not only did he fail to pay sufficient heed to the open door (also an Alianza goal), but his egalitarian reformism was as unacceptable to U.S. corporate investors as it was to the local oligarchic sectors. In addition, he failed to fully respect military autonomy. Hence, his elected government was deposed on September 25, 1963. According to Lieuwen, "the United States-trained police joined the Army in ousting President Bosch, following which both the police and the anti-guerrilla units, trained during 1963 by a forty-four man United States Army Mission, were used to hunt down Bosch's non-Communist partisans in the name of anti-Communism."[48]

The military installed a so-called triumvirate government to replace Bosch who had fled into exile. What of its policies?

The subsequent Dominican government was headed by Donald Reid Cabral, former auto dealer, close friend of U.S. business and member of the oligarchy. Realizing his power rested with the military, Reid turned his back on the officers' contraband and smuggling operations, thereby winning their tacit support.

The Reid government put an end to economic and social development. With the shelving of the 1963 Constitution, the Bosch reform legislation, and the Zurich consortium credit, large numbers of United States businessmen turned again to the Dominican Republic. George Walker (of the Mellons' Koppers Company), a close Reid friend, had made a visit to the Dominican Republic during the Consejo period (Reid was one of the seven Consejo members) on behalf of the Businessmen's Council on International Understanding. Walker "had brought in high-level U.S. industrialists to study the former Trujillo properties and advise the Consejo what to do with them." It was from these operations that Bosch had hoped to partially finance his social reforms. But Walker recommended gradually selling or leasing the properties to private investors.

The Midland Cooperatives of Minneapolis, Minnesota, landed an oil refinery contract. Falconbridge Nickel Mines, a Canadian-American concern, announced plans to build a $78 million refinery. The Inter-American Development Bank and the AFL-CIO launched a joint housing project. The World Bank gave $1.7 million for a hydroelectric study, replacing the one Bosch intended to finance with European capital.

And there was more. At the same time, plagued by plunging world sugar prices, Reid continued Bosch's austerity program and received International Monetary Fund (IMF) and AID credit and loans as well as a $30 million loan from six U.S. commercial banks.

During the Bosch regime the 18,000 workers at South Puerto Rico's La Romana had won a 30 percent increase in their meager wages. Under Reid, that contract was broken and in February, 1965, one much more favorable to the company was negotiated with a parallel or "ghost" union set up by La Romana. Since low wages are crucial for profits in this marginal industry requiring surplus labor, the company greatly benefited from Reid's antilabor policy. Reid had previously declared a state of seige in Santo Domingo to offset a general strike called by the Sindicato Independiente de Choferes. By spring of 1965 the country was $200 million in debt; there were rumors of a Reid campaign to sell the former Trujillo properties, now in the hands of the state, to private investors; the armed forces were carrying on increasingly flagrant and publicly known contraband operations; and Santo Domingo itself was facing a serious water shortage—Reid had discontinued the renovation and expansion of the city's water system initiated by Bosch. One other persistent rumor was that Reid was planning to rig the promised elections to insure his victory. More than one commentator was led to observe that only U.S. support kept Reid in power.[49]

While *de jure* (but not *de facto*) relations were suspended between the U.S. and Santo Domingo for eighteen weeks after Bosch was deposed, this was largely a *pro forma* exercise in public relations as can be seen from the economic aid provided to that regime by the United States and international organizations over which it exercises considerable leverage.

There have also been "domino effect" cases when U.S. corporate interests were little more than symbolically "threatened" by nationalistic Third World regimes. As in Greece, when other neocolonial Western powers appeared incapable of maintaining control, the United States has opted to "carry the ball." Thus four years after the Truman Doctrine was proclaimed, an Iranian nationalist movement led by Premier

Mohammed Mossadegh had undercut the power of the Shah and proceeded to nationalize the Anglo-Iranian Oil Company. The British government had received more taxes from the company than the Iranian government had received for its own natural resource, and the company consequently provided a very convenient target in an impoverished land where five hundred babies died out of every thousand births. The British demanded payment for the confiscated holdings, a demand that the Iranians could not meet without binding themselves to foreign lenders. With oil exports at a standstill, the Iranian economy began to sink, since income from oil provided thirty percent of its total income and sixty percent of its foreign exchange. When Eisenhower entered office, the United States, in spite of extensive efforts by Acheson, had not been successful in acting as a mediator.

After a three-week trip through the Middle East in May 1953, Dulles reached some disturbing conclusions. Western power had "deteriorated" in the area, he believed, and unless drastic action was taken, the Arab nations would become "outright" neutrals in "the East-West struggle." Israel and intra-Arab squabbles accounted for some of the problems, but Dulles also wondered about the British. "They interpret our policy as one which in fact hastens their loss of prestige in the area. To some extent," the Secretary admitted, ". . . this may be true," but Great Britain's loss of power was also due to "altered world power relationships." Dulles decided that he would have to convince the Middle East that the United States had little to do with British and French colonialism. The opportunity came within the next two months when the State Department concluded that Mossadegh was moving into the Soviet orbit. Rumors of a Soviet-Iranian loan began to circulate, and in August, Mossadegh received 99.4 percent of the votes in a plebiscite, a percentage which Eisenhower later used as proof of increased Communist influence. Having earlier refused to help Mossadegh rebuild the Iranian economy, the United States now cut off all aid.

In August the Shah staged a successful *coup* to regain power. The United States provided guns, trucks, armored cars, and radio communications for the Shah's forces. The new government quickly undertook discussions with representatives of the oil company, but the representatives were not those of the year before. Since the turn of the century the United States had been trying to get into the Iranian oil fields only to be constantly repulsed by the British. Now the breakthrough occurred by the grace of the Shah and under the guidance of the State Department official Herbert Hoover, Jr., who had gained wide experience in the complexities of the international oil problem as a private businessman. A new international consortium was established giving the British 40 percent, five American firms (Gulf, Socony-Vacuum, Standard Oil of California, Standard Oil of New Jersey, and Texaco) 40 percent, and Dutch Shell and French Petroleum the remaining 20 percent of Iranian oil production. Profits would be divided equally between the consortium and Iran. Iranian oil once more freely flowed into the international markets, the Shah's government was securely within the Western camp and the British monopoly on the oil fields had been broken. For Dulles and Eisenhower, it was one revolution with a happy ending. More accurately, it was not a revolution at all.[50]

Our fourth case involves the most populous and potentially rich nation of South-East Asia. Before the overthrow of Sukarno, Indonesia vigorously opposed neo-colonialism and had expropriated resources which had formerly been exploited by Western multinational corporations. Although military training for the army was continued, the United States terminated virtually all economic aid after 1963.

When Indonesia's General Suharto seized state power after the Army-directed program against the Communist party in 1955-6 and the 1966 overthrow of Sukarno, he brought with him a group of American-trained economists who have since launched a 'New Economic Policy' that has opened Indonesia's once nearly closed economy to heavy investment by multinational Corporations. A consortium of Indonesia's capitalist creditors has rescheduled the nation's debts and granted it big new loans. But the heart of the government's new economic policy is an attempt to attract foreign investment. Companies may invest in any area of the Indonesian economy except national defense, and foreign capital may own one hundred percent of any enterprise except in the areas of shipping, railroads, utilities, and the mass media. Indonesia levies no duties on imported fixed assets, raw materials, or semi-finished goods.

As if these conditions were not attractive enough to foreign investors, the Indonesian government last fall liberalized the tax laws, reducing corporate income taxes from a maximum of sixty percent to a maximum of forty-five percent. The Law on Foreign Capital Investment of 1967, which allows a tax holiday up to five years, may be extended up to six years for some companies. Foreign investors may choose between the tax holiday and an investment allowance deduction on profits before taxes.[51]

As one would suspect, consortia and multinational corporations have moved quickly to assume control over the most dynamic and potentially profitable Indonesian resources.[o]

These cases have been discussed only because there is available some concrete evidence indicating MAAG involvement. We doubt that they are atypical of many rightist coups. Three of them involved not only an opening of the door to foreign investor interests but not coincidentally a shift to a pro-Western (Indonesia) or pro-American (Brazil and Iran) foreign policy.[p] Since the Dominican Republic had not followed an "independent" foreign policy under

[o]Two years before the military in-fighting which resulted in the overthrow of Sukarno's Nationalist-Communist coalition government, Congressman Barry and an AID assistant administrator for the Far East agreed that Jakarta needed the "free enterprise system." Following the consolidation of the new right-wing military dictatorship, CINPAC Commander in Chief Admiral McCain testified that "as far as Indonesia is concerned, General Suharto has brought a stable government to that country. In addition, there are millions of dollars in private investment going into Indonesia including some from our own country." *Foreign Assistance Act of 1963, Hearings*, House, p. 775. *Foreign Assistance Act of 1969, Hearings* House, p. 613. Cf., "Indonesia: The Making of a Neo-Colony," *Pacific Research and World Empire Telegram*, 1, no. 1 (1969), pp. 6-16.

[p]Indonesia forced a rupture in relations with China, deactivated and purged its Soviet trained Air Force and Navy, and terminated its struggle against Malaysia. Brazil broke relations with Cuba and joined the U.S. invasion of the Dominican Republic after radical nationalists attempted to restore Bosch to the presidency. Iran became a formal military ally of the United States and joined the U.S. sponsored CENTO.

Bosch, the principal changes involved domestic policy. All four illustrate the criteria used in classifying coups as rightist for the quantitative tables that appear in the first section of this chapter. They are also consonant with the MAP foreign policy goals delineated in the first two chapters of this study. In each country non-Communist radicals as well as Marxist-Leninists have been systematically suppressed.

It is probable that the MAP has been one of the more effective instruments of American foreign policy. Although its efficacy would seem to be affected by the proportion of a particular country's officers who had been trained and to a lesser degree upon the length of time provided for the structuring and reinforcement of intermilitary "rapport," it seems reasonable to conclude that it has played a moderate role in limiting the number of nonaligned and socialistic regimes in the Third World. In saying this we do not underestimate the difficulties in manipulating the political behavior of foreign officer corps, nor do we depreciate the multiplicity of factors which condition their behavior. These along with trends in military assistance will be examined in the final chapter.

8 Conclusions

> The doctrine of nonintervention was developed during an earlier period. In recent decades those circumstances have changed markedly. The relative inviolability of nations has declined and intervention, in various forms, has become common place. Powerful nations do not observe the principle of nonintervention, and cannot, for the exercise of leadership [*sic*] often involves intervention—economic aid, technical assistance, alliance activities, military training missions, cultural exchange programs and so on.[1]
>
> A.M. Scott, 1968

> There is a danger that fashionable terms like 'self-sufficiency' and 'Nixon doctrine' will become rhetorical tokens which may communicate vastly different things to different people, or, in fact, have no real meaning at all.[2]
>
> Staff Report for the Subcommittee on National Security Policy, 1971

> Overall, the critics of military aid policy remain weak, scattered, and inconsistent, and there is no single line of opposition.[3]
>
> L. Siegel, 1971

Trends in Military Training

Unlike equipment, the cost of training foreign officers has remained almost static since the mid-fifties at approximately $2,500 per man.[4] During the same 1955 to 1970 period, the share of total MAP costs allocated to training has risen from 2 to about 4 percent.[5] Hence, despite the paucity of uncensored testimony clearly demonstrating a high level of attainment in securing foreign policy goals, the relatively low cost of training may explain congressional readiness to act upon faith in supporting a continued expansion of the program. That something more than "faith" is involved, however, is implied by the admittedly inconclusive evidence recounted in the preceding chapter. In any case, while equipment grants and economic aid have been curtailed in recent years, the provision of training has actually increased (see Figure 8-1).[a] During this same period, an effort has been made to channel more CONUS officer trainees into "profession-

[a]For FY 1971, in excess of 16,000 FMTs were slated for CONUS training spaces. This represents a 50 percent increase over 1970 as indicated in Table 8-1. The latter in turn omits Thailand and Indochina. Drew Middleton, "Thousands of Foreign Military Men Studying in U.S.," *New York Times*, November 1, 1970, p. 18.

Figure 8-1. Military Assistance Program

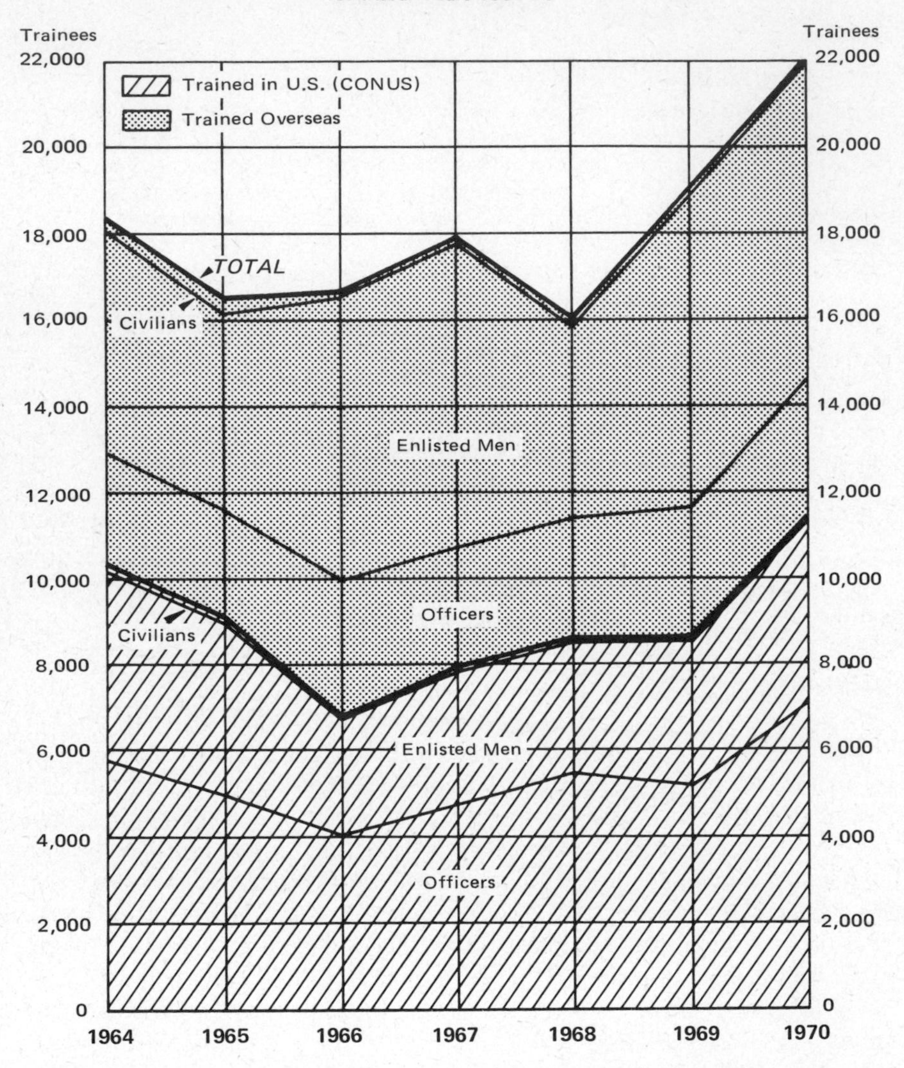

Source: U.S., Congress, House, Committee on Foreign Affairs, *Military Assistance Training, Report,* of the Subcommittee on National Security Policy and Scientific Developments to the Committee on Foreign Affairs, House of Representatives, April 2, 1971, p. 8. Based upon Defense Department figures.

[a]Includes MASF trainees.

al" rather than technical specialty courses. At the present time, "the orientation/influence and generally professional courses attract over half the students trained in the United States."[6] Among officers then, as many as two-thirds may be exposed to a high degree of ideological indoctrination at CONUS installations. Enlistees, it will be recalled, are virtually always programmed for technical—usually unit—training.[b]

It should be borne in mind that the new Defense Department "Information Program" was fully systematized and institutionalized at the training installation level only during 1966-70 period. Given this fact and the tendency to increase the officer proportion of FMTs being brought to the United States, it is likely that the program's impact upon the political roles abroad will grow during the mid and late seventies. The Nixon Doctrine, it should be noted, envisages further increases in the number of foreign officers trained. As one administration spokesman recently put it, "the people and Congress must come to understand that they cannot have it both ways—cut back overseas troop commitments and cut MAP. As the first goes down, the second must come up. . . ."[7] This assumes, as we do, an inflexible foreign policy commitment to oppose the rise of radical nationalist and socialist regimes in the Third World."[c] Similarly, it is based upon the premise that China, other Communist nations, and several radical Third World regimes will in varying degrees encourage such tendencies and movements.

American policy-makers have only recently become cognizant of the need to specially tailor nontraining aspects of the MAP to *an increasingly nationalist environment.*[d] Thus MAAG contingents are being given a low profile through reductions in size "particularly where anti-American sentiment is prevalent."[8] Similarly, both the Rockefeller Report and the Subcommittee on National Security Policy and Scientific Developments have urged other changes in order to make MILGRPs *less conspicuous*. The former recommended that the United States

[b]Thus, a recent U.S. Army Regulation which implements DOD Directive 5410.17. "An Informational Program for Foreign Military Trainees and Visitors in the United States, January 15, 1965," stipulates that "the Informational Program essentially applies to foreign officer personnel. The inclusion of enlisted personnel in Informational Program activities is encouraged when considered appropriate by the installation commander." U.S. Army, *Foreign Countries and Nationals: Training of Foreign Personnel by the U.S. Army* (Washington, D.C.: Headquarters, Dept. of the Army, Army Regulation AR 550-50, October 1970, Effective 1 December 1970), pp. 1-1, 2-18.

[c]This applies to Nixon and quite probably to a hypothetical McGovern administration. Instructive sources include: U.S., Congress, House, Committee on Foreign Affairs, *Foreign Assistance Act of 1972, Hearings*, before the Committee on Foreign Affairs, 92nd Cong., 2nd sess., March 1972, pp. 3, 49, 134; Suzannah Lessard, "What's Wrong with McGovern," *Washington Monthly*, June 1972, pp. 6-16.

[d]Derived in great measure from an ever more widespread unwillingness to accept the subordination and underdevelopment which flows from transnational corporate exploitation of Third World resources, the drive for national sovereignty constitutes the basic tension or dynamic psychocultural variable in the Western neocolonial system. Demographic trends and the failure of non-Socialist societies to close the technological gap reinforce nationalistic appeals at elite and counterelite levels.

should provide, on request, military and technical training missions but should no longer maintain the permanent military missions in residence in other nations which too often have constituted too large and too visible a United States presence.[9]

Similarly:

By physically removing themselves from the host military commands, the MAAGs and Milgroups can contribute to several practically and psychologically important objectives, including (1) enhancing the self-confidence of the host military, (2) discouraging over-identification of U.S. personnel with their foreign military counterparts, and (3) lowering the U.S. 'profile' generally in the military affairs of MAP recipient countries.[10]

The low profile has also been implemented through renaming MAAGs as "Defense Liaison Groups" (Indonesia), "Military Equipment Development Teams" (Cambodia), or as the "Morocco-United States Liaison Office" in Rabat. Along similar semantic lines, the Rockefeller Report advocated that the MAP be retitled the "Western Hemisphere Security Program."[11] This recalls the renaming of the U.S. Air Force School of the Americas as the Inter-American Air Force Academy. While a low profile has long characterized the "soft-sell" approach to CONUS political indoctrination of trainees, it is now being extended to overseas military representation.

Although it can hardly be described as innovative, new priority will be assigned to inducing third countries to carry out Washington's strategic objectives. Under the Nixon Doctrine this will involve the substitution of foreign armies for American casualties at the cost of increased military assistance in the form of grants. The latter will be partially offset by eliminating all grant aid to allies with viable economies. Additional training both in the United States and friendly allies will be mandatory to maintain neocolonial control and to enhance surrogate military proficiency.[e] Thus Japan, France, New Zealand, Australia, and Britain are currently contributing to the United States effort in Laos and a similar group including Indonesia are endeavoring to save the rightist Lon Nol regime in Cambodia.[12] One prominent member of the House Foreign Affairs Committee has characterized this as another dimension of the low profile,[13] which itself mirrors the approach of the CIA.

Ironically, in both Cambodia and Laos the CIA has been intimately involved

[e]Hence: "Consideration should be given to establishing in Thailand, with U.S. support, a common facility for training military personnel from allied and friendly nations in Asia. . . . The proposed regional training facility in Thailand would appear to further several American aims. It would permit the United States to adopt a 'lower profile' in training Asians, especially in the operation and maintenance of MAP-supplied equipment. It would promote Asian regionalism and military cooperation. And, finally, it would promise some long-term financial savings when compared with costs of CONUS training for comparable numbers of troops." *Military Assistance Training in East Asia, Report*, House, p. 26.

in military assistance efforts and the new emphasis would seem to imply an even more prominent role for that functionally specialized agency. In a slightly different context, former Deputy CIA Director Richard Bissell articulated the essence of the Nixon Doctrine's third-country, low-profile orientation several years ago when he counseled a "special study group" of the quasi-official Council on Foreign Relations:[f]

The United States should make increasing use of non-nationals, who, with effort at indoctrination and training, should be encouraged to develop a second loyalty, more or less comparable to that of the American staff. As we shift our attention to Latin America, Asia, and Africa, the conduct of U.S. nationals is likely to be increasingly circumscribed. The primary change recommended would be to build up a system of unofficial cover; to see how far we can go with non-U.S. nationals, especially in the field. The CIA might be able to make increasing use of non-nationals as 'career agents' that is with a status midway between that of the classical agent used in a single compartmented operation perhaps for a limited period of time and that of a staff member involved through his career in many operations and well informed of the Agency's capabilities. Such career agents should be encouraged with an effort at indoctrination and training and with a prospect of long-term employment to develop a second loyalty and they could of course never be employed in ways that would conflict with their primary loyalties toward their own countries. This still leaves open, however, a wide range of potential uses. The desirability of more effective use of foreign nationals increases as we shift our attention to Latin America, Asia, and Africa where the conduct of United States nationals is easily subject to scrutiny and is likely to be increasingly circumscribed.

These suggestions about unofficial cover and career agents illustrate and emphasize the need for continuing efforts to develop covert action capabilities even where there is no immediate need to employ them. The central task is that of identifying potential indigenous allies—both individuals and organizations—making contact with them, and establishing the fact of a community of interest.

There is some room for improvement, Mr. Bissell thought, in the planning of covert action country by country. Covert intervention is probably most effective in situations where a comprehensive effort is undertaken with a number of separate operations designed to support and complement one another and to have a cumulatively significant effect.[14]

---

[f]Thus a former member of CIA Director Richard Helms' staff has recently commented upon the linkage between the CFR and White House decision-making: "the reorganization of the U.S. intelligence community late last year in no way altered the CIA's mission as the clandestine action arm of American foreign policy. . . . The extensive review conducted by the White House staff in preparation for the reorganization drew heavily on advice provided by the CIA and that given by former agency officials through such go-betweens as the influential Council on Foreign Relations. Earlier in the Nixon Administration, the Council had responded to a similar request by recommending that in the future the CIA should concentrate its covert pressure tactics on Latin American, African and Asian targets, using more foreign nationals as agents and relying more on private U.S. corporations and other institutions as covers." Victor Marchetti, "CIA: The President's Loyal Tool," *Nation*, April 3, 1972, p. 431. On the relationship of the CFR to the corporate owning American upper class, see G. William Domhoff, "Who Made American Foreign Policy, 1945-1963?" in *Corporations and the Cold War*, David Horowitz, ed. (New York: Monthly Review Press, 1969), pp. 25-69.

The cumulative separate operation approach to affecting the "internal power balance" parallels the American strategy of simultaneously endeavoring to shape a variety of civilian as well as military institutions within penetrated Third World countries.

It is unlikely then, that the Nixon Doctrine's emphasis upon "self-sufficiency" can amount to more than a symbolic facade if the MAP is to continue as an effective foreign policy instrumentality. The problem is epitomized by the success of the Japanese in attaining technical military "self-sufficiency." Because they now see no need to buy training from the United States, a recent congressional report warns that unless it is given to Tokyo gratis, the United States will inevitably lose its leverage over Japanese military leaders and ultimately that country may follow a truly independent foreign policy. This in turn would adversely affect American political, economic, and strategic goals for Japan.[15] Ultimately then the Nixon Doctrine cannot be reconciled with the universally acknowledged political goals of the military aid program. If viewed as a public relations device for obtaining higher military aid appropriations from the American Congress, however, it might be functional for the MAP. Furthermore, this need not become generally apparent because the socioeconomic and political regimes prevalent in most Third World countries are unlikely to attain the degree of self-sufficiency which characterizes Japan. Thus South Korea and Taiwan—two of the four "forward defense" countries which have over the years received about 70 percent of American military aid—are so dependent upon the United States for training that self-sufficiency is not even envisaged.[16] Ironically, neither North Korea nor China have been dependent upon the Soviet Union for some years.

Although the Vietnamese imbroglio, crass deceit, and overextension of American resources in recent years have catalyzed congressional criticism of military aid, no serious effort has been made to curtail the training program. While official MAP figures indicate a decline in the late sixties and early seventies, this is spurious because trainee authorizations for various countries simply have been transferred to the Armed Services from the Foreign Affairs Committees. As we have seen, the number of FMTs has actually increased.[17] Although the 1969 Fulbright Amendment to section 510 of the Foreign Assistance Act limited the number of MAP grant aid trainees to the number foreigners brought to the U.S. during the preceding year under the State Department's Cultural Exchange Program, this provision was intended to increase the latter category rather than reduce the former.[18] The restriction moreover

does not apply to civilians trained under this authorization, nor does it apply to the military trainees of South Vietnam, Laos, and Thailand, an exception that is not insignificant. Training of students from these three countries in the United States under the Military Assistance Service Funded (MASF) program has been much more generously funded than the MAP training. In fiscal 1970,

$14,857,000 was spent under MAP for student training in the United States compared to $44,975,000 under MASF. In fiscal 1971, $13,537,000 was spent under MAP for the same MAP programs and $26,936,000 for similar MASF training.[19]

While there have been some across-the-board cuts in MAP authorized training, it would seem that temporary budget reallocations necessitated by the Cambodian intervention were more influential than section 510.[20] Thus the FY 1970 reduction of the CONUS $250/trainee MAP Information Program expenditure was fully restored the following year.[21]

Criticism by the congressional "dove" minority has prompted the Foreign Affairs Committee to undertake an investigation of the MAP. The committee's report from which we have liberally quoted, does not advocate a curtailment let alone a reduction in military training. On the contrary, its recommendations are as follows:

1. Section 510 of the Foreign Assistance Act which limits the number of foreign military students to be trained in the United States should be eliminated.
2. Lengthen tours for MAP personnel abroad from 2 years with an option for a third, to 3 years with an option for a fourth.
3. U.S. military missions abroad should be operated on as integrated a basis as possible, emphasizing interservice cooperation.
4. Co-location of U.S. military missions in the military headquarters of the host country should be ended.
5. Orientation tours should be given to foreign officers on an individual basis only.[22]

Like section 510, other restrictions governing the military aid program since 1967 have been designed primarily to limit presidential or executive autonomy by institutionalizing greater responsiveness to congressional oversight and concomitant vulnerability to congressional criticism.[g] Although most of these

---

[g]These limitations introduced during FY 1968 and FY 1969 encompassed: (1) ending the use of repayment-based revolving fund to finance military sales; (2) setting a $190 million maximum for sales; (3) termination of DOD guarantees for Eximbank loans; (4) the Conte-Long Amendment to the Foreign Assistance Act of 1967 which terminates economic aid to most underdeveloped countries should they purchase "sophisticated" weapons without a presidential determination of "security" need; (5) the Symington Amendment to the Foreign Assistance Act of 1967 directs the president to terminate economic aid if an underdeveloped country is diverting funds "to unnecessary military expenditures to a degree which materially interferes with its development." In general, during the 1960s an increasing proportion of military "aid" has been in the form of credit sales rather than grants. Ironically, this trend has been dysfunctional for "development" goals. See Harold M. Hochman and C. Tait Ratcliffe, "Grant Aid or Credit Sales: A Dilemma of Military Assistance Planning," *Journal of Developing Areas*, 4 (July 1970), 461-76.

enactments have involved funding for weapons systems and supplies, only one in addition to 510 was directly related to training. Thus an amendment to the Foreign Assistance and Related Appropriations Act of 1969 stated

> that the military assistance program for any country could not be increased beyond the amount justified to the Congress unless there was a determination by the President that an increase was essential to U.S. national interests. Previously if a country failed to use all of its MAP training dollars, they could be transferred to other countries which wished additional training slots at the Inter-American Air Force Academy or other training schools. . . . Subsequent to the delegation's return to the United States, the restrictive provision was amended by Congress to provide an increased flexibility by permitting a 20-percent deviation in the congressional presentation level for military assistance programs.[23]

The fact that congressional "doves" have opted to symbolically reduce executive autonomy rather than to materially undermine the program reflects their basic consensus insofar as America's counterrevolutionary goals are concerned. For Senator Fulbright the tragedy of Vietnam is not America's attempt to frustrate a thirty-year movement for national independence which happens to be Communist led. His grounds for condemning the executive are based upon the "pragmatic" conclusion that the war as conducted was "an unwise, unprofitable and disastrous political misjudgment . . . an intellectual error."[24]

To be fair, it must be admitted that Senate criticism of military aid did antedate the failure of policy in Vietnam. Commencing in the late fifties, reservations began to be voiced concerning: (1) whether Congress should strengthen antipopulist repressive regimes; (2) whether arms aid contributed to local arms races; (3) whether the MAP stimulated buildup of armed forces in Third World countries unduly taxed resources needed for development and thus actually enhanced the appeal of revolutionaries; (4) whether military aid really does "win friends and influence people," and (5) whether the congressional oversight function was being given sufficient deference by the executive. The failure of these liberal critics to act in unison and their limited legislative backing by colleagues rendered nugatory much of their well-intentioned criticism. The executive has simply reallocated funds, utilized waiver authority, marshalled impressive research reports to rebut some of the charges and assigned higher priority to a *low domestic* profile![h] Hence the Deputy Assistant Secretary of Defense for *Military Assistance* and Sales has been retitled the Deputy Assistant Secretary of Defense for *Security Assistance*, while new high-level civilian State Department positions have been created with at least symbolic authority in

---

[h]The principal effects of domestic criticism have been: (1) to slow the rate of increase in military aid; (2) force a new emphasis upon sales rather than grants; (3) result in more funds being channeled through economic "security supporting assistance." For an incisive assessment of recent trends, see Siegel's "Future of Military Aid," pp. 1-3, 7.

this area. And the aura of innovation had been cast by such new slogans as the Nixon Doctrine and "total force planning."[i] In short, the momentum of American militarism all but vitiates criticism which fails to transcend the parameters of necessity that are so intertwined with an inherently expansionist corporate-dominated social order.

Theoretical Perspectives

Jose Nun has proposed that we begin an analysis of military intervention in Third World countries by distinguishing structural from circumstantial factors.[25] With respect to the former, Nun believes certain historical events have neutralized interventionist propensities in West Europe.[26] Both Nordlinger and Nun premise their explanations of intervention in underdeveloping areas upon the following factors: (1) middle-class reference groups for officers; (2) institutional interests that are threatened in common with middle-class privileges by lower-class mobilization; (3) antipopulist professional values such as hierarchy, order (Nordlinger); (4) antiradical ideological beliefs (Nun). In relatively stagnant economies, consumption is a zero-sum question. Thus, (3) and (4) are reinforced by the first two and condition a military propensity to view lower-class demands as a potential threat to their share of the budget. Insofar as Communists are concerned, the threat is even more far reaching since such regimes have for the most part completely eliminated military autonomy by politicizing (i.e., integrating) the officer corps.[j] According to Nun, "anti-Communism" may also *function* as a *rationalization* for less defensible antipopulist (radical) interventions.[27] Within this context simplistic anti-Communist indoctrination is likely to be especially effective in promoting such tendencies.[k] So

[i] "Total force planning" involves improved programming to reduce duplication. The United States will not maintain units which can be financed from willing allies through "security assistance." This quasi-mercenary approach—recently tested with Thai and Korean units in Indochina—mirrors the shift from the citizens' to a highly paid professional army within the United States. On TFP, see the *Foreign Assistance Act of 1972, Hearings*, House, pp. 48, 158-59.

[j] Both enforced austerity and political subordination constitute blockages of major importance to the upward mobility status aspirations of many officers from lower middle-class backgrounds. For senior officers who in a sense have "already made it" or themselves come from the more privileged classes, an acute sense of relative deprivation is likely to be occasioned by prospective policy changes of this nature. The significance of these status aspirations in relatively stagnant economies where ascriptive criteria and particularism limit upward mobility prospects in civilian sectors is so great according to Merle Kling that they account for many nonideological coups through which ambitious officers can win access to highly remunerative positions in the public sector. See Kling's excellent "Toward, a Theory of Power and Political Instability in Latin America," *Western Political Quarterly*, 9 (March 1956), 21-35.

[k] This hypothesis could be operationalized by correlating intensities of anti-Communism expressed by foreign officers with their breadth of definition of what groups or leaders are furthering the expansion of Communism. The recruitment of officers from lower status and less well educated backgrounds during recent decades would almost certainly strengthen this tendency as would their entry into adolescence in the cold war era.

for that matter would be *the indoctrinational emphasis* upon linking corporate capitalism ("free enterprise") with "liberty." Socialism and "exploitive" theories of capitalism were conversely identified with Satanic totalitarianism. For officers who have received "professional" command and staff courses and multiple training tours—regular or orientation—little more than a word of encouragement from embassy attachés or MILGRP officers would be necessary to catalyze a rightist plot to intervene. The utility of interface "rapport" becomes apparent in such circumstances. Unsolicited "advice" to counterpart officers might be as unwise as it would be unnecessary.[l]

The expert opinion and official testimony examined in Chapter Seven support an estimate of moderate effectiveness as does *the decline in the number of radical interventions*. This trend is significant not only for Latin America (Needler), but is especially notable for countries with a high proportion of officers who were CONUS trained (1 percent of armed force category). It is somewhat weaker but still observable for regimes which have received U.S. or Western training for more than eight years. The more moderate relationship here is probably explained by the fact that a number of these recipients did not send as large a proportion for training as did others.

MAP's effectiveness in reinforcing rightist ideological tendencies and cohesion among foreign officer corps is manifested by the reduction in the number of *successful* radical interventions. Not coincidentally, official testimony quoted in Chapter Seven quantified the number of (radical) military rebellions that had been suppressed by "loyal" forces as evidence of MAP efficacy. Although we do not possess it, data confirming a secular decline in the incidence of radical plots would be particularly germane here.[m]

Interestingly, the data provided in Chapter Seven do not indicate an increase in rightist interventions as the proportion trained rises or the time period lengthens. An attempt to explain this apparent MAP "failure" requires that we return to Nun's dichotomy between structural and circumstantial factors. There is no reason why military establishments whose rightist propensities have been reinforced should launch a rightist coup against such pro-American (Western, capitalist) rulers as the Shah of Iran or President Balaguer

[l]While it is reasonable to suppose that this data is on file in Washington, without it our conclusions must remain inconclusive. References to rightist interventions in Iran, Guatemala, Vietnam, and Brazil which appear in Chapters Six and Seven suggest that in an indeterminate number of cases mission personnel may be no more than junior partners in the operation. Thus we have a cast of high ranking officials such as Generals Schwartzkopf, Walters, and Taylor along with Ambassadors Puerifoy and Lodge who were apparently invested with special authority to supervise the conspiratorial process in conjunction with the CIA.

[m]So for that matter might be other possible causal determinants such as: (1) failure of past radical plots; (2) fear of U.S. invasion (Lebanon 1958, Dominican Republic 1965 as bases for the fear); (3) the deposition of existing radical regimes; and (4) the demonstrated incapacity of some non-communist radical regimes to close the per capita GNP gap with the advanced nations.

in the Dominican Republic. Obviously *there must be a radical threat* which can assume at least three forms: (1) possession of civil authority (Arbenz in Guatemala, Goulart in Brazil); (2) inability or unwillingness of the government to suppress a growing radical mobilization (Diem in Vietnam, Sihanouk in Cambodia); or (3) a move to the left by a formerly moderate government (Tanzania). Given the fact that many long-term and large-scale recipients of military training also open their societies to other forms of intervention (propaganda, cultural exchange, peace corps, CIA subsidization of antiradical "community development, organizing efforts, etc.), it may well be that radical potential is diminished by mobilizing institutional and mass attitudinal bias against such aspiring counterelites. Thus the "penetrated society" may exhibit fewer radical provocations to an increasingly rightist military.[n]

Situational factors probably also play an important part in explaining the lack of variance in the number of rightist coups. While environmental restructuring oriented towards suppressing leftist counterelites, coopting them, preempting their mass constituencies by increasing the prevalence of middle-class consciousness may tend to diminish radical provocations from the civilian sector, this "mobilization of bias" is often quite fragile and imperfectly attained in many countries. Tactical overconfidence and errors as in Chile during 1970 may more than offset the long-range stagnation of leftist electoral constituencies. Or the mass misery and inability to transcend seemingly perpetual underdevelopment may attract some centrist politicians to radical policies. Hence because the masses have been incorporated into the political process *prior to successful economic "take-offs,"* the situation remains difficult to fully control and inherently unstable.[28] For this reason–even in "penetrated" societies–some radical elements continue to emerge as "threats" to the United States and indigenous military leaderships and perhaps more will in the future. Continued surplus drainage and structural dependence thus partially neutralize intervention in nonmilitary sectors and ensure that the incidence of rightist military coups will not markedly decline as the number of radical coups. If the former do in fact decline–and this is conceivable–it may be as a consequence of an increasing prevalence of rightist regimes which have systematically suppressed and physically liquidated left-wing elements. Thus we can see no reason why there should be *rightist* coups in the foreseeable future in such countries as Brazil, Iran, the Dominican Republic, Indonesia, Sudan, Bolivia, or Uganda.

To the extent that military institutional or corporate interests are *defined* as

---

[n]This hypothesis would be supported if we found that countries with say a high proportion (more than 1 percent) of CONUS trained officers also were substantially higher than other nations on such indices as: leader grants to labor leaders and political elites; percentage of news/feature electronic media programming originating from USIS and other North American sources; percentage of population within the constituencies of subsidized U.S. nongovernment organizations, etc.

encompassing nonpuritanical middle-class privileges (luxurious clubs, servants, high salaries, automobiles, etc.), they *reinforce* antiradical propensities and are consistent with Nordlinger's class model.[o] Nevertheless, too much emphasis upon structural factors such as middle-class interaction, economic zero-sum pies, etc., can obscure the countervailing role of imposed values and ideological interpretive frameworks. Thus even in countries with long term U.S. or Western training relationships, there is no significant decline in radical coups *when training has been taken from Eastern or neutralist donors.*[p] It is unlikely that officers indoctrinated in Eastern countries were recruited from the lower classes. Similarly, one of the key taproots of the current radical nationalist regime in Peru was an indigenous war college *which had not been Americanized.* On the contrary, it was staffed in part by Peruvian intellectuals of a radical bent.[q]

As the Egyptians have demonstrated, corporate privileges are at best partially compatible with radical developmental policies that end surplus drainage by foreign corporations. Hence, resentment against neocolonial domination and the perception of the growth potential inherent in radical economic and mobilizational policies *can be reconciled* with both traditional as well as new definitions of institutional interest.[r] This is not to say that Eastern trained officers are transmuted into Marxist-Leninists–only that their own perspectives are broadened in a radical and socialistic direction. No ex-trainees from such countries have attempted to introduce a Communist regime with any success, but a fair number have established radical nationalist ones. Given continued underdevelopment and surplus expropriation by Western corporations, officers who consider themselves as guardians of the nation–even a few trained in the United States–may be quite vulnerable to leftist ideological appeals. Peru's generals claim that some of them were deeply impressed by captured guerrillas' propaganda tracts.[29] Even more determinative, however, was their exposure to leftist intellectuals who lectured at the Centro de Altos Estudios Militares–a national war college that had not been completely Americanized (de-nationalized).[30]

We are not arguing that military officers–who constitute one of the more privileged and status insecure classes in their societies–are as open to radical

[o]So for that matter does the expansion of urban service sector based middle-class strata which intermarry with military officers. This of course is a pervasive phenomenon in the underdeveloping areas.

[p]The motivational effects of transnational reference groups are explored in Robert M. Price, "A Theoretical Approach to Military Rule in New States: Reference-Group Theory and the Ghanaian Case," *World Politics*, 23 (April 1971), 401ff.

[q]There has been little if any systematic investigation of Western or American influence upon staffing and ideological curricula in Third World military academies, war colleges, and command and staff colleges. Kaplan's previously cited findings on the Brazilian ESG along with the Peruvian case suggest the possible value of such a study. Because few university graduates find military careers appealing, the military educational institutions perform important youth and adult political socialization roles.

[r]In even greater measure so can military autonomy where as in Libya or Peru, the non-Communist but nationalist regime is actually a creation of the Armed Forces.

ideological appeals as to rightist propaganda. As force-oriented and conservative personality types who tend to view intellectuals and all civilians with chauvinistic disdain, they are naturally more receptive to American Western indoctrination.[s] But *residual nationalism* and the failure of their countries to "take-off" remain as ever present "problems" for the United States. The "danger" that some patriotic junior officers may come to intensely resent their neocolonial subordination is ever present. This explains the *increasing priority* assigned to systematic indoctrination, orientation tours, and the political ends of training by American officialdom. It also casts light upon the determination that Third World nations–new or old–should never be permitted to balance Eastern against Western "aid." Perhaps national humiliation when leavened with radical perspectives will for an indeterminate number neutralize or erode profession-related conservatism and reference group identifications with the United States and domestic privileged classes. Future research then should not neglect the sources, content, and functional significance of transnational military political socialization.[t] We can be confident that governmental policy-makers will not!

[s]The consistent denigration of civilian supremacy by MAP may be counterposed to the party-dominant military subordination that is a hallmark of all "non-Western" (i.e., Marxian) regimes. Hence, by stressing the legitimacy of "civil" governments, military missions and talents as "nation-builders," and the illegitimacy of opposition parties which are not perceived as "loyal" (defined as those based upon religious and especially socialist principles), MAP political socialization operates to legitimize a praetorian systemic role for Third World military establishments. Officer corps function then–to use Vagts' conception–as militaristic sovereigns rather than as guardians of state sovereignty. And in constituting a materially privileged social stratum, they have a direct interest in wielding their sovereignty to extract forced savings from the underlying population through control over the government. Regardless of whether one prefers autonomy or sovereignty to classify the institutional position of military establishments, it seems that on balance they would be most comfortable with and hence receptive to American efforts to reconcile counterrevolutionary goals with legitimized militarism. The Communist military aid donors cannot approximate this easy fit between ideological values and military privileges including the assurance of long-term leverage via civilian subordination. On the other hand for reasons to be explicated in our text, this role strain–particularly for junior officers–would be considerably diminished by training tours in such radical *military-led* nations as Algeria, Somalia, Congo (Brazzaville), and Syria. Ironically, with the qualified exception of Algeria, these regimes have neither the resources nor the military expertise to act as donors of aid.

[t]And it should attempt to "tell us not that political socialization somehow does affect the political system, but how it produces particular, substantively significant effects." Or as another political scientist has put it, the "so-called revolutionary transformations in comparative politics of the last two decades have involved, among other things, a proliferation of macro-theories of politics, political systems and political behavior. Many, perhaps most, such 'theories' are of extremely dubious value. . . . There has occurred an alarming neglect of the political process, itself, and of key institutions directly involved in the policy-making process." T.G. Harvey, "Political Socialization Research: Problems and Prospects," (paper prepared for presentation at the annual meeting of the Canadian Political Science Association, McGill University, Montreal, June 1972), p. 6. J. LaPalombara, "Macro-theories and Micro-applications in Comparative Politics: A Widening Chasm," *Comparative Politics*, 1 (October 1968), 52-78, as abstracted in *International Political Science Abstracts*, 19, no. 2 (1969), 129-30.

Appendixes

Appendix 1: Key Defense Officials

Bogart, Maj. Gen. Theodore F.
Commanding General, U.S. Army, Caribbean (1961)

Cunha, Capt. George M.
U.S. Navy, Staff Officer, Joint Staff, Joint Chiefs of Staff (1962); Staff Officer, Commander-in-Chief, Caribbean (1959-1961)

Fuqua, Brig. General Stephen O., Jr.
Director, Near East, South Asia, Africa Region, Officer of the Assistant Secretary of Defense for International Security Affairs (1963)

Heinz, Vice Admiral L.C.
Director of Military Assistance, Office of the Assistant Secretary of Defense for International Security Affairs (1966)

Hughes, Lt. Col. David R.
U.S. Army, Assistant for Counterinsurgency, Officer of the Secretary of Defense (1967)

Jordon, Col. Amos A.
Professor and Head of the Department of Social Sciences, U.S. Military Academy (1970). Former Director of the Near East and South Asia Region in the Office of the Assistant Secretary of Defense for International Security Affairs. Author of *Foreign Aid and the Defense of Southeast Asia* (1962)

Laird, Melvin R.
Secretary of Defense (1969-

McCain, Admiral John S., Jr.
U.S. Navy, Commander-in-Chief, Pacific (1969)

McNamara, Robert S.
Secretary of Defense (1961-1967)

Mather, Gen. George R.
Commander-in-Chief, Southern Command (1969) (1970)

Newsom, David D.
Assistant Secretary for African Affairs, Department of State (1970)

Nutter, Hon. G. Warren.
Assistant Secretary of Defense for International Security Affairs (1970)

O'Meara, Lt. Gen. Andrew P.
Commander-in-Chief, Southern Command (1961)

Porter, General Robert W.
Commander-in-Chief, Southern Command (1968)

Rockefeller, Hon. Nelson A.
Governor of New York. Chairman of a United States Presidential Mission for the Western Hemisphere (1969)

Seignious, Lt. Gen. George M., II,
U.S. Army. Deputy Assistant Secretary of Defense (International Security Affairs) for Security Assistance and Director, Defense Security Assistance Agency (1972)

Spiers, Ronald I.
Director, Bureau of Politico-Military Affairs, Department of State (1970)

Warnke, Paul C.
Assistant Secretary of Defense for International Security Affairs (1968)

Warren, Lt. Gen. Robert H.
Director of Military Assistance (1968); Deputy Assistant Secretary of Defense (ISA) for Military Assistance and Sales (1969)

Wheeler, Gen. Earle G.
U.S. Army, Chairman, Joint Chiefs of Staff (1965) (1970)

Wood, Gen. Robert J.
U.S. Army, Director of Military Assistance, Office of the Assistant Secretary of Defense for International Security Affairs (1965)

Appendix 2: Congressional Stipulations on Expropriation

The Foreign Assistance Act of 1961, As Amended:
Sec. 601. ENCOURAGEMENT OF FREE ENTERPRISE AND PRIVATE PARTICIPATION–

(a) The Congress of the United States recognizes the vital role of free enterprise in achieving rising levels of production and standards of living essential to economic progress and development. Accordingly, it is declared to be the policy of the United States to encourage the efforts of other countries to increase the flow of international trade, to foster private initiative and competition, to encourage the development and use of cooperatives, credit unions, and savings and loan associations, to discourage monopolistic practices, to improve the technical efficiency of their industry, agriculture, and commerce, and to strengthen free labor unions; and to encourage the contribution of United States enterprise toward economic strength of less developed friendly countries, through private trade and investment abroad, private participation in programs carried out under this Act (including the use of private trade channels to maximum extent practicable in carrying out such programs), and exchange of ideas and technical information on the matters covered by this subsection.

(b) In order to encourage and facilitate participation by private enterprise to the maximum extent practicable in achieving any of the purposes of this Act, the President shall–

  (1) make arrangements to find, and draw the attention of private enterprise to, opportunities for investment and development in less developed friendly countries and areas;

  (2) accelerate a program of negotiating treaties for commerce and trade, including tax treaties, which shall include provisions to encourage and facilitate the flow of private investment to, and its equitable treatment in, friendly countries and areas participating in programs under this Act;

  (3) seek, consistent with the national interest, compliance by other countries or areas with all treaties for commerce and trade and taxes, and take all reasonable measures under this Act or other authority to secure compliance therewith and to assist United States citizens in obtaining just compensation for losses sustained by them or payments exacted from them as a result of measures taken or imposed by any country or area thereof in violation of any such treaty;

  (4) to the maximum extent practicable carry out programs of assistance through private channels and to the extent practicable in conjunction with local private governmental participation, including loans under the

authority of section 201 to any individual, corporation, or other body of persons;

(5) take appropriate steps to discourage nationalization, expropriation, confiscation, seizure of ownership or control of private investment and discriminatory or other actions having the effect thereof, undertaken by countries receiving assistance under this act, which divert available resources essential to create new wealth, employment, and productivity in those countries and otherwise impair the climate for new private investment essential to the stable economic growth and development of those countries; and

(6) utilize wherever practicable the services of United States private enterprise (including, but not limited to, the services of experts and consultants in technical fields such as engineering).

(c) (1) There is hereby established an Advisory Committee on Private Enterprise in Foreign Aid. The Advisory Committee shall carry out studies and make recommendations for achieving the most effective utilization of the private enterprise provisions of this Act to the head of the agency charged with administering the program under part I of this Act, who shall appoint the Committee.

(2) Members of the Advisory Committee shall represent the public interest and shall be selected from the business, labor and professional world, from the universities and foundations, and from among persons with extensive experience in government. The Advisory Committee shall consist of not more than nine members, and one of the members shall be designated as chairman.

(3) Members of the Advisory Committee shall receive no compensation for their services but shall be entitled to reimbursement in accordance with section 5 of the Administrative Expenses Act of 1946 (5 U.S.C. 73b-2) for travel and other expenses incurred in attending meetings of the Advisory Committee.

(4) The Advisory Committee shall, if possible, meet not less frequently than once each month, shall submit such interim reports as the Committee finds advisable, and shall submit a final report not later than June 30, 1965, whereupon the Committee shall cease to exist. Such reports shall be made available to the public and to the Congress.

(5) The expenses of the Committee, which shall not exceed $50,000, shall be paid from funds otherwise available under this Act.

(d) It is the sense of Congress that the Agency for International Development should continue to encourage, to the maximum extent consistent with the national interest, the utilization of engineering and professional services of United States firms (including, but not limited to, any corporation, company, partnership, or other association) or by an affiliate of such United States firms in connection with capital projects financed by funds authorized under this Act.

Sec. 602. SMALL BUSINESS.–

(a) Insofar as practicable and to the maximum extent consistent with the accomplishment of the purposes of this Act, the President shall assist American small business to participate equitably in the furnishing of commodities, defense articles, and services (including defense services) financed with funds made available under this Act–

(1) by causing to be made available to suppliers in the United States, and particularly to small independent enterprises, information, as far in advance as possible, with respect to purchases proposed to be financed with such funds;

(2) by causing to be made available to prospective purchasers in the countries and areas receiving assistance under this Act information as to such commodities, articles, and services produced by small independent enterprises in the United States; and

(3) by providing for additional services to give small business better opportunities to participate in the furnishing of such commodities, articles, and services financed with such funds.

(b) There shall be an Office of Small Business, headed by a Special Assistant for Small Business, in such agency of the United States Government as the President may direct, to assist in carrying out the provisions of subsection (a) of this section.

(c) The Secretary of Defense shall assure that there is made available to suppliers in the United States, and particularly to small independent enterprises, information with respect to purchases made by the Department of Defense pursuant to part II, such information to be furnished as far in advance as possible.

Sec. 603. SHIPPING ON UNITED STATES VESSELS.–

The ocean transportation between foreign countries of commodities and defense articles purchased with foreign currencies made available or derived from funds made available under this Act or the Agricultural Trade Development and Assistance Act of 1954, as amended (7 U.S.C. 1691 et seq.), and transfers of fresh fruit and products thereof under this Act, shall not be governed by the provisions of section 901 (b) of the Merchant Marine Act of 1936, as amended (46 U.S.C. 1241), or any other law relating to the ocean transportation of commodities on United States flag vessels.

Sec. 604. PROCUREMENT.–

(a) Funds made available under this Act may be used for procurement outside the United States only if the President determines that such procurement will not result in adverse effects upon the economy of the United States or the industrial mobilization base, with special reference to any areas of labor surplus or to the net position of the United States in its balance of payments with the rest of the world, which outweigh the economic or other advantages to the United States of less costly procurement outside the United States, and only if the price of any commodity procured in bulk is lower than the market price prevailing in the United States at the time of procurement, adjusted for differences in the cost of transportation to destination, quality, and terms of payments.

(b) No funds made available under this Act shall be used for the purchase in bulk of any commodities at prices higher than the market price prevailing in the United States at the time of purchase, adjusted for differences in the cost of transportation to destination, quality, and terms of payment.

(c) In providing for the procurement of any surplus agricultural commodity for transfer by grant under this Act to any recipient country in accordance with its requirements, the President shall, insofar as practicable and when in furtherance of the purposes of this Act, authorize the procurement of such surplus agricultural commodity only within the United States except to the extent that such surplus agricultural commodity is not available in the United States in sufficient quantities to supply emergency requirements of recipients under this Act.

(d) In providing assistance in the procurement of commodities in the United States, United States dollars shall be made available for marine insurance on such commodities where such insurance is placed on a competitive basis in accordance with normal trade practice prevailing prior to the outbreak of World War II: *Provided,* That in the event a participating country, by statute, decree, rule, or regulation, discriminates against any marine insurance company authorized to do business in any State of the United States, then commodities purchased with funds provided hereunder and destined for such country shall be insured in the United States against marine risk with a company or companies authorized to do a marine insurance business in any State of the United States.

. . . .

Sec. 620. PROHIBITIONS AGAINST FURNISHING ASSISTANCE.–

(a) (1) No assistance shall be furnished under this Act to the present government of Cuba; nor shall any such assistance be furnished to any country which furnishes assistance to the present government of Cuba unless the President determines that such assistance is in the national interest of the United States. As an additional means of implementing and carrying into effect the policy of the preceding sentence, the President is authorized to establish and maintain a total embargo upon all trade between the United States and Cuba.

(2) Except as may be deemed necessary by the President in the interest of the United States, no assistance shall be furnished under this Act to any government of Cuba, nor shall Cuba be entitled to receive any quota authorizing the importation of Cuban sugar into the United States or to receive any other benefit under any law of the United States, until the President determines that such government has taken appropriate steps according to international law standards to return to United States

citizens, and to entities not less than 50 per centum beneficially owned by United States citizens, or to provide equitable compensation to such citizens and entities for property taken from such citizens and entities on or after January 1, 1959, by the Government of Cuba.

(3) No funds authorized to be made available under this Act (except under section 214) shall be used to furnish assistance to any country which has failed to take appropriate steps, not later than 60 days after the date of enactment of the Foreign Assistance Act of 1963–

(A) to prevent ships or aircraft under its registry from transporting to Cuba (other than to United States Installations in Cuba)

(i) any items of economic assistance.

(ii) any items which are, for the purposes of title I of the Mutual Defense Assistance Control Act of 1951, as amended, arms, ammunition and implements of war, atomic energy materials, petroleum, transportation materials of strategic value, or items of primary strategic significance used in the production of arms, ammunition, and implements of war, or

(iii) any other equipment, materials, or commodities, so long as Cuba is governed by the Castro regime; and

(B) to prevent ships or aircraft under its registry from transporting any equipment, materials, or commodities from Cuba (other than from United States installations in Cuba) so long as Cuba is governed by the Castro regime.

(b) No assistance shall be furnished under this Act to the government of any country unless the President determines that such country is not dominated or controlled by the international Communist movement.

(c) No assistance shall be provided under this Act to the government of any country which is indebted to any United States citizens or person for goods or services furnished or ordered where (i) such citizen or person has exhausted available legal remedies, which shall include arbitration, or (ii) the debt is not denied or contested by such government, or (iii) such indebtedness arises under an unconditional guaranty of payment given by such government, or any predecessor government, directly or indirectly, through any controlled entity: *Provided*, That the President does not find such action contrary to the national security.

(d) No assistance shall be furnished under section 201 of this Act for construction or operation of any productive enterprise in any country where such enterprise will compete with United States enterprise unless such country has agreed that it will establish appropriate procedures to prevent the exportation for use of consumption in the United States of more than twenty per centum of the annual production of such facility during the life of the loan. In case of failure to implement such agreement by the other contracting party, the President is authorized to establish necessary import controls to effectuate the agreement. The restrictions imposed by or pursuant to this subsection may be waived by the President where he determines that such waiver is in the national security interest.

(e) (1) The President shall suspend assistance to the government of any country to which assistance is provided under this or any other Act when the government of such country or any government agency or subdivision within such country on or after January 1, 1962–

(A) has nationalized or expropriated or seized ownership or control of property owned by any United States citizen or by any corporation, partnership, or association not less than 50 per centum beneficially owned by United States citizens, or

(B) has taken steps to repudiate or nullify existing contracts or agreements with any United States citizen or any corporation, partnership, or association not less than 50 per centum beneficially owned by United States citizens, or

(C) has imposed or enforced discriminatory taxes or other exactions, or restrictive maintenance or operational conditions, or has taken other actions, which have the effect of nationalizing, expropriating, or otherwise seizing ownership or control of property so owned, and such country, government agency, or government subdivision fails within a reasonable time (not more than six months after such action, or, in the event of a referral to the Foreign Claims Settlement Commission of the United States within such period as provided herein, not more than twenty days after the report of the Commission is received) to take appropriate steps, which may include arbitration, to discharge its obligations under international law toward such citizen or entity, including speedy compensation for such property in convertible foreign exchange, equivalent to the full value thereof, as required by international law, or fails to take steps designed to provide relief from such taxes, exactions, or conditions, as the case may be; and such suspension shall continue until the President is satisfied that appropriate steps are being taken, and no other provision of this Act shall be construed to authorize the President to waive the provisions of this subsection.

Upon request of the President (within seventy days after such action referred to in subparagraphs (A), (B), or (C) of paragraph (1) of this subsection), the Foreign Claims Settlement Commission of the United States (established pursuant to Reorganization Plan No. 1 of 1954, 68 Stat. 1279) is hereby authorized to evaluate expropriated property, determining the full value of any property nationalized, expropriated, or seized, or subject to discriminatory or other actions as aforesaid, for purposes of this subsection and to render an advisory report to the President within ninety days after such request. Unless authorized by the President, the Commission shall not publish its advisory report except to the citizen or entity owning such property. There is hereby authorized to be appropriated such amount, to remain available until expended, as may be necessary from time to time to enable the Commission to carry out expeditiously its functions under this subsection.

(2) Notwithstanding any other provision of law, no court in the United States shall decline on the ground of the federal act of state doctrine to make a determination on the merits giving effect to the principles of international law in a case in which a claim or title or other right to property is asserted by any party including a foreign state (or a party claiming through such state) based upon (or traced through) a confiscation of other taking after January 1, 1959, by an act of that state in violation of the principles of international law, including the principles of compensation and the other standards set out in this subsection: *Provided*, That this subparagraph shall not be applicable (1) in any case in which an act of a foreign state is not contrary to international law or with respect to a claim of title or other right to property acquired pursuant to an irrevocable letter of credit of not more than 180 days duration issued in good faith prior to the time of the confiscation or other taking, or (2) in any case with respect to which the President determines that application of the act of state doctrine is required in that particular case by the foreign policy interests of the United States and a suggestion to this effect is filed on his behalf in that case with the court.

(f) No assistance shall be furnished under this Act, as amended (except section 214 (b)), to any Communist country. This restriction may not be waived pursuant to any authority contained in this Act unless the President finds and promptly reports to Congress that: (1) such assistance is vital to the security of the United States; (2) the recipient country is not controlled by the international Communist conspiracy; and (3) such assistance will further promote the independence of the recipient country from international communism. For the purposes of this subsection, the phrase "Communist country" shall include specifically, but not be limited to, the following countries:

Peoples Republic of Albania,
Peoples Republic of Bulgaria,
Peoples Republic of China,
Czechoslovak Socialist Republic,
German Democratic Republic (East Germany),
Estonia,
Hungarian Peoples Republic,
Latvia,

Lithuania,
North Korean Peoples Republic,
North Vietnam,
Outer Mongolia-Mongolian Peoples Republic,
Polish Peoples Republic,
Rumanian Peoples Republic,
Tibet,
Federal Peoples Republic of Yugoslavia,
Cuba, and
Union of Soviet Social Republics (including its captive constituent republics).

(g) Notwithstanding any other provision of law, no monetary assistance shall be made available under this Act to any government or political subdivision or agency of such government which will be used to compensate owners for expropriated or nationalized property and, upon finding by the President that such assistance has been used by any government for such purpose, no further assistance under this Act shall be furnished to such government until appropriate reimbursement is made to the United States for sums so diverted.

(h) The President shall adopt regulations and establish procedures to insure that United States foreign aid is not used in a manner which, contrary to the best interests of the United States, promotes or assists the forieign aid projects or activities of the Communist-bloc countries.

(i) No assistance shall be provided under this or any other Act, and no sales shall be made under the Agricultural Trade Development and Assistance Act of 1954, to any country which the President determines is engaging in or preparing for aggressive military efforts directed against–

(1) the United States,
(2) any country receiving assistance under this or any other Act, or
(3) any country to which sales are made under the Agricultural Trade Development and Assistance Act of 1954, until the President determines that such military efforts or preparations have ceased and he reports to the Congress that he has received assurances satisfactory to him that such military efforts or preparations will not be renewed. This restriction may not be waived pursuant to any authority contained in this Act.

(j) No assistance under this Act shall be furnished to Indonesia unless the President determines that the furnishing of such assistance is essential to the national interest of the United States. The President shall keep the Foreign Relations Committee and the Appropriations Committee of the Senate and the Speaker of the House of Representatives fully and currently informed of any assistance furnished to Indonesia under this Act.

(k) Until the enactment of the Foreign Assistance Act of 1965 or other general legislation, during the calendar year 1965, authorizing additional appropriations to carry out programs of assistance under this Act, no assistance shall be furnished under this Act to any country for construction of any productive enterprise with respect to which the aggregate value of such assistance to be furnished by the United States will exceed $100,000,000. No other provision of this Act shall be construed to authorize the President to waive the provisions of this subsection.

(l) No assistance shall be provided under this Act after December 31, 1966, to the government of any less developed country which has failed to enter into an agreement with the President to institute the investment guaranty program under section 221(b) (1) of this Act, providing protection against the specific risks of inconvertibility under subparagraph (A), and expropriation or confiscation under subparagraph (B), of such section 221(b) (1).

(m) No assistance shall be furnished on a grant based under this Act to any economically developed nation capable of sustaining its own defense burden and economic growth, except (1) to fulfill firm commitments made prior to July 1, 1963, or (2) additional orientation and training expenses under part II hereof during each fiscal year in an amount not to exceed $500,000.

(n) In view of the aggression of North Vietnam, the President shall consider denying assistance under this Act to any country which has failed to take appropriate steps, not later than sixty days after the date of enactment of the Foreign Assistance Act of 1965

(A) to prevent ships or aircraft under its registry from transporting to North Vietnam

(i) any items of economic assistance,

(ii) any items which are, for the purposes of title I of the Mutual Defense Assistance Control Act of 1951, as amended, arms, ammunition and implements of war, atomic energy materials, petroleum, transportation materials of strategic value, or items of primary strategic significance used in the production of arms, ammunition, and implements of war, or

(iii) any other equipment, materials or commodities; and

(B) to prevent ships or aircraft under its registry from transporting any equipment, materials or commodities from North Vietnam.

(o) In determining whether or not to furnish assistance under this Act, consideration shall be given to excluding from such assistance any country which hereafter seizes, or imposes any penalty or sanction against, any United States fishing vessel on account of its fishing activities in international waters. The provisions of this subsection shall not be applicable in any case governed by international agreement to which the United States is a party.

According to the President's Annual Report to Congress on the Foreign Assistance Program for FY 1969, "Countries or Areas with Investment Insurance and Guaranty Agreements" were:

(June 30, 1969)

Convertibility	Expropriation	War, Revolution and Insurrection	Extended Risk Guaranties
Afghanistan	Afghanistan	Afghanistan[a]	
Antigua	Antigua	Antigua	Antigua
Argentina	Argentina	Argentina	Argentina[b]
Barbados	Barbados	Barbados	Barbados
Bolivia	Bolivia	Bolivia	Bolivia
Brazil	Brazil	Brazil	Brazil
British Honduras	British Honduras	British Honduras	British Honduras
Burundi	Burundi	Burundi	Burundi
Cameroon	Cameroon	Cameroon	Cameroon
Central African Rep.	Central African Rep.	Central African Rep.	Central African Rep.
Ceylon	Ceylon	Ceylon	Ceylon
Chad	Chad	Chad	Chad
Chile	Chile	Chile	Chile
China, Rep. of	China, Rep. of	China, Rep. of	China, Rep. of
Colombia	Colombia	Colombia	Colombia
Congo (Brazzaville)[c]	Congo (Brazzaville)[c]	Congo (Brazzaville)[c]	Congo (Brazzaville)[c]
Congo (Kinshasa)	Congo (Kinshasa)	Congo (Kinshasa)	Congo (Kinshasa)
Costa Rica	Costa Rica	Costa Rica[d]	Costa Rica[d]
Cyprus[c]	Cyprus[c]	Cyprus[c]	Cyprus[c]
Dahomey	Dahomey	Dahomey	Dahomey
Dominica	Dominica	Dominica	Dominica
Dominican Republic	Dominican Republic	Dominican Republic	Dominican Republic
Ecuador	Ecuador	Ecuador	Ecuador
El Salvador	El Salvador	El Salvador[d]	El Salvador[d]
Ethiopia	Ethiopia	Ethiopia	Ethiopia
Gabon	Gabon	Gabon	Gabon
Gambia	Gambia	Gambia	Gambia
Ghana	Ghana	Ghana	Ghana
Greece	Greece	Greece	Greece
Grenada	Grenada	Grenada	Grenada

Guatemala	Guatemala		
Guinea	Guinea	Guinea	Guinea
Guyana	Guyana	Guyana	Guyana
Haiti	Haiti		
Honduras	Honduras	Honduras	Honduras
India	India	India	India
Indonesia	Indonesia	Indonesia	Indonesia
Iran	Iran		
Israel	Israel	Israel	Israel
Ivory Coast	Ivory Coast	Ivory Coast	Ivory Coast
Jamaica	Jamaica	Jamaica	Jamaica
Jordan	Jordan	Jordan	Jordan
Kenya	Kenya	Kenya	Kenya
Korea	Korea	Korea	Korea
Laos	Laos	Laos	Laos
Lesotho	Lesotho	Lesotho	Lesotho
Liberia	Liberia	Liberia	Liberia
Malagasy	Malagasy	Malagasy	Malagasy
Malawi	Malawi	Malawi	Malawi
Malaysia	Malaysia	Malaysia	Malaysia
Mali	Mali	Mali	Mali
Malta[c]	Malta[c]	Malta[c]	Malta[c]
Mauritania[b]	Mauritania[b]	Mauritania[b]	Mauritania[b]
Morocco	Morocco	Morocco	Morocco
Nepal	Nepal	Nepal	Nepal
Nicaragua	Nicaragua	Nicaragua	Nicaragua
Niger	Niger	Niger	Niger
Nigeria	Nigeria		
Pakistan	Pakistan	Pakistan	Pakistan
Panama	Panama	Panama[a]	
Paraguay	Paraguay	Paraguay	Paraguay
Peru[b]			
Philippines	Philippines	Philippines	Philippines
Portugal	Portugal		
Rwanda	Rwanda	Rwanda	Rwanda
St. Christ. Nevis-Ang.	St. Christ. Nevis-Ang.	St. Christ. Nevis-Ang.	St. Christ. Nevis-Ang.
St. Lucia	St. Lucia	St. Lucia	St. Lucia
Senegal	Senegal	Senegal	Senegal
Sierra Leone	Sierra Leone	Sierra Leone	Sierra Leone
Singapore	Singapore	Singapore	Singapore
Somali Rep	Somali Rep	Somali Rep	Somali Rep
Sudan[b]	Sudan[b]	Sudan[b]	Sudan[b]

Swaziland	Swaziland	Swaziland	Swaziland
Tanzania (excl. Zanzibar)	Tanzania (excl. Zanzibar)	Tanzania (excl. Zanzibar)	Tanzania (excl. Zanzibar)
Thailand	Thailand	Thailand	Thailand
Togo	Togo	Togo	Togo
Trinidad-Tobago	Trinidad-Tobago	Trinidad-Tobago	Trinidad-Tobago
Tunisia	Tunisia	Tunisia	Tunisia
Turkey	Turkey	Turkey	Turkey
Uganda	Uganda	Uganda	Uganda
U.A.R. (Egypt)[b]	U.A.R. (Egypt)[b]	U.A.R. (Egypt)[b]	U.A.R. (Egypt)[b]
Upper Volta	Upper Volta	Upper Volta	Upper Volta
Uruguay[b]	Uruguay[b]		
Venezuela	Venezuela	Venezuela	Venezuela
Vietnam	Vietnam	Vietnam	Vietnam
Yugoslavia[c]	Yugoslavia[c]		
Zambia	Zambia	Zambia	Zambia

[a]Includes only insurance against loss due to damage from war.

[b]Program inoperative.

[c]Restricted availability.

[d]Although application will be accepted, guaranties will not be processed until agreement is ratified by country's legislative body and in force.

Although economically developed countries are excluded from the Investment Insurance Program, insurance may be available for some of the less developed dependencies of France, the Netherlands, Portugal, and the United Kingdom.

Appendix 3: Memorandum of the Secretary of Defense, 9/13/63

The Secretary of Defense,
Washington, September 13, 1963.

Memorandum for: The Secretary of the Army.
The Secretary of the Navy.
The Secretary of the Air Force.
The Assistant Secretary of Defense (International Security Affairs).
The Assistant Secretary of Defense (Manpower).
Commanders of unified and specified commands.

Subject: An informational program for foreign military trainees and visitors in the United States.

References:

(a) Military Assistance Manual, "Part I–Objectives"; section B, "Training Objectives and Guidance" (S).

(b) Memorandum Secretary of Defense, subject: "A Program for Developing and Maintaining Contact with Foreign Military Personnel," dated April 6, 1959 (S) (to services only).

(c) DOD Directive 5132.3, "Department of Defense Policy and Responsibilities Relating to Military Assistance," dated July 8, 1963 (U).

I. PURPOSE

The purpose of this memorandum is to prescribe the objectives of an informational program for foreign military trainees and visitors in the United States to complement their formal training courses or military orientation and to assign responsibilities for the attainment of these objectives.

II. APPLICABILITY

This memorandum is applicable to the Departments of the Army, the Navy, and the Air Force, to unified Commands and to the Director of Military Assistance. It applies to all U.S. Department of Defense personnel dealing with foreign military personnel training in CONUS. Additionally, it applies to all U.S. Department of Defense personnel dealing with military-sponsored visitors to this country except those visiting by personal invitation of a service Secretary or

Chief of Staff. Although this memorandum is not applicable to foreign trainees at or visitors to U.S. installations abroad, commanders of those installations will be informed of its contents so that they may follow its guidance where they deem it appropriate.

III. POLICY

It is the policy of the Department of Defense to give foreign military trainees and military-sponsored visitors, in addition to their military training or orientation in the United States, a balanced understanding, insofar as possible within the limits of the time available, of U.S. society, institutions, and ideals. Steps undertaken to give such persons a perspective on American life are intended to complement and strengthen the strictly military side of their experience while in the United States.

IV. INFORMATIONAL OBJECTIVES

In furtherance of this policy, the following specific informational objectives are established. Each objective bears on a significant area of American life, an understanding of which contributes to a sound grasp of our institutions and purposes. In devising a program to attain these objectives, each commander responsible for foreign trainees or visitors is expected to supplement or modify the objectives, where necessary, to fit the character and background of the foreigners involved and the time and resources available for such purposes. Successful attainment of these objectives and full exposure of the foreign trainee to the nonmilitary aspects of American life are considered to be of importance to the military assistance program second only to the strictly military training objectives of that program.

(a) Governmental institutions.–In order that foreign trainees may understand the structure and operation of the Federal system, they should be acquainted directly or indirectly with our political institutions on the local, State, and Federal levels and with the relationships between them. Additionally, they should be made aware of the principle of checks and balances and its effects upon the initiative of government.

(b) The judicial system and the doctrine of judicial review.–In order that foreign trainees may grasp that this is a government of law, before which all men are equal, they should be acquainted with the constitutional position of the courts, the relationships between them on their various levels and the practical effects of the doctrine of judicial review both of statute law and of decisions of lower courts, by which successive Supreme Court interpretations of constitutionality have adapted our institutions to meet new requirements. They should be acquainted with the constitutional and legal status of the Armed Forces, with special emphasis on their non-political character.

(c) The role of the opposition in a two-party system.–In order that they may grasp the working of our electoral institutions foreign trainees should be acquainted with the basis of party political organization, with the role which the opposition plays as a check on and alternative to the administration of the day and with the relations between government and opposition. They should be acquainted with the range and frequency of elections in order to appreciate the extent of popular control over government at all levels and the importance of the right to vote.

(d) The role of press, radio, and television.—Foreign trainees should be acquainted with the ways in which a free press works and how its owners and managers define for themselves their responsibilities to the public. Trainees should be aware of the U.S. belief that a free people with full access to factual information and to differing opinions will tend in the long run to seek the general good.

(e) The diversity of American life.—Foreign trainees should be given an idea of the diversity of American life, geographically, regionally, ethnically, religiously, and socially, and the ways in which recent technological and demographic changes have modified this historic trait.

(f) The position of minority groups in the United States, with special reference to the Negro problem.—Foreign trainees should be made aware of the constitutional rights of minorities in this country, of their history and present position in American life, of the means open to them both politically and legally for seeking redress of grievances and of the organizations and procedures through which they have done and are doing so.

(g) Agriculture.—Foreign trainees should be acquainted with American agriculture and farmers to emphasize the principles on which they operate and the factors underlying their productiveness. They should understand the importance and magnitude of education, research, and governmental assistance in improving productivity.

(h) The economy.—Foreign trainees should be acquainted with American business and industrial enterprise to underline the diversity of enterprise and ownership in our mixed economy, the principles and procedures on which capital investment is based and the range of personal and mercantile credit available to the average American. They will be acquainted with the extent of Government-owned enterprise and Government assistance to private enterprise and the importance of Federal and State regulation of economic activities.

(i) Labor.—Foreign trainees should be acquainted with the labor union movement to emphasize the scope of its organizations, the quality of its leadership and the economic power and independence.

(j) Education.—Foreign trainees should be acquainted with our institutions of secondary and higher education to emphasize their purpose, their range, and their affiliations. Trainees should understand the connection between education and a responsible citizenry.

V. RESPONSIBILITY

The military departments, the unified commands and the Director of Military

Assistance are responsible for carrying out the policy established in this memorandum. They will instruct subordinate commanders and chiefs of MAAG's/missions to use the informational objectives outlined above as a guide for developing informational programs.

VI. IMPLEMENTATION

The military departments will submit two copies of their directives promulgating policy outlined above to the Director of Military Assistance within 90 days. Two copies of any subsequent changes will likewise be submitted. The Director of Military Assistance will make such changes in the military assistance manual as this policy requires.

VII. EFFECTIVE PERIOD

This memorandum will remain in force until superseded by a Department of Defense directive.

Roswell Gilpatric,
Deputy

Enclosures:

1. Orientation in home country.
2. Reception and travel arrangements.
3. Requirements for officers escorting foreign visitors on tours.
4. Informational programs at training establishments.
5. Suggested trips and visits.
6. Tours to Washington.

ORIENTATION IN HOME COUNTRY

1. The informational objectives set forth in Section IV of the basic directive will guide the orientation given a foreign visitor or student before his departure for the United States and also subsequent contacts with him after his return. In conjunction with the country team the chief of MAAG/mission or attaché where appropriate will plan predeparture orientation, debriefing upon return, and followup contacts in conjunction with the country team so as to make full use of available resources, including USIS material and facilities, to attain the informational objectives. It is especially important to assure insofar as possible that the future student or visitor be designated sufficiently well in advance that full advantage may be taken of his natural interest in his forthcoming trip to the United States.
2. The predeparture orientation and administrative processing should particularly emphasize the following:
   (a) Informational materials for individual study to include USIS material on the United States, information about the school courses or special trip planned, and administrative requirements for the trip.
   (b) Personal interviews with U.S. officers who can explain the purpose of the

trip and of any courses to be taken who can explain or answer questions on the study material furnished.

(c) Encouragement and assistance in English language training. Material for use in this training should further the informational objectives. Where language training facilities are not sufficient, full use of USIS bilingual centers will be made.

(d) Advice and assistance with travel arrangements. Adequate advance notice of each individual's arrival must reach the port and school or other appropriate authorities in the United States. Individual sponsorship by a MAAG officer may be helpful in some cases and letter contacts part way through the U.S. trip in others.

(e) Social contacts with U.S. personnel prior to the trip where this is needed to prepare individuals for associations in the United States.

3. After a foreign visitor or trainee returns to his home country from the United States a U.S. official should take any opportunity to discuss with him the results and impressions of his trip and to determine any improvements which could be made for subsequent similar trips. Full use should be made of this meeting and subsequent contacts with selected individuals to further the basic informational objectives.

RECEPTION AND TRAVEL ARRANGEMENTS

1. Reception and travel arrangements for foreign military visitors and trainees coming to the United States must contribute to the accomplishment of the informational program for foreigners. Officers in charge of such arrangements must be acquainted with the informational objectives and be prepared to take advantage locally of any opportunities that may be present themselves to contribute to their attainment while the foreign trainee is in their charge.
2. The following points will be stressed in order to give favorable first impressions:

(a) An atmosphere of welcome, courtesy, efficiency, and consideration for the foreign officer.

(b) Due care and formality in dealing with foreign officers who often are more sensitive than American officers in matters of rank and protocol.

(c) Simple and slow English in speaking to foreigners.

(d) Travel arrangements for all personnel designed to shield them from affronts and indignities arising from their race.

(e) Arrangements to meet problems arising from religious or national dietary requirements.

(f) Ready availability of information on the United States and on specific travel arrangements and problems, especially during long layovers.

REQUIREMENTS FOR OFFICERS ESCORTING FOREIGN VISITORS ON TOURS

1. Military departments in charge of orientation tours for distinguished or senior foreign visitors will plan them with an eye to the informational objectives

contained in Section IV of the basic memorandum. Because of the short duration and specific purposes of most such tours, special care will be required to devise an effective informational program. Chiefs of MAAG's/ missions or attaches, when proposing a tour or recommending particular individual, for inclusion, should recommend to the appropriate military department the informational objectives to be stressed and the contribution they can make to the total program.

2. Military departments should instruct the escort officer on the informational objectives contained in the basic memorandum on the specific informational program for each tour and on his responsibility for that program and methods to be used to attain objectives.

INFORMATIONAL PROGRAMS AT TRAINING ESTABLISHMENTS

1. Commanders of training establishments bear the prime responsibility for acquainting foreign military trainees within their cognizance with the main areas of American life set forth in Section IV of the basic memorandum. Taking into account any orientation previously given trainees in their home country local commanders will devise a comprehensive and integrated program to achieve the aforementioned informational objectives.
2. This program should include (but not be limited to) the following:
   (a) A precourse orientation at training establishment.
   (b) Provision of materials to foreign students for individual study.
   (c) Revision of content of formal military courses where appropriate.
   (d) Sponsors and social contacts.
   (e) Special lectures and visiting speakers.
   (f) Visits and trips.
   (g) Material used in English language training, when given.
3. Within the guidance laid down in Section IV of the basic memorandum, for trainees whose total period of training will be 8 to 13 weeks, military departments will arrange for an orientation of a minimum of 10 lessons. For training of 13 weeks or longer and for such shorter courses as the military departments may designate, an orientation of 20 lessons will be provided for foreign students. The minimum amount of time devoted to achieving informational objectives, exclusive of language training, will be:
   (a) Ten hours of lecture, conference or discussion for the 10-lesson program.
   (b) Twenty hours of lecture, conference or discussion for the 20-lesson program.

   Short trips or visits should be substituted for several of the lecture-conference periods, where local arrangements permit the attainment of certain informational objectives by this method.
4. The Assistant Secretary of Defense (Manpower) will prepare and distribute a model program for 10 lessons, conferences or discussions and a model program for 20-lesson series. Background material, bibliography and training aids will be included with each program.

5. At every installation responsible for training foreign students, either by means of formal courses or by observer and on-the-job training, the commander will designate an individual primarily responsible for establishing the program to achieve the informational objectives which is directed in paragraph 1 above.

   The individual so designated may be either an officer or a civilian. In either case he will be selected with an eye to his experience and his ability to insure attainment of the informational objectives. Additional personnel will be assigned as required by both the number and the character of the foreign trainees present. Military departments will periodically review staffing patterns at service schools and training installations to assure that personnel support is adequate and that individuals with the necessary prerequisites for the successful discharge of these duties have been provided.

   Conferences of persons charged with training, administration, and orientation of foreign military trainees will be conducted by each military service for the purpose of exchanging information, ideas, and methods. Regional conferences among the services are encouraged and will be sponsored on a rotational basis as directed by the Department of Defense.

SUGGESTED TRIPS AND VISITS

1. Listed below are typical trips or visits which might appropriately be scheduled for foreign military trainees to acquaint them with various aspects of American life. They are designed to meet overall information objectives for orientation of foreign students. Local commanders will use this list as a guide, programming actual visits with an eye to their own means and to local conditions and in consonance with the other means employed to meet the same objectives, including visiting speakers. Some of the suggested visits may be combined and others may be repeated, such as visits to more than one type of farm (ranch) or industrial plant. For full value, commanders must plan trips carefully, provide knowledgeable and well-briefed escorts and insure that persons who speak to the group are made aware of its level of English language comprehension.
   (a) Local governments.—Commanders should introduce foreign military trainees to agencies and principal personnel of local governments on the city, township, or county level at the earliest opportunity. This may best be accomplished at the time foreign military trainees are formally presented to local officials, in case they are. The purpose of such an introduction is to make a point many if not most foreigners misapprehend, that is, local government officials are locally elected and responsible, within broad limits, to local people rather than to the central authorities.
   (b) State government.—At some time during their stay here as many trainees as possible should be taken to the State capital to be presented to the Governor, etc., and have an opportunity to observe selected operations of the State government. The purpose of this visit, like those outlined above,

is to stress the autonomy of State governments and the independence of Governors and legislatures. Obviously, the State supreme court should always be included in such visits. One of the best additional means for making this point is to take trainees to the headquarters of the State police or rangers.

(c) The opposition.–On the occasion of the above mentioned visits or, if convenient, at other times, commanders should arrange for foreign military trainees to meet and talk with leaders of opposition parties, preferably office holders rather than party personnel. Such a visit should be designed to show trainees that there is a "loyal" opposition, that its leaders perform official duties and have official status (limousines, for example) and that administration and opposition in fact are united by more issues than divide them.

(d) The party system.–An understanding of the "grassroots" character of American party organization is best gained by bringing foreign trainees in touch with representatives of one or the other party–sometimes with both–to give them some idea of the problems of local party organization, of means by which candidates are chosen, of the use of publicity and other means to convince voters and of the relationships between local and national party organization. Since in certain areas such visits might embroil the military and military installations in controversy that will detract from their missions, final decision in this matter is left to the discretion of the local commander. If he judges such visits to be imprudent, other means should be adopted to get this point across.

(e) The newspaper press, radio, and TV.–As a free press is one of the American institutions foreigners find most difficult to grasp, visits to newspaper editorial offices should be arranged in order to underline through personal contact how a free press works and the ways in which editors and publishers define for themselves their responsibility to the public. Radio and TV stations are interesting from a technical point of view but do not make this point quite as firmly.

(f) Minority groups.–Foreign trainees who interest themselves in the affairs of American minority groups should be put in touch with responsible leaders of minorities in order to give them an idea of the quality of such people and the character of their organization and purposes, procedures, and objectives.

(g) Farms.–Commanders should arrange tours to modern "scientific" farms to show foreign trainees the character of American agriculture. In such trips, it may be advisable to match the interests and regional background of foreigners with certain specialized types of farming operations in the vicinity. Especially worth underlining in such visits are the handling of crop credits and marketing and the question of the kinds of aid farmers receive from Federal, State, and other agricultural services in combating pests and diseases, controlling breeding stock, etc.

(h) Agricultural experiment stations.—Such trips will permit foreign military trainees to view development of new and hybrid plants, animal and fish stock, experiments in controlling local soil conditions, pests and diseases, etc. The financing of the station and the means it uses to make information available to farmers are worthy of emphasis.

(i) Business.—The following four kinds of trips are designed to suggest the scope and diversity of American business enterprise:

1. Visits to industrial enterprises should be designed to give foreign trainees an idea of the range of different kinds of industrial enterprises in the American "mixed economy," including Government-operated dams and hydroelectric institutions, local affiliates of large national corporations and smaller locally owned industries. Among other matters which company officials should be encouraged to discuss are: the relation between ownership and management of the company; management-union relationships; decisionmaking procedures in the fields of product research and development, production scheduling, marketing and cost controls; and character and effect of governmental controls over operations.
2. Visits to banks, personal loan agencies, savings and loan associations, FHA offices and agricultural cooperative credit facilities will underline the range and ease of credit facilities available to the average American.
3. Visits to local brokerage houses and discussion with brokers will emphasize the principles on which American speculative investment is based and the procedures through which it is undertaken. On these occasions obtaining a "bid-and-asked" quotation over the New York Stock Exchange wires will undoubtedly impress the foreigners.
4. Visits to large transportation centers for rail, air, water, truck, or pipeline will give foreign military trainees an opportunity to discuss the problems of management, maintenance, scheduling, and interconnection with transport officials.

(j) Labor unions.—In addition to putting interested foreign trainees in personal touch with local union officials where appropriate, tours to regional and national union headquarters will serve the useful purpose of emphasizing the scope of such organizations, the character of their leadership and their political and financial independence. In addition, trainees should be introduced to plant union officials during visits to industrial plants where that seems indicated.

(k) Schools and colleges.—Visits to nearby schools and colleges should be undertaken to show foreign trainees classes, laboratories, research facilities, extension course programs, agricultural experiment stations and, where available, cultural activities such as symphony performances, drama workshops, etc. "Area study programs," where they exist, will be of special interest to foreigners. Such visits should seek to underline the role of our schools and universities–to teach and learn, not to function as political instruments–and to show the diversity of our educational institutions, including privately endowed colleges, State or city colleges, land-grant universities and church-affiliated institutions.

(l) Housing development.–Visits to model homes, apartments, and publicly supported housing developments designed for low-and middle-income groups will be of particular interest to foreign military trainees. Visits to rural housing areas should be included where appropriate.

(m) Historic sites and National or State parks.–In such trips emphasis should be laid on battlefield parks of our various wars to underline the care we have taken to preserve and commemorate the military side of American history.

(n) Sporting events.–Visits to baseball and football games, golf matches, and other sporting events such as rodeos, regattas, and automobile races will show the foreign military trainee multiplicity of American athletic interests.

TOURS TO WASHINGTON

1. Selected foreign officers in senior, career, post-graduate and other significant courses which the military departments will designate, including officer or cadet training courses, will at some time during their stay in this country make a visit to Washington, D.C. The purpose of this visit is to round out their acquaintance with U.S. political institutions by showing them the work of the Federal Government as it goes forward in its Washington environment and by introducing them to some of the key officials, both civil and military, who bear its burdens.
2. The Washington tour is therefore an integral part of informational programs. In planning it, commanders of training establishments will coordinate the timing and itinerary of this visit with other aspects of their informational programs to assure that it achieves the fullest possible effect.
3. Responsibility for overseeing the necessary arrangements for foreign military trainees and visitors in the Washington area lies with the Director of Military Assistance. Military departments will therefore keep that office informed of their schedules for Washington visits and will routinely provide it with the detailed information necessary for the discharge of that responsibility.

Appendix 4: U.S. Service Schools Attended by Foreign Military Students in FY 1963

Department	School	Location	Annual Enrollment		Average Tour (Weeks)
			Officers	Enlisted Men	
Air Force	Lackland AFB[a] (English language)	San Antonio, Tex	831	584	15
Army	Air Defense School	Fort Bliss, Tex	565	482	19
	Infantry School	Fort Benning, Ga	391	–	16
	Special Warfare School	Fort Bragg, N.C.	340	–	6
	Engineer School	Fort Belvoir, Va	258	128	14
	Armor School	Fort Knox, Ky	252	45	15
Air Force	Keesler AFB	Biloxi, Miss	240	729	30
Army	Signal School	Fort Monmouth, N.J.	236	89	24
	Artillery and Missile School	Fort Sill, Okla	209	26	16
	Ordnance School	Aberdeen Proving Ground, Md.	199	71	1
Air Force	Air University, Maxwell AFB	Montgomery, Ala	199	–	20
	Chanute AFB	Illinois	187	272	19
Army	Transportation School	Fort Eustis, Va	162	99	13
	Command and General Staff College	Fort Leavenworth, Kans.	154	–	39/18
	Aviation School	Fort Rucker, Ala	147	29	12
Air Force	Amarillo AFB	Texas	143	214	9
	Sheppard AFB	Wichita Falls, Tex	140	250	16
	Randolph AFB	San Antonio, Tex	138	31	9
Army	Ordnance Guided Missile School	Redstone Arsenal, Ala	123	471	14

Department	School	Location	Annual Enrollment		Average Tour (Weeks)
			Officers	Enlisted Men	
Navy	Naval Intelligence School (English Language training)	Washington, D.C.	122	123	6
	Amphibious Training Command	San Diego, Calif	121	100	15
	Antisubmarine Warfare School	do	119	39	8
Army	Quartermaster School	Fort Lee, Va	117	31	12
Air Force	Moody AFB	Valdosta, Ga	111	–	42
Navy	Marine Corps Schools	Quantico, Va	101	13	b
	Naval Postgraduate School	Monterey, Calif	89	–	39/104
	Gunnery School	San Diego, Calif	97	–	8
	U.S. Naval Schools Com.	Norfolk, Va	92	17	4
	U.S. Naval Training Center	San Diego, Calif	88	142	4/10+
	U.S. Naval Schools, Mine Warfare	Charleston, S.C.	8	71	9
	Fleet AntiAir Warfare School	San Diego, Calif	80	44	1
Air Force	Luke AFB	Arizona	80	6	14
	Craig AFB	Alabama	75	23	50
Army	Southeast Signal School	Fort Gordon, Ga	71	49	16
Air Force	Reese AFB	Texas	70	23	50
	Brooks AFB	Texas	68	16	9
	Lowry AFB	Denver, Colo	65	136	20
Navy	Fleet Training Center	San Diego, Calif	65	647	6
Army	Chemical School	Fort McClellan, Ala	62	2	10
Navy	Fleet AntiAir Warfare Training Center	Virginia Beach, Va	62	105	1

	Naval Air Basic Training Center	Pensacola, Fla	57	–	78
Army	Civil Affairs School	Fort Gordon, Ga	54	–	8
Navy	Submarine School	Groton, Conn	51	–	6
	Naval Training Center	Great Lakes, Ill	50	264	20
	Marine Corps Base	Camp Lejeune, N.C.	37	47	12
Air Force	McClellan AFB	California	36	18	14
	Williams AFB	Arizona	36	39	52
Navy	Supply Corps School	Athens, Ga	35	4	12
	Sonar School	Key West, Fla	35	31	15
	Damage Control Training Center	Philadelphia, Pa	34	21	7
Air Force	Laughin AFB	Texas	32	3	45
	Webb AFB	Texas	27	16	55
Navy	Explosive Ordnance Disposal School	Indian Head, Md	27	16	10
	Naval Air Advanced Training Command	Corpus Christi, Tex	26	–	16
	Naval War College	Newport, R.I.	26	–	40
	Naval Schools Command	Treasure Island, Calif	25	54	8
	Naval Hospital	Oakland, Calif	24	3	34
Air Force	Laredo AFB	Texas	23	17	39
Navy	U.S. Navy Medical School	Bethesda, Md	20	–	11
	Naval Training Center	Bainbridge, Md	17	79	6
Air Force	Perrin AFB	Texas	17	2	26
Navy	Naval Hospital	Philadelphia, Pa	15	2	52
	Marine Corps Base	Camp Pendleton, Calif	14	16	b
	Civil Engineer Corps	Port Hueneme, Calif	13	–	8
	Undersea Weapons School	Key West, Fla	12	29	12
	Pacific Fleet Training Command	San Diego, Calif	12	–	1/2

Department	School	Location	Annual Enrollment		Average Tour (Weeks)
			Officers	Enlisted Men	
Air Force	George AFB	California	11	–	6
	Military Air Transport Service	(Various locations)	8	–	52
	Air Force Academy (Instructors)	Colorado	5	–	52
Navy	USN Aviation Medical Center	Pensacola, Fla	44	2	24
	Naval Amphibious Base	Norfolk, Va	4	–	12
Air Force	Wilford Hall Hospital Lackland AFB	San Antonio, Tex	4	–	13
Navy	Guided Missile School	Virginia Beach, Va	54		4
	Marine Corps Recruit Depot	San Diego, Calif	39		b
Air Force	Robins AFB	Georgia	28		b
	Nellis AFB	Nevada	b	b	14
Totals					
Top 10 (officer enrollment annually			3,521	2,154	5,675
Top 20			5,036	3,643	8,679
Top 30			6,031	4,087	10,118
All 76 includes 30 civilians, 235 officers and enlisted men not differentiated).			7,440	5,757	13,462

[a]All Lackland trainees proceed to other schools for special training.

[b]Unknown.

Source: U.S., Congress, House, Committee on Foreign Affairs, *Winning the Cold War: The U.S. Ideological Offensive, Hearings*, before the subcommittee on International Organizations and Movements of the Committee on Foreign Affairs, House of Representatives, 88th Cong., 2d sess., pp. 1040-41.

Appendix 5: Excerpts from "Know Your Communist Enemy: Communism in Red China" (1955)

"The Chinese people have one of the world's oldest civilizations built on the arts of peace. But today they are under the domination of Communist rulers who head the Chinese People's Republic. This Government is militantly spreading communism among its neighbors. It uses internal subversion to weaken and undermine free governments; if this is insufficient, open force is used.

After years of conspiracy against the Republic of China (Nationalist), the Reds achieved control of the Chinese mainland in 1949. . . .

When the Communists emerged as rulers of the Chinese mainland, there were diverse opinions as to what would happen. Some believed that Chinese Communists were different from Russian Communists, that they would–as they had promised–work to unify and modernize China under Mao Tse-tung, their chief. Others warned–Communists are Communists, wherever you find them. Their aim is to overthrow free governments and work for Communist domination of the world.

The years since 1949 have shown that Chinese Communists are an integral part of the international Communist conspiracy. Mao Tse-tung has developed in China the same ruthless and aggressive imperialism we have seen in Soviet Russia.

Inside China itself, the Communists have brought about many drastic changes in the life of the people. They have done this with a terror and ruthlessness that easily matches that of the Soviet Communists."

Appendix 6: Excerpts from "Que Es El Comunismo?" and "Democracia Contra Comunismo" (Early 1960s)

"No es fácil definir el comunismo; es una palabra antigua que ha significado diferentes cosas para pueblos que existieron mucho antes del establecimiento del Gobierno Soviético en Rusia. Algunas veces el comunismo ha significado el sueño de un paraíso en la tierra. A veces ha significado la revolución contra la pobreza y la miseria, con engendros de violencia y destrucción. Actualmente, comunismo significa, en general, la forma de gobierno de la Rusia Soviética ye de la China Comunista. Significa el gobierno de un pequeño grupo de hombres que han obtenido el poder mediante una revolución violenta. Ellos alegan que gobiernan en nombre de los trabajadores y campesinos, pero, en realidad, utilizan la fuerza y el engaño para mantenerse en el poder. Su gobierno es el unico dueño y administrador de todas las propiedades. Estos dictadores hablan de una sociedad ideal, pero no permiten que el pueblo tome parte alguna en las decisiones que han de tomarse para producir esa sociedad ideal. Los dictadores comunistas prometen libertad, pero ellos han destruido las libertades de p palabra y prensa, así como otras muchas libertades personales.

Una de las creencias básicas de los comunistas es la de que la propiedad, es decir, la tierra, minas, fábricas y los astilleros deben ser propiedad del Estado y operados por éste. De acuerdo con esta idea, el gobierno debe administrar las haciendas, funcionar las minas y operar las fábricas y ferrocarriles. El comunismo ha significado la drástica reglamentación por el gobierno de la vida privada de todo el pueblo, llegando al extremo de dictar normas respecto a cómo debe pensar el individuo y como ha de vivir, casarse, trabajar y distraerse. . . ."

En una ciudad de los Estados Unidos llamada Bristol, Vd. puede situarse en la acera de una calle y estar en el Estado de Virginia y con solo cruzar la calle, se traslada Vd. al Estado de Tennessee. Sin embargo, al hacer esto no se observa diferencia alguna en el Estado de Tennessee, porque la gente allí trabaja y se divierte de la misma manera que lo hace al otro lado de la calle. En cambio, si en la ciudad de Berlín Vd. cruza cierta calle, se encuentra, de repente, en un mundo distinto.

La ciudad de Berlín está dividida. Una parte está dominada por los dictadores comunistas. La otra, está bajo la protección de las naciones occidentales, cuyos ciudadanos disfrutan de libertades que no tienen las personas que se encuentran bajo el dominio comunista. Otros folletos de esta serie tratarán de cómo viven los pueblos de los países comunistas sin estas libertades.

¿Podemos Albergar Esperanzas de que se Logre la Coexistencia Pacífica? En los dos grandes bandos de nuestro mundo, hombres y mujeres por igual desean la paz. Y la paz es algo tan grande, tan sublime, que no parece haber frustración alguna capaz de ahogar la esperanza de los pueblos del mundo en poseerla.

Sin embargo, ya en 1961 hasta las personas más optimistas tuvieron que reconocer que nuestro mundo estaba dividido en tres grupos. Por una parte, el mundo comunista cuenta con más de 13 millones de millas cuadradas de territorio y una población ascendente a un billon de personas. En contraste, el Mundo Libre tiene más de 37 millones de millas cuadradas y una población de casi dos billones de personas. Pero, más de 800 millones de personas del Mundo Libre no toman parte activa en la labor de contrarrestar las amenazas de la Rusia Comunista y la China Roja. Esos 800 millones de personas en las llamadas naciones neutrales, que tienen una superficie le 10 millones de millas cuadradas. Las estadísticas señalan que aquellas naciones que están aliadas para protegerse y proteger a las demás naciones libres contra los ataques comunistas, tienen una población de 1.122 millones de personas con 26 millones de millas cuadradas. Militarmente, el Mundo Comunista es muy poderoso, ya que de acuerdo con cifras del año 1961, sus fuerzas armadas ascenian a 7,100,000 hombres, mientras que el Mundo Libre contaba con unos 4,500,000 hombres. Es indudable que cantidades tan elevadas de hombres armados y equipados, representan una amenaza a la paz del mundo entero.

Aunque la Unión Sovietica utiliza parte de sus fuerzas para vigilar a satélites como Hungría y Alemania Oriental, la mayor parte de esas fuerzas las mantiene en reserva. Es bien sabido, sin embargo, que la Unión Soviética ha alentado y dirigido guerras en otros países mediante la subversion y propaganda. Subversión significa el derrocamiento o destrucción de un gobierno por medio de acciones internas en vez de ataques realizados desde el exterior. Propaganda significa la difusión de ciertas ideas y creencias. Además, hay pruebas de que pilotes soviéticos operaban aviones MIG durante la guerra de Corea en contra de las tropas de las Naciones Unidas.

El cacareo de la "coexistencia pacífica" no ha sido más que un truco de la propaganda soviética. Todos los dirigentes comunistas, empezando por el propis Lenín, lo utilizaron. Ellos aspiran a que el Mundo Libre crea que se puede vivir en paz con los comunistas. La realidad, sin embargo, es otra muy distinta, pues lo que ellos esperan y estan planeando es la dominacion del Mundo Libre y para realizar su anhelo utilizian la fuerza si ellos lo consideran necesario o conveniente. Por ejemplo, el mismo año en que los comunistas anunciaron la coexistencia pacífica, la China Roja completó su invasión del Tibet.

Así es como se ha ido desarrollando la guerra fría. Hacia los finales de la década 1950-1960, agentes comunistas se introdujeron en Guatemals, nación a la cual planeaban conquistar. Al mismo tiempo, al sureste de Asia fuerzas comunistas chinas invadieron a Viet-Nam y empezaron una revolución en Laos, cuyas dos naciones son relativamente pequeñas. Estas fuerzas comunistas ocuparon el Tibet e invadieron el norte de la India. En el Oriente Medio agentes comunistas organizaron entonces disturbios en Irak, Jordania y el Líbano (donde fuerzas norteamericanas desembarcaron a peticion del Gobierno libanés). Los agentes comunistas prosiguieron también sus trabajos de subversión en Argelia, y disfrazados de técnicos, entraron en dos nuevas naciones del Africa: Ghana y la República del Congo, y para hacer más grave su amenaza, la China Comunista anunció que estaba casi lista para detonar su primera bomba atómica.

Además, cuando los trabajadores de Hungría protestaron en 1956 contra sus dictadores comunistas, Rusia mandó 100,000 soldados armados y 5,000 tanques para aplastarlos, con el resultado de que dieron muerte a más de 30,000 patriotas húngaros. Los rusos también cerraron las fronteras de Hungría para impedir la huída de refugiados a Austria, mataron a mansalva a todos húngaros que huían y se robaron los suministros que la Cruz Roja envió a Hungría para aliviar la desgracia del pueblo húngaro. Cuando en la Asamblea General de las Naciones Unidas se pidió que fuera autorizada la entrada de observadores de la Organización en Hungría para investigar lo que estaba ocurriendo, la petición fué bruscamente rechazada por el gobierno comunista de Hungría. Acciones como esta en Hungría y en otras partes del mundo, han convertido el postulado comunista de la coexistencia pacífica en palabras huecas de todo sentido y en letra muerta.

1. ¿En qué países han tenido éxito las Naciones Unidas ayudando a zanjar desavenencias?
2. ¿Cómo ha obstaculizado la Unión Soviética los esfuerzos de las Naciones Unidas para promover la paz?
3. ¿Qué fué la Conferencia de la Cumbre y qué logró?
4. ¿Qué significa la palabra neutral?
5. ¿Por qué tantas personas no podrían aceptar como hecha de buena fe, la propuesta soviética para la coexistencia?

Peligros Futuros. Muchos hombres de Occidente consideran que el peligro para las naciones nuevas y débiles del Mundo Libre radica en la creciente penetración de estos países por la Unión Soviética y la China Comunista, mediante tratados comerciales, ayuda económica y la propaganda y actividades subversivas. Mientras en estas naciones se conducen como amigos y difunden propaganda de odio contra Occidente, los comunistas planean y se preparan para apoderarse de sus respectivos gobiernos. Al parecer, los comunistas prefieren una victoria sin guerra (arrebatándole las riendas del poder a gobiernos débiles e inestables, como lo han hecho en Cuba, para aislar entonces de Occidente a esos países y muy en especial para aislarlos de los Estados Unidos.

La Indiferencia Puede Ser Nuestro Enemigo. Un patriota irlandés, John Philpot Curran, dijo una vez que "La vigilancia eterna es el precio de la libertad". La gran verdad que encierra esta frase ha sido demostrada frecuentemente en la historia. En la antigua Grecia, los ciudadanos de Atenas perdieron sus libertades, a pesar de que el orador Demóstenes les había advertido sobre la amenaza que representaba el Rey Filipo de Macedonia. Y otra vez, en la década de 1930-1940, antes de la segunda guerra mundial, los cuidadanos alemanes se encontraron el férreo yugo de una dictadura que los llivó a la guerra, la devastación y la derrota, únicamente porque no rechazaron la falsa propaganda de Adolfo Hitler y de su Partido Nazi.

Hasta en los Estados Unidos, donde la libertad ha alcanzado una posición prominente en la historia, la ignorancia y la indiferencia de los ciudadanos han tenido a veces graves consecuencias. Cuando los ciudadanos descuidan su vigilancia, permiten la corrupción en los gobernantes de sus ciudades. La indiferencia nacional en lo que respecta a la conservación de nuestras libertades sería desastrosa. Ninguna nación puede exponerse al riesgo de tratar la libertad como si fuera una planta silvestre, que vive y florece sin ser cuidada.

Donde nos Encontramos Actualmente. Si se observa el mapa del mundo, aun cuando no esté dividido para mostrar los dos grandes campos, es decir, las naciones comunistas y las naciones libres, incluyendo entre las últimas a las naciones neutrales, se pueden ver, no obstante, estas divisiones. Las naciones occidentales, la Rusia Soviética y la China fueron aliadas durante la segunda guerra mundial. Pero, ahora, los que una vez fueron aliados están divididos y mantienen una paz rara, que no es paz, en realidad, sino la guerra fría. Conocemos del más pequeño grupo de naciones neutrales que no se han unido a las naciones libres en alianzas defensivas contra la agresión, hemos visto los esfuerzos realizados por las Naciones Unidas en pro de un acercamiento entre los dos bandos y del mantenimiento de la paz; y hemos conocido de las conferencias que han fracasado a pesar de los llamamientos que se han hecho a favor de soluciones pacíficas y algunos de los medios que usa el comunismo para reclutar adictos aun dentro de las mismas democracias.

Aunque gran parte de este relato puede inquietar tal vez a muchas personas, son alentadores en cambio los esfuerzos que muchos hombres realizan en distintas naciones y en las Naciones Unidas, para encontrar puntos de avenecia entre los bandos en que está dividido nuestro mundo."

Appendix 7: Installation Hospitality Program for Foreign Naval Students at The Great Lakes Naval Training Center*

SERVICE SCHOOL COMMAND
U.S. Naval Training Center
Great Lakes, Illinois

16 November 1957

SERVICE SCHOOL COMMAND INSTRUCTION 5710.1

From: Commanding Officer, Service School Command
To: Service School Command
Subj: Hospitality program for foreign trainees

1. *Purpose*. The purpose of this Instruction is to outline detailed procedures for conducting a comprehensive hospitality program for foreign naval trainees.
2. *Discussion*. In addition to conducting technical training for a large number of foreign naval trainees this command is afforded the opportunity to help acquaint these men with America and Americans, thus planting the seeds of real friendship and understanding. When the trainees return to their homes they will carry with them impressions of the United States, formed for the most part from their experiences while attached to this command. It is vital, therefore, that they be given sufficient opportunity to absorb as much of American life and customs as possible—but in a manner which will not detract from classroom time.
3. *Responsibility*.
   a. The Officer in Charge, Journalist School, is assigned responsibility for conducting the foreign student hospitality program. He will serve as chairman of the Foreign Student Hospitality Committee and will insure maintenance of files, host lists, and contacts required to conduct the program.
   b. The Officer in Charge of each school training foreign students will designate one officer to serve on the Foreign Student Hospitality Committee. The Assistant Officer in Charge of Journalist School will serve as permanent secretary of this committee.
   c. It shall be the duty of the Foreign Student Hospitality Committee to arrange programs described in paragraph 4, to disseminate appropriate information to foreign students and to furnish escorts for trips away from the command as required.

*This reproduction of the Service School Command directive establishing the hospitality program for foreign trainees at Great Lakes, Illinois, is quoted verbatim from John M. Dunn, "Military Aid and Military Elites: The Political Potential of American Training and Technical Assistance Programs," (unpublished Ph.D. dissertation, Princeton University, 1961), pp. 358-61.

d. Hospitality for our foreign students is a continuing "All Hands" responsibility. Initiative and enthusiastic support is expected.

4. *Program*. The Hospitality Program shall include, but not be limited to, the following features:

a. *Commanding Officer's Reception*. The Committee shall arrange a reception at least every two months at which the Commanding Officer will be host to all officer students, including U.S. officers, who have reported to the command since the previous such reception. If practical, receptions will be held at the Gunnery Officer's School on first Tuesday of every other month.

b. *Home Hospitality*. A Major share of the program should be devoted to local home hospitality. Efforts should be made to encourage local Navy and civilian families to invite foreign students to their homes for visiting and, if possible, informal dining. Brochures outlining the program should be mailed to church, civic, and professional groups inviting their participation in this program. Invitations will be received from such groups or from individuals. Information required to complete arrangements will be recorded and retained on file. A general invitation or invitations to special categories of personnel, as desired by the host, will be extended through the appropriate school's representative of the hospitality committee. The chairman or secretary of the committee will insure that the host is informed of acceptances and that the student is provided memorandum confirmation of arrangements and appropriate transportation by the host or adequate directions to arrive by public conveyance. Hosts calling for foreign students by private automobile will be cleared through the 22nd Street gate and directed to Building 300. The chairman or secretary of the Hospitality Committee shall confirm such arrangements in memorandum to the Service School Command duty officer and the Provost Marshal. School representatives will obtain an informal oral report from each trainee after return from such visits and will report significant matters to the chairman of the committee.

c. *Tours*. A tour to Chicago, Milwaukee, or another nearby community will be arranged approximately once every six weeks. The tour should include a visit to an industrial, business or cultural centers, a call on the mayor or other officials as appropriate, a luncheon sponsored by a civic or industrial group, entertainment and home hospitality arranged through a civic group if practicable. Information on such tours should be disseminated ten days or more in advance and spaces allotted on an equitable and rotational basis.

d. *Athletics*. The game of soccer is as popular in many foreign countries as baseball and football are in the United States. Two teams of foreign students were organized in the spring of 1957 and contests arranged between them and other teams. This command has purchased 24-pairs of soccer shoes for this purpose. Equipment is available through the Special Services Department, Administrative Command, which has custody of the soccer shoes. This, and similar sporting programs appropriate to the nations represented, are to be encouraged.

e. *Special Events and Entertainment*. Free admission to sporting events, concerts, and other entertainment can sometimes be obtained for groups of this nature. An effort shall be made to do so wherever possible.

f. *Recorded Christmas Messages*. To the extent that funds permit, all foreign trainees who will be here over Christmas shall be offered an opportunity to record Christmas greetings in their own languages for forwading to their homes. Recordings will be processed by the Journalist School and forwarded to Military Assistance Advisory Groups, Naval Missions or Naval Attaches in countries concerned, no later than 1 December.

g. *Christmas Hospitality*. A special effort shall be made to secure invitations for all foreign trainees who desire them to spend Christmas in an American home. Naval training center chaplains have been of assistance in these arrangements.

5. *Funds*. Limited official funds available to support this program will be requested and administered by the Material and Fiscal Officer.
6. *Transportation*. Transportation, when required, will be requested from the Special Services Officer, Administrative Command.
7. *Publicity*. The Officer in Charge, Journalist School, in his capacity as Service School Command Public Information Officer, will publicize this program as appropriate in accordance with applicable directives.
8. *Letters of Appreciation*. Appropriate letters of appreciation to appropriate civilian hosts will be prepared by the secretary of the Hospitality Committee for the signature of the Commanding Officer. When practicable, photographs of the event will be included. The MDAP Clerk, Bldg 300, will provide clerical assistance in all matters pertaining to the Hospitality Program.
9. *Reports*. A memorandum report shall be made to the Commanding Officer by the Chairman, Foreign Students' Hospitality Committee by the last day of each month informing him of events arranged and enclosing clippings if any publicity has been obtained.

/s/ ______________________________
Commanding

Appendix 8: The Allied Sponsor Program at Fort Leavenworth*

U.S. ARMY COMMAND AND GENERAL STAFF COLLEGE
Fort Leavenworth, Kansas

7 July 1959

SUBJECT: Allied Officer Sponsor Program
TO: See Distribution

1. The purpose of this letter is to provide information to state policies concerning the Allied Officer Sponsor Program.
2. The program applies to Allied *student* officers only.
3. General:
   a. Each year approximately 140 Allied officers from 40 to 45 nations of the free world attend courses at the USACGSC. Allocations normally are 80 in the Regular Course and 30 in each Associate Course. Occasionally special additional courses are conducted when specifically directed by the Department of the Army. Normally about half the Allied officers attending the Regular Course are accompanied by their dependents–or are joined later by their families. Associate Course students rarely bring dependents.
   b. Selection of students is made by the country concerned in coordination with the appropriate MAAG, mission or attache based on criteria set forth in current regulations (AR 551-50). Those selected to attend are highly motivated officers, many of whom presently occupy or are destined to occupy influential positions in their armies or governments. Former Allied graduates have become chiefs of staff, cabinet ministers and even presidents in their countries.
   c. Aside from the purely military value, the presence of these outstanding representatives of the free nations of the world at the College affords a unique opportunity for promoting improved Allied relations by close contact between US personnel and the Allied students. These relations, both official and social can assist materially in achieving understanding and respect on the part of our Allies toward the United States, and of US personnel toward our Allies. An important element in promoting effective relationships is the Allied Officer Sponsor Program. The Commandant places great importance on the program and considers that officers who

*This reproduction of the installation directive establishing the sponsor program for foreign officer students at the U.S. Army Command and General Staff College, Fort Leavenworth, Kansas, is quoted verbatim from John M. Dunn, "Military Aid and Military Elites: The Political Potential of American Training and Technical Assistance Programs," (unpublished Ph.D. dissertation, Princeton University, 1961), pp. 352-54.

act as sponsors perform a valuable service to the College and to the United States.

4. Participants–US officers assigned to the Staff and Faculty and the Post are eligible to participate in the program by sponsoring Allied students.
5. Responsibility–The Office of the Allied Personnel Supervisor is responsible for implementation of the Allied Officer Sponsor Program.
6. Objectives–The objectives of the sponsor program are to:
   a. Promote cultural and social integration of Allied students and their families into Post life.
   b. Assist the Allies in adapting themselves to our culture and way of life.
   c. Make the Allies feel "accepted."
   d. Create friendship and good will for the United States.
   e. Provide an opportunity for US officers and their families to broaden their understanding of our Allies.
7. Procedures:
   a. Upon notification by Department of the Army (generally in March of each year) of spaces accepted by foreign countries for the following year's courses, the Office of the Allied Personnel Supervisor furnishes a list of countries scheduled to send students to the next Regular Course and Fall Associate Course to current sponsors offering them an opportunity to continue their participation in the program. This list is followed by a College letter offering sponsorships to all officers assigned to the College and Post for countries not spoken for by present US participants in the program. In October of each year a similar procedure is followed to afford opportunity for US officers to sponsor Spring Associate Course Allied students.
   b. After approval of the sponsors by the Assistant Commandant, notification and informational material is furnished by the Allied Personnel Supervisor to each sponsor, and to the Allied sponsoree upon arrival. This is followed by a letter from the Commandant to each sponsor acknowledging his selection as sponsor.
   c. There are no assigned duties for sponsors; however, they are expected to maintain close contact with their sponsorees throughout the academic year and in general to act as personal friends, providing a personal touch to supplement the sponsoree's official association with the College. Written guidance to sponsors is provided by the Office of the Allied Personnel Supervisor at time of assignment of sponsors.
8. Relationship to Women's Club activities–The International Group of Fort Leavenworth Women's Club is organized to assist in the social integration of Allied students, with particular attention to dependents. In view of similarity of objectives, opportunity for participation in the International Group activities is extended to sponsors and their wives.
9. Efficiency reports–Mention will be made in item 14 of efficiency reports

of the fact that an officer and country concerned will be included (Supplement Nr. 5, USACGSC Faculty Memorandum 22, 21 January 1959 and Memorandum Nr 3, Hq, Fort Leavenworth, Kansas, 16 January 1959).

FOR THE COMMANDANT:

/s/________________________

Colonel, Infantry
Secretary

Appendix 9: Extracurricular Activities for Foreign Military Students at Lowry Air Force Base, Colorado*

A. *Civic and Community Relationships*

1. Foreign students are made honorary members of the International House and of the Pan-American Club, which cater to students from countries all over the world. Many prominent civic leaders and businessmen are members of these two organizations. Daily social and educational events are conducted for the benefit of the students.
2. *Kiwanis and Rotary Clubs*
   The foreign students, escorted by a member of the USAF, regularly attend the luncheons and social functions of these organizations where they are introduced and are able to make acquaintances.
3. *Off Duty Education*
   Foreign students are encouraged and most of them take advantage of the Opportunity School, free of charge, conducted by the City of Denver. This school teaches classes in English, other languages and many of the social science subjects.
4. *Private Parties and Social Invitations*
   Civic and business leaders and others interested in the foreign student program hold social functions at their homes and at clubs for the students. Regular invitations for dinner or a social evening are constantly being received. Church organizations particularly take an interest in our foreign students and invite them to the many functions which they sponsor. During the Thanksgiving and Christmas holidays, invitations were received to more than adequately insure that each and every student had an invitation.
5. *Formal Dining-In*
   Foreign officers participate in our Formal Dining-In functions. A Formal Dining-In to be conducted at the Officers' Club on 10 April 1959 will provide an occasion for fellowship and to honor the foreign officers attending technical school at Lowry Air Force Base.
6. *Publicity*
   Pictures and articles pertaining to the foreign students and their activities are an integral part of the city and base newspapers. Pictures of the

---

*This reproduction of an unpublished memorandum submitted by the Commander, Lowry Air Force Base to the Office of the Assistant Secretary of Defense, International Security Affairs, in response to a request for a summary of the extracurricular activities provided for foreign military students is quoted verbatim from John M. Dunn, "Military Aid and Military Elites: The Political Potential of American Training and Technical Assistance Programs," (unpublished Ph.D. dissertation, Princeton University, 1961), 355-57.

students are provided to them as a memoir of their enjoyable stay at Lowry Air Force Base and Denver, Colorado. Television appearances have been made on several occasions to honor our foreign students.

7. *United Nations Day*
   All foreign students who desired to do so participated in the United Nations Day ceremonies in downtown Denver and afterwards attended a social event held by the civic and business leaders of the city.
8. *Guest Speakers*
   Requests are regularly being received for foreign students to act as guest speakers at many of the church and civic organizations in the city. Topics are limited, of course, to appropriate subjects.
9. *Tours*
   Frequent and regular tours to include the following are made: State Capital, U.S. Treasury Department, USAF Academy, Fitzsimons General Hospital, Radio Stations, TV Stations, Gates Rubber Company, Coors Brewery, Stock Show and Rodeo, Colorado Springs, Boulder, Colorado, University of Colorado. On two occasions during the visit to the State Capital, the students were granted an audience of one half hour to meet and talk with the Governor of the State.
10. *Beauty Contest*
    A beauty contest was held on 6 December 1958 to select "Miss Foreign Intrigue." Students from different countries acted as judges. The title winner was selected at a holiday party attended by all of the foreign students and many of the Center and Technical School Staff and prominent civic leaders. The Officers' Club was reserved for this event at which a smorgasbord dinner, refreshments and dancing to one of the better bands were available. Publicity on television and in the base and city newspapers was given to this contest.
11. *Las Vegas, Nevada*
    Ten foreign students were taken on a two-day trip at no expense to the government to Las Vegas, Nevada. The Chamber of Commerce and hotel owners cooperated to the fullest in showing these students a most enjoyable time.
12. *Picnic*
    During the summer months frequent picnics are held for the foreign students in the beautiful mountains of Colorado. Many guests, including civic leaders, are always invited to attend these outings and to meet with and to help create a relationship of cordiality and good will.

B. The above activities are typical of those that have been conducted and will be continued for the benefit of the students and to maintain our excellent civic relationships in the surrounding community. Because of the importance of this program, every effort will be made to make it a most successful one and to allow the students, when they return to their homeland, to take with them fond memories and pleasantries of their stay at Lowry Air Force Base and the City of Denver and the State of Colorado.

C. The addition of a program similar to that being utilized by the Air University and the City of Montgomery and incorporating the presentation of a booklet to foreign students is now in the process of being developed at Lowry Air Force Base. Our Foreign Liaison Officer and Information Services Officer plan not only to prepare a greeting booklet for the students but also to arrange welcoming meetings with the leading civic representatives to include both the Government of the State and the Mayor of the City of Denver.

Appendix 10: Foreign Aid and Military Interventions

The tables which appear in Chapter Seven are based upon the following data. Sources on military aid for each country are listed in end notes on p. 290. Not included are annual Defense Department *Military Assistance Facts* and the president's *Annual Report to the Congress* on foreign aid. These have been freely drawn upon for the 1966-70 period.

Our country universe is divided into two categories: those which experienced conservative or radical interventions and those which did not. The definition presented in Chapter One establishes the criteria for classification, and the *New York Times* is the principal source for the period from January 1945 until January 1970. In many instances it was difficult to ascertain whether ideological goals were prominent among the motives for military intervention. At times we have used the explicit declarations or broadcasts of the plotters. Frequently, postcoup behavior or commentator opinion were relied upon. While it cannot be doubted that some of our classifications may be arbitrary, it is equally certain that such bias was not in one direction. The limited size of the universe and the inherent difficulty of imputing motives require that our conclusions and determinations be viewed as no more than tentative in nature.

In general we have assumed the continuance of military aid from the date of its first inception. There are two reasons for this. First, reliable information on the extent to which relationships were maintained is exceedingly difficult to obtain. Second, it is probable that in most cases liaisons of some sort continued to exist. Furthermore, in the absence of wholesale purges some effects of prior foreign training would still be experienced.

Ideological Interventions

Nation	Interventions		Military Aid	
Algeria[1]	Left:	July 1962	Egypt	1956-
	Other:	June 1965	USSR	1962-
			Jordan	1962-
			France	1964-
Burma[2]	Right:	Sept. 1958	Britain	1948-
	Other:	March 1962	U.S.A.	1953-
			France	1965-
			Australia	
			India	
Cambodia[3]	Right:	March 1970	U.S.A.	1954-
			China	

Nation	Interventions		Military Aid	
			USSR	
			France	
			Japan	
Congo (B)[4]	Left:	Aug. 1963	France	1960-
	Other:	May 1968	USSR	1964-
			China	1964(?)-
Congo (K)[5]	Right:	Sept. 1960	Belgium	1960-
	Other:	Nov. 1965	U.S.A.	1962-
			Britain	1963-
			Italy	1963-
			Israel	1963-
Egypt[6]	Other:	1952	Britain	1964-
	Left:	March 1954	U.S.A.	1948-
			USSR	1955-
			CSR	1955-
			W. Germ.	1960-
Ghana[7]	Right:	February 1966	Britain	1957-
			Israel	1958-
			USSR	1960-
			India	1960-
			Canada	1961-
Greece[8]	Right:	April 1967	Britain	1945-
			U.S.A.	1947-
Indonesia[9]	Right:	March 1966	U.S.A.	1950-
			Britain	1954(?)-
			France	1954(?)-
			USSR	1958-
			CSR	1958-
Iran[10]	Right:	August 1953	U.S.A.	1946-
			Britain	1965(?)-
			USSR	1967-
Iraq[11]	Left:	July 1958	Britain	1948-
	Right:	February 1963	U.S.A.	1954-
			USSR	1958-
Jordan[12]	Right:	April 1957	Britain	1946-
			U.S.A.	1957-
Laos[13]	Right:	January 1960	U.S.A.	1954-
	Left:	August 1960	USSR	1960-
	Right:	April 1964		
Libya[14]	Left:	Sept. 1969	Britain	1952-
			U.S.A.	1954-
			France	

Nation	Interventions		Military Aid	
Mali[15]	Right:	Nov. 1968	USSR	1960-
			U.S.A.	1961-
Pakistan[16]	Right:	October 1958	Britain	1948
	Other:	March 1969	U.S.A.	1948-
			France	1965-
			China	1965-
			USSR	1968-
Somalia[17]	Left:	October 1969	Britain	1959(?)-
			U.S.A.	1963-
			USSR	1964-
			China	1964(?)-
South Yemen[18]	Left:	May 1968	Britain	1967-
	Other:	June 1969	USSR	1968-
Sudan[19]	Other:	Nov. 1958	Britain	1956-
	Left:	October 1964	U.S.A.	1957-
	Left:	May 1969	USSR	1968-
			Yugoslavia	
Syria[20]	Right:	March 1949	Turkey	1949-
	Other:	May 1949	USSR	1954-
	Other:	Nov. 1951	U.S.A.	1963-
	Left:	Feb. 1954		
	Right:	Sept. 1961		
	Other:	March 1962		
	Other:	March 1963		
	Left:	Feb. 1966		
	Other:	March 1969		
	Other:	November 1970		
Thailand[21]	Other:	Nov. 1947	U.S.A.	1950-
	Other:	April 1948		
	Right:	Nov. 1951		
	Other:	Sept. 1957		
	Other:	Oct. 1958		
Zanzibar[22]	Left:	January 1964	Britain	1963-
			USSR	1964-
			E. Germany	1964-
			China	1964-

Nonideological Interventions

Nation	Interventions: Other	Military Aid	
Afghanistan[23]	1953	USSR	1954-
	1963	U.S.A.	1958-

Nation	Interventions: Other	Military Aid	
Burundi[24]	Sept. 1966	Belgium	1962-
	Nov. 1966		
Cambodia[25]		U.S.A.	1955-
		USSR	1963-
Cameroons[26]		France	1962-
		U.S.A.	1962-
Central African Rep.[27]	Jan. 1966	France	1960-
		U.S.A.	1962-
Ceylon[28]		Britain	1948-
		U.S.A.	1967-
Chad[29]		France	1960-
		Israel	?
Cyprus[30]		Britain	1960-
		USSR	1964-
Dahomey[31]	October 1963	France	1960-
	November 1965	U.S.A.	1961-
	December 1967		
	June 1968		
	November 1969		
Ethiopa[32]		U.S.A.	1953-
		Sweden	1934-
		Britain	
		France	
		India	1960-
Gabon[33]	Feb. 1964	France	1960-
Gambia		Britain (?)	1965-
Guinea[34]		USSR	1959-
		CSR	1959-
		U.S.A.	1963-
		W. Germ.	1963-
India[35]		Britain	1948
		U.S.A.	1948-
		USSR	1963
		France	1965-
Ivory Coast[36]		France	1960-
		U.S.A.	
		Israel	1963-
Japan[37]		U.S.A.	1952-
Kenya		Britain	1963-
Korea[38]	March 1960	U.S.A.	1948-
	May 1961		
Kuwait[39]		Britain	1961-

Nation	Interventions: Other	Military Aid	
Lebanon[40]		U.S.A.	1957-
		France	
Lesotho		Britain (?)	
Liberia[41]		U.S.A.	1959-
Malagasy[42]		France	1960-
		W. Germ.	1963-
Malawi[43]		Britain	1964-
Malaysia[44]		Britain	1957-
		Australia	1962-
		U.S.A.	1964-
Mauritania[45]		France	1960-
Morocco[46]		France	1956-
		U.S.A.	1956-
		USSR	1961-
		Britain	
		Spain	
Nepal[47]		Britain	196?-
		U.S.A.	1963-
		India	
		USSR (?)	
		China (?)	
Niger[48]		France	1960-
		U.S.A.	1961-
Nigeria[49]	January 1966		
	August 1966	Britain	1960-
		U.S.A.	1961-
		USSR	1966-
		W. Germ.	1963-
		France	1965-
		Canada	
		India	
		Pakistan	
		Ethiopia	
Philippines[50]		U.S.A.	1946-
Rwanda[51]		Belgium	1962-
Saudi Arabia[52]	Nov. 1964	Britain	1947-
		U.S.A.	1948
Senegal[53]		France	1960-
		U.S.A.	1961-
Sierra Leone[54]	March 1967	Britain	1961
	April 1968		
Taiwan		U.S.A.	1945-

Nation	Interventions: Other	Military Aid	
Tanzania[55]		Canada	1964-
		China	1964-
		W. Germ.	1965-
Togo	January 1963	France	1960-
	January 1965	U.S.A.	1968(?)-
Tunisia[56]		U.S.A.	1957-
		Britain	1957-
Turkey[57]	May 1960	U.S.A.	1947-
Uganda[58]		Britain	1962-
		Israel	1964-
		USSR	
		China (?)	
Upper Volta[59]	January 1966	France	1960-
		U.S.A.	1961-
Yemen[60]	Sept. 1962	Egypt	1954-
	Nov. 1967	USSR	1957
		U.S.A.	1963-
		Bulgaria	
Zambia[61]		Britain	1964-
		China	1968-

Appendix 11: Excerpts from Army Regulation AR550-50*

Chapter 2–School Training

Section I. Prerequisites

2-1. National prerequisites. *a.* Training assistance will be provided in response to specific requests presented through appropriate channels by an authorized representative of the foreign government or foreign international organization concerned. The MAAG may encourage the foreign country to submit requests for needed training which is known to be available from U.S. sources; however, care should be taken to assure that no U.S. commitment is made or implied by such encouragement.

b. Requests for MAP training are presented to the MAAG. Prior to submission of foreign government requests for U.S. Army training, the MAAG should screen the requests to assure conformity with the following general policies:

(1) Training is in support of approved programs, plans, and objectives.

(2) Qualified students will be available for the training requested.

(3) Recipient country is making optimum use of personnel previously trained under MAP.

(4) Recipient country is making maximum use of its own training resources.

(5) Emphasis is placed on the training of foreign instructor personnel and career personnel.

(6) Training is in skills in which actual deficiencies exist, is in furtherance of overall objectives, and is clearly beyond the capability of the recipient nation.

(7) Assigned MAAG personnel are being fully utilized to provide in-country training assistance.

c. Except in the advanced weapons area, special classes normally will not be conducted in CONUS Army service schools. Exceptions must be fully justified and must permit sufficient leadtime to the school concerned to provide the desired training. Interpreters and translation required will be provided by the recipient country. When special foreign classes for a single nation are conducted at U.S. Army service schools through the use of interpreters, the invitational travel orders will designate a chief interpreter and/or class commander. In this case the class commander should not be a student, and his command authority must be clearly delineated.

d. Training benefits must warrant the high cost of the transocean travel involved. When transocean travel costs in connection with training in the United

---

*Foreign Countries and Nationals, Training of Foreign Personnel by the U.S. Army, Effective 1 December 1970. Headquarters, Department of the Army, October 1970.

States are borne by the MAP, such training normally will be arranged only when the total training in formal school course(s) or a combination of formal school and observer training is a minimum of 8 weeks, including CONUS travel time.

2-2. **Student prerequisites**. Foreign personnel to be trained in U.S. Army service schools and installations must satisfy the following:

a. Rank. The grade and position of foreign military personnel nominated for attendance at U.S. Army service school courses should be commensurate with the grade or experience of the U.S. personnel attending these courses. Any foreign officer who holds a position equivalent to the U.S. Army Chief of Staff or higher will not be scheduled for training at U.S. Army service schools and installations.

b. Health.

(1) Prior to the issuance of invitational travel orders (ITO) the MAAG will require the presentation of a signed statement from a competent medical authority designated by the recipient government that the student selected for training has received within the 3-month period preceding departure for training a thorough physical examination including a chest X-ray and complete immunization prescribed by the U.S. Public Health Service, as approved by the World Health Organization (WHO). This statement will further certify that the student or orientation tour visitor is free from communicable disease and/or any disease or condition likely to require medical care during the period of training scheduled or any condition which would preclude participation in the regular curriculum.

(2) The procedures in (1) above will be followed for dependents who accompany or join the trainee.

(3) If the medical facilities of the recipient country are unable to accomplish the medical requirements in (1) above, the MAAG will have the medical examination for student and dependents conducted by local U.S. medical facilities. Costs of transportation of the trainee for medical examination at U.S. facilities will be borne by the recipient government.

(4) Trainees who do not meet the medical standards for training after arrival in CONUS will be returned home.

(5) School and installation foreign training officers (FTO) will insure that immunization requirements of the WHO are met prior to the student's departure for the port of embarkation (POE) for return to his home country.

c. Technical qualification. The student must meet the prerequisites, except U.S. service retainability for the course for which selected, as prescribed for U.S. personnel in DA Pam 350-10, the appropriate oversea school catalog, or prerequisites established by the U.S. Army component commander providing the training.

d. Language qualifications.

(1) Students selected for training will be required to pass the prescribed

Defense Language Institute (DLI) English examination and meet the English comprehension level (ECL) for each course as indicated in the MASL. Students invitational travel orders (ITO) will include a statement of date of completion of the test, ECL score achieved, and ECL test form used. Exceptions to this policy are—

(a) Courses conducted in the native tongue of students in oversea U.S. Army schools or elsewhere.

(b) Special courses conducted with the aid of interpreters.

(c) When English is the native tongue.

(2) To confirm validity of the ECL test scores, direct entry students will be retested not earlier than 10 days after the student's arrival at the first training activity (other than Defense Language Institute, English Language School (DLIELS). The currently authorized ECL test will be administered using a different DLI test form than the one previously administered in-country. Students certified by DLIELS, following language training at that activity, will not be retested at the first CONUS training activity.

(3) In the event CONUS testing reveals a serious deficiency in the student's ECL, the training activity will make appropriate recommendations to DA concerning additional English language training.

e. Motivation. The student should be motivated for the training course for which selected. This is particularly essential for air-borne, ranger, special warfare, and aviator courses.

2-3. Special prerequisites for flight and air-borne training. *a.* Applicants for the ranger, airborne, U.S. Army aviation training, and pathfinder courses will be rigidly screened and only students who meet prerequisites outlined in paragraph 2-2 will be selected. Invitational travel orders of personnel scheduled for training involving aerial flights or parachute training (e.g., special warfare, ranger, airborne, pathfinder courses) must indicate that students are qualified parachutists and authorized to participate in aerial flights as required by the training course.

b. Students taking other than airborne training, but who are airborne qualified, may be placed on airborne status for the duration of CONUS training to maintain proficiency, if such status is approved by the student's government and specific authority is included in the basic ITO. Implementation of this authority will be contingent upon school capability.

c. Approval for use of Army aircraft for the maintenance of flight proficiency by students taking other than flight training will be determined under the provisions of AR 95-1. The following additional requirements must also be met:

(1) Specific authority to participate in flights for the purpose of maintaining flight proficiency must be approved by the student's government and such authority included in the student's basic ITO.

(2) Supporting documentation is provided to the major Army commander

by the ITO issuing authority concerning the student's professional and physical qualifications to participate in flight proficiency activities.

(3) No cost to the U.S. Government will be involved.

d. Prospective flying students are required to have proven adaptability for flight training, to include solo flight in light aircraft when possible prior to reporting to the primary fixed wing or rotary wing courses. Army Fixed Wing Aptitude Battery (AFWAB) or Army Rotary Wing Aptitude Battery (ARWAB) as outlined in AR 611-110 will be administered. Invitational travel orders should include a statement that the test battery was completed, give the date of completion, and should indicate the composite battery score attained.

e. Individuals who are scheduled for flight training in U.S. Army service schools will be required to meet class I, IA, or II medical standards as appropriate, in accordance with chapter 4, AR 40-501. Personnel who have received a current valid aviator rating in the Armed Forces of their respective countries will be accorded the same consideration as U.S. Army aviators and will be required to pass a class II flight medical examination.

f. A U.S. Army aviation medical examination will be given to individuals selected for pilot training by a qualified U.S. Army Aviation Medical Officer or Flight Surgeon, U.S. Air Force Flight Surgeon, or U.S. Naval Flight Surgeon prior to the student's departure from his home area for the United States. If the country does not have a U.S. Armed Forces Aviation Medical Officer, the foreign student candidate, upon approval of the Chief, MAAG, will report to the closest U.S. Armed Forces Aviation Medical Officer for medical examination. The appropriate MAAG will issue the necessary travel orders. The examining officer will determine the individual's physical qualification for the flying course and approve or disapprove his application. When waiver of a medically disqualifying condition is considered appropriate, the results of the medical examination will be referred by the examining officer to The Surgeon General, ATTN: MEDPSSP, Department of the Army, Washington, D.C. 20314, for advice and recommendation.

g. Flight physical examination records will accompany the individual throughout the period of his aviation training.

h. Foreign students undergoing physical reexamination in the United States prior to initiation of flight training will be required to meet class II medical standards for flying as prescribed in AR 40-501.

2-4. Prerequisites for U.S. Army Command and General Staff College (USACGSC). The following governs the attendance of foreign nationals at the Command and General Staff College:

a. All foreign officers selected for Course 1-250-C2 are required to attend the appropriate phases of the Allied Officer Preparatory Course at USACGSC. The preparatory course is divided into two phases—

(1) *Phase I (5 weeks)* is a course of instruction to enhance knowledge of

English and to provide a better appreciation of political, social, and economic factors which have a bearing on U.S. people, traditions, and way of life.

(2) *Phase II (3 weeks)* is a course in U.S. military terminology, organization, and tactics. Foreign students from non-English speaking countries will be required to attend both Phase I and Phase II of the preparatory course. Foreign students from English-speaking and other selected countries are exempted from attending Phase I; however, they are required to attend Phase II of the preparatory course. Foreign students attending Phase I must report at the college on or about 20 June; those attending Phase II only must report on or about 1 August.

b. U.S. Army Command and General Staff College Special Extension Courses as outlined in DA Pamphlet 350-60 and as indicated below are available to MAAG for use of foreign students. In the event there is no MAAG in a particular country, these courses are available to the U.S. Defense Attaché (USDAO). Upon request, USACGSC will forward all lessons, solutions, and summaries for these extension courses to the MAAG or USDAO, as appropriate. The lessons and their solution are to be administered by the U.S. Army element of the MAAG or the USDAO. Lesson solutions and summaries may be released to students upon completion of each lesson. There are no examinations as a part of these extension courses. Direct communication is authorized between MAAG or USDAO and USACGSC for the purpose of administering this program.

(1) *Special Extension Course–Preparatory.* This extension course is for the use of those students selected to attend the resident Command and General Staff Officer Course. It should be used to the maximum extent possible by personnel selected or potential nominees to attend the resident course of instruction.

(2) *Special Extension Course–Graduate Refresher.* This course provides a means for graduates of resident and nonresident courses to bring themselves up to date on the latest College instruction on Army organization, doctrine, and procedures. The applicant must have successfully completed the resident or nonresident Command and General Staff Officer Course offered by the College.

2-5. Prerequisites for civilian personnel. *a.* Training of foreign civilian personnel is permissible only if they are civil service employees (or the equivalent thereof) of host country armed forces.

b. The recipient country must agree to the same administrative control over civilians in training as currently applies to military personnel. Civilians will be afforded the same status as officers attending training courses, and those bearing grant aid invitational travel orders will be authorized living allowance rates applicable to officers.

c. Training of foreign civilian personnel in civilian schools or installations other than under MAP auspices is not arranged by Headquarters, Department of the Army. Request for this type of assistance should be submitted through appropriate country representatives.

2-6. Prerequisites for civilian schools. The Department of the Army will consider graduate and undergraduate level training in U.S. civilian schools under the auspices of the MAP for foreign military personnel and certain selected foreign civilian personnel from foreign military services provided—

a. Personnel are destined for assignment in the host country's military establishment, in regional security organizations, in high governmental positions, or will be assigned as instructors at a national military academy or at a military school of command and staff level or above.

b. They are academically qualified to pursue the course selected.

c. They are fluent in reading, writing, speaking, and understanding the English language.

d. The training is not available in the student's home country or in a U.S. Army School or installation.

e. Personnel to be trained are career personnel.

2-7. General. This section outlines procedures for the administration of the foreign training program. For matters of a purely administrative nature concerning foreign students, i.e., nonreceipt of ITO, personal problems concerning dependents' welfare, nonreceipt of pay and similar-type matters, CONUS training installation commanders at the discretion of CGCONARC and CG-USAMC, are authorized direct communication with MAAG. Information copies will be furnished DA, CGCONARC, or CGUSAMC as a minimum.

2-8. Academic reports. *a.* DA Form 3288-R (Academic Report—Foreign Students Attending Service Schools) (fig. 2-1), will be prepared for each foreign student upon completion of each course of instruction (except preparatory courses). DA Form 3288-R (image size: 7 4/10 x 9 5/6") will be reproduced locally on 8-inch by 10½-inch paper. If a student cross-trains at two or more installations, a copy of the academic report for the previous training will be forwarded to the second installation for information and guidance.

b. The academic report is designed not only to provide a record of the trainee's academic achievements but also to furnish an account of his participation in extra-curricular and community activities. In order that the trainee may receive adequate recognition, it is essential that all information requested be furnished in as much detail as possible.

c. Numerical grades and/or information as to an individual student's class standing will not be released by CONUS schools to foreign students except for the following:

(1) Students first in class standing among the foreign students.

(2) Foreign students from the same country participating in special all foreign classes (releasable to liaison officers or senior foreign officer of country concerned).

(3) Individual foreign students, provided grade and class standing are not

published in such manner that the grades of foreign students may be compared with each other.

d. Academic reports on foreign students will be distributed as follows:

(1) MAAG concerned, original and 2 copies *(except as indicated in (5) below)*.

(2) Defense Intelligence Agency, ATTN: DIACO-3, Dissemination Center, 1 copy.

(3) U.S. Defense Attaché concerned, 1 copy.

(4) Commander, U.S. Army Forces, Southern Command, Fort Amador, Canal Zone, for students from USCINSCO area, 3 copies.

(5) Office of Public Safety (AID), Building CT, 3600 M St., N.W., Washington, D.C. (for students under auspices of the Agency for International Development in lieu of MAAG), 1 copy.

(6) Commanding General, U.S. Army Intelligence Command, ATTN: ICDS, Fort Holabird, MD 21219, 1 copy.

(7) Special addressees as indicated in student's ITO, 1 copy.

e. In the case of special classes of students from a single country, and at the discretion of the school commandant concerned, an academic report may be rendered on the class as a whole rather than on each student. As an exception, individual reports will be submitted on students who do not successfully complete the course.

f. Release of academic reports will be at the discretion of the MAAG.

2-9. Predeparture requirements. *a.* Prior to his departure from home country, each trainee will be thoroughly briefed by the MAAG on the specific items outlined in paragraph F-5, appendix F. Other items considered appropriate based on country requirements will be included.

b. Not later than 15 days prior to the student's expected time of arrival in CONUS, the authority issuing the invitational travel order will send a cable to the U.S. terminal at which the student will arrive with an information copy to all installations at which he will receive training, stating his name, the course number, a remark that security requirements in accordance with paragraph 2-40 have been met, country security clearance is appropriate, and his expected time of arrival. The message will also indicate the number of dependents accompanying student, when appropriate. Whenever possible students should be scheduled to arrive at CONUS installations during the week–(Mon-Fri).

c. Students arriving by commercial transportation at other than regularly established ports of entry will be advised of their responsibility to inform the receiving installation 48 hours in advance of their estimated time of arrival (ETA) and mode of travel.

2-10. CONUS arrival and departure. *a. Port.* Particular emphasis is laid on initial reception at the port of entry/departure (aerial or surface, commercial or military). Commanders of the Army area in which the port is located will be responsible within their area of control for the coordination of all arrival and departure information of foreign trainees and visitors at ports of embarkation/debarkation through appropriate protocol officers. Personnel assigned to meet foreign visitors must be acquainted with the Informational Program (app J) and be prepared to take advantage locally of any opportunities to contribute to its attainment while the foreign trainee or visitor is in their charge. The following points will be stressed:

(1) Foreign trainees and visitors should be considered guests of the United States. An atmosphere of welcome, courtesy, efficiency, and consideration for foreign trainees and visitors is therefore essential.

(2) Wherever possible, personnel of equal rank should meet arriving personnel, particularly in cases of general and flag rank, and applicable protocol procedures should be followed.

(3) Expedited assistance through customs and currency exchange should be furnished.

(4) Information and instructions should be given in simple, easily understood English.

(5) Prior arrangements should be made to meet religious or national dietary requirements, e.g., list of local restaurants with type and price of food served.

(6) General information should be available on such items of local interest as special events, bus schedules, taxi rates, hotels, local community organizations established for assistance of foreign personnel.

(7) Assistance to dependents should be given.

The port will inform the receiving installation by appropriate means and sufficiently in advance of the student's ETA, the mode of travel, flight number, number of dependents, and all other pertinent travel information to insure proper reception.

b. Training installation. Commanders of CONUS installations and commandants of U.S. Army service schools where foreign military personnel receive training will–

(1) Request port calls for foreign students returning to home country in accordance with the provisions of chapter 3, AR 55-28.

(2) Inform the appropriate Army headquarters of the departure of all general or flag officer students. This notification will include the name, grade, country, date and hour of arrival, name of carrier, flight or train number, or the information that the student is traveling at his own expense.

(3) Insure that each student departing from the United States through a CONUS port of debarkation has in his possession a copy of the last pay voucher and complete orders, including basic invitational travel orders in a minimum quantity as indicated below:

(a) For Transportation via MAC–3 copies.

(b) For final travel and final living allowance–9 copies.

2-11. Attrition. *a.* Foreign students who fail to meet the training standards set for U.S. personnel through lack of interest, application, or other reasons will be relieved and returned to their homeland. When it is apparent that a foreign student should be relieved from training, the Deputy Chief of Staff for Military Operations, ATTN: OPS-IA-MA, Department of the Army, Washington, DC 20310, will be advised immediately of the full particulars of the case with information copy to appropriate intermediate commands. Under no circumstances will the student be relieved for cause without authority from Headquarters, Department of the Army. Pending receipt of this authority, temporary suspension is authorized at the discretion of the school commandant. Headquarters, Department of the Army, will advise the MAAG and Unified Commander concerned and, through the Assistant Chief of Staff for Intelligence, the appropriate foreign attaché in Washington, D.C., when authority has been given for relief of a foreign student.

b. In addition to matters relating to the relief of foreign students for academic deficiency, the types of incidents involving foreign students which should be reported in accordance with procedures outlined in *a* above, include but are not limited to—

(1) Hospitalization of trainee to include date of hospitalization, diagnosis, prognosis, and probable date of release. Minor illnesses of foreign students need not be reported. Information furnished in accordance with AR 40-3 on medical services provided foreign students does not fulfill the notification requirements contained in this regulation.

(2) Relief of foreign student for disciplinary reasons, to include basis for request, report of investigation, and recommendations for disposition of student, if appropriate.

(3) Requirement to recycle student due to academic deficiency.

(4) Accident reports involving foreign trainees or their dependents where damage to property and injury to other persons results.

(5) Emergency leave or other significant items of student welfare.

(6) Absence of student without official leave.

(7) Any event involving a foreign student that may have intergovernmental implications. This will include any complaint by a foreign student, or behavioral attitude indicated or reported, which reveals the foreign student's dissatisfaction with his environment or social acceptance therein.

c. Following the dispatch of initial notification, HQDA and intermediate commands will be kept informed and detailed written reports will be provided when appropriate.

d. Foreign students attending oversea U.S. Army service schools who fail to meet the training standards through lack of interest, application, or other reasons will be released and returned to their homeland upon authority of the U.S. oversea commander. The MAAG concerned will be advised of the full particulars of the case.

2-12. Biographical data. *a.* The MAAG will furnish biographical data for each foreign officer student 30 days prior to his reporting date. Information may be provided on DD Form 1396-1, as modified by appendix L, or on a form completed by the country concerned and attached to a DD Form 1396.

b. Distribution of biographical data will be made to the following activities (distribution by activity will be indicated on the format used):

(1) Each school or installation the officer will attend–1 copy.

(2) Defense Intelligence Agency, ATTN: DIACO-3, Dissemination Center, Arlington Hall Station, Arlington, VA 20301–1 copy.

(3) CG, U.S. Army Intelligence Command, ATTN: ICDS, Fort Holabird, MD 21219–1 copy.

c. Submission of biographical data for enlisted personnel is not required.

2-13. Briefing and orientation for students. *a.* Prior to students' departure from home country for the United States, CONUS training installations will forward through the appropriate MAAG all necessary information, e.g., school brochures, maps of local area, living costs, type of clothing required, housing facilities, and other pertinent data which should be furnished prospective students. Further, a special text containing the terminology peculiar to the course concerned should be provided to assist in preparation of the students. The MAAG should follow up to make certain that this information is furnished by the CONUS training installation. MAAG will contact schools for additional copies as required.

b. Commandants of schools or activities concerned will provide for necessary orientation of students upon arrival at their installation. The appropriate points stressed for reception at the POE (para 2-10*a*) will also apply to training establishments.

2-14. Clothing and equipment. *a. Organizational clothing and equipment.* Organizational clothing and equipment required by foreign personnel for a prescribed training course are authorized for issue during the extent of the course only, and no reimbursement is required from the individual. However, maintenance costs of such equipment will be included in the course costs. The basis for issue to such personnel will be as authorized for officers and enlisted personnel of the U.S. Army in accordance with applicable tables of allowance, tables of organization and equipment, or applicable tables of distribution and allowances published by the Department of the Army. Lost, damaged, or destroyed property will be accounted for in accordance with AR 735-11. Collection from foreign personnel held pecuniarily liable will be in accordance with AR 735-11. Accounting procedures are prescribed in AR 735-35.

b. Uniform requirements. Foreign military trainees will report to U.S. installations in uniform. Trainees will wear the prescribed uniform when traveling to and from CONUS, unless the wearing of civilian clothing is specifically authorized in the basic ITO. This authorization will be held to a minimum and granted for cogent reasons only. While undergoing training in the U.S. Army service schools and installations, students will wear seasonable uniforms prescribed and furnished by the country concerned.

c. Wearing of U.S. uniforms. In the event that the country concerned does not provide a uniform suitable for the climatic conditions of the United States, an Army uniform other than distinctive U.S. uniforms is authorized for wear provided insignia of the country represented is placed thereon. Sale of distinctive uniforms or items of uniforms listed in AR 670-5 and AR 670-30 is prohibited. MAAG will bring this fact to the attention of the host country representatives. Authorized uniforms may be purchased by the country or by individual students and will be worn as prescribed by the country concerned.

d. Individual apparel and equipment.

(1) *MAP, MASF, FWMAF, and FMS students.* Individual apparel and equipment required for prescribed training courses, except as authorized by *a* above for issue on an organizational basis, will be made available to students as required, and expenses incurred will be included in the course costs. Except for items which are nonrecoverable for hygienic or aesthetic reasons, issued individual clothing will be withdrawn from foreign students on completion of their training at a particular installation. Items withdrawn will be processed and maintained for reissue. Nonrecoverable items may be retained by the students.

(2) *Non-MAP students.* Expenses in connection with individual apparel and equipment required for prescribed training courses will be defrayed by the country or the individual student concerned.

2-15. Commissary and Army exchange privileges. *a.* Commissary, Army exchange, and similar privileges ordinarily available to officers and enlisted men of the U.S. Army in CONUS will be extended to foreign military personnel coming within the meaning of this regulation, in accordance with AR 31-200, AR 210-65, and AR 60-20.

b. Privileges extended to foreign civilian students within CONUS are limited to those authorized for DOD civilian employees under the provisions of paragraph 3-8, AR 60-20, and appendix A, AR 31-200.

c. Privileges extended to foreign military and civilian students in oversea areas must be in accordance with applicable international agreements. When agreements between the United States and the host nation do not expressly authorize the United States to accord these privileges to such personnel, they may be accorded such privileges only with the consent of the host government.

2-16. Death/burial. *a.* In the event of the death of a foreign student under DA auspices (MAP, non-MAP, MASF, FMS) the U.S. Army activity at which the incident occurs will immediately notify DCSOPS (MA), by telephone, of the occurrence.

b. A casualty report will be furnished in accordance with the provisions of

AR 600-10. DCSOPS (IA-MA) will be included as an action addressee to the casualty report. The following additional information will be included in the Remarks item of the casualty report:

(1) Student's ITO number.

(2) Request for disposition of remains.

(3) Request for permission to perform autopsy, if required.

c. The MAP/MASF fund citation in the student's ITO will be used to defray burial preparation expenses and costs for transportation of the body to the deceased's home country. Oversea return transportation costs will be paid from MAP funds only for deceased personnel from countries for which transocean costs are defrayed from MAP funds.

d. Burial or shipment of the remains and personal effects of dependents and of deceased FMS or non-MAP personnel will be arranged and expenses paid by the government concerned. If the assistance of the installation mortuary officer is desired by the foreign government, that officer will, without charge, negotiate with a civilian mortuary on behalf of the foreign government for the preparation of the remains for burial or shipment. No payment for services rendered locally will be made by the U.S. Government. All charges incident thereto are the responsibility of the foreign government.

e. Unless otherwise directed, effects (under U.S. Army control) of deceased personnel (*a* above) will be inventoried and forwarded to the appropriate MAAG.

2-17. Dependents. *a.* Foreign students will be discouraged from bringing dependents to the United States during their training periods. Housing on and around most Army posts is either scarce or unavailable. Travel and maintenance of dependents cannot be supported by the MAP, and no other U.S. Government funds are available for this purpose.

b. Should the commander of the training facility determine that the length and nature of the course and the availability of housing and other amenities make suitable the presence of dependents, he may forward a request for approval for a specific course through channels to HQDA for consideration. If approved, the exception will apply to all trainees for a specific course, and HQDA will authorize an additional living allowance for those students accompanied by dependents based on the availability of Government quarters for the student and his dependents.

c. As an exception to the policy in *a* above dependents may accompany students attending USACGSC Course 1-250-C2; however, transportation of dependents at MAP expense is not authorized. Students basic invitational travel orders will include authorization for dependents to accompany student at no expense to the U.S. Government. The living allowance rate for the student will be based on the availability of Government quarters for the student and his dependents. Students cross training from course 1-250-C2 are authorized like living allowance for the follow-on course when accompanied by dependents.

d. Schools will be notified 1 month in advance of the arrival of students accompanied by dependents.

e. In keeping with the purposes of the Informational Program, the use of Government-owned vehicles in the reception and departure of bona fide dependents of students is authorized.

2-18. Diplomas and certificates of attendance. The following policies govern the award of diplomas and certificates to foreign students attending U.S. Army service schools in CONUS.

a. Diplomas issued foreign students will be identical to diplomas issued to U.S. students. The notation "Foreign Course of Instruction" will not appear thereon. When practicable, the diplomas should not differ in size or format, and they should be signed by the school commandant or assistant commandant. Similarly, individual letters of commendation should be signed by the school commandant or assistant, and forwarded to the student through the MAAG with information copy of letter furnished the appropriate unified commander.

b. Diplomas for graduation from a U.S. formal course of instruction will be given foreign students only when such students have met the training standards established for U.S. military personnel. It is not the intent of this policy that numerical grades be used entirely in determining whether the foreign student has achieved the standards set for U.S. military personnel. Determination that the student can accomplish to a satisfactory degree that for which he was trained is the controlling factor. This determination will be influenced by the factors of aptitude, application, practical effort, and demonstrated understanding, as well as by numerical grades.

c. Certificates of attendance in a U.S. formal course of instruction will be given foreign students only when such students do not meet the training standard set for U.S. military personnel but have been diligent and sincere in their training efforts.

d. It is recognized that in some courses the number of classified hours of instruction which are not available to foreign students is high. However, if the foreign student successfully completes the portions of the course available to him, based on the grading standards used for U.S. students, the training meets the requirements contained in *b* above.

e. If a foreign student is eligible for early graduation, the school or installation concerned will notify the Deputy Chief of Staff for Military Operations, ATTN: OPS-IA-MA, Department of the Army, Washington DC 20310.

f. For pay purposes certain foreign governments require their embassies to report the actual training period (beginning date and graduation or termination date) for all students sent by those governments to U.S. military service schools and installations. This requirement will be included in the student's ITO. Accordingly, U.S. Army schools and installations concerned will furnish each foreign student a certificate (affixing the adjutant's seal thereto) which contains such information, when the requirement is so indicated in the student's ITO.

2-19. Extension courses. *a.* Foreign student participation in the Army Extension Course Program will be encouraged by service schools and MAAG to the extent deemed appropriate for the country concerned.

b. Applications for enrollment will be forwarded through, approved, and serviced by the appropriate MAAG of the country concerned, without referral to HQDA (see AR 350-60).

c. DA Pamphlet 350-60 outlines those courses available. Within security limitations, copies of extension course subcourses, including instructional material, tests, and answer sheets can be furnished to the MAAG. The MAAG will distribute the lesson material to the foreign student and, except as indicated in *d* below, grade the lesson answer sheets completed by the student. The school will grade the subcourse examinations and maintain the academic record of the student.

d. All lessons and examinations of the USACGSC General Staff Officer Extension Course are graded by the College.

e. All correspondence with the foreign student in the administration of the Extension Course Program will be accomplished through the MAAG who will monitor the progress of the student.

2-20. Identification cards. Identification cards will be furnished foreign students by the first training facility in CONUS in accordance with AR 606-5. Identification cards will be recovered prior to the students' departure from the last CONUS school en route to the port. The foreign students may use their invitational travel orders if identification is necessary on leave en route to the port of embarkation. All foreign students and orientation tour participants who have been admitted to the United States on official visas are exempt from the provisions of the Immigration and Nationality Act pertaining to registration, fingerprinting, and reporting of address unless the status of the students changes and they are no longer pursuing the training prescribed in their invitational travel orders.

2-21. Laundry. Laundry service will be furnished to officer and enlisted personnel on a cash basis. Collection will be made by the local laundry officer. An exception is made for Latin American students who are provided this support in the Canal Zone.

2-22. Legal status of foreign trainees. *a. Jurisdiction.*

(1) Military and civilian foreign trainees and their dependents, while in the United States, are subject to the jurisdiction of the U.S. courts, both State and Federal, unless they are expressly exempted by treaty or have diplomatic immunity.

(2) Currently there are no treaties or agreements to which the United States is a party which exempt any foreign trainees or their dependents from the jurisdiction of U.S. courts.

(3) The NATO Status of Forces Agreement (NATO SOFA) is the only agreement which limits the jurisdiction of U.S. courts over members of foreign forces and their dependents in the United States. Under this agreement the United States and the NATO SOFA country concerned exercise concurrent jurisdiction over the members of the forces of that NATO SOFA country, and their dependents, who are present in the United States for official duty purposes. The treaty states the circumstances when the authorities of the United States or the NATO country will have the primary right to exercise jurisdiction. Trainees and their dependents from countries which have ratified this agreement will be accorded all the rights guaranteed under this treaty.

(4) The following countries have ratified the NATO-SOFA: Belgium, Canada, Denmark, France, Federal Republic of Germany, Greece, Italy, Luxembourg, Netherlands, Norway, Portugal, Turkey, United Kingdom, and the United States. The only other authority for the exercise of jurisdiction by a foreign country over its personnel in the United States is contained in a U.S. statute (22 U.S.C. 701-706), which provides that the service courts of friendly foreign forces within the United States may exercise limited jurisdiction over their personnel when its provisions are made operative by Presidential Proclamation. (See AR 27-51 for U.S. Army implementation of the statute.) At present Australia is the only foreign country to which the statute has been made operative by Presidential Proclamation. Thus, at the present time only Australia and those nations which are parties to the NATO SOFA have the right, in appropriate cases and under specified conditions, to try their personnel by military courts sitting in the United States.

(5) No other foreign forces are authorized to try their personnel by any type of tribunal sitting in the United States. All questions concerning the jurisdictional status of foreign trainees or their dependents should be referred to the local staff judge advocate.

b. Diplomatic immunity. Foreign trainees who believe themselves entitled to diplomatic immunity or other special status should convey this information to the local staff judge advocate. Staff judge advocates will contact the Deputy Chief of Staff for Military Operations, ATTN: OPS-IA-MA, Department of the Army, Washington, DC, 20310, for information as to the diplomatic or other status of the individual concerned. As a general rule a sponsor's diplomatic immunity extends to his dependents as well; however, foreign trainees usually do not have diplomatic immunity.

c. Control of trainees. Foreign trainees are not subject to the Uniform Code of Military Justice, and, generally, there is no authority under which U.S. military authorities may place foreign trainees in confinement. Under the Federal statute discussed in *a* above, however, Australian military authorities in the United States may request the assistance of U.S. military authorities in the apprehension and confinement of members of Australian forces in the United States (AR 27-51). U.S. civil authorities, State or Federal, may also apprehend and confine foreign trainees for breaches of State or Federal law. Except in the case of express authorization by treaty or agreement (such as NATO SOFA), or by statute, Executive order, or Presidential proclamation (such as in the case of Australia), foreign military attachés or commanders stationed in this country have no authority to arrest, detain, or confine members of their forces within the United States. Neither can they empower U.S. military authorities to arrest, detain, or confine members of their forces. When warranted by urgent circumstances, the installation commander may authorize temporary restraint to prevent bodily harm to the trainee or to other persons, pending arrival of civilian authorities.

d. Claims. For information concerning claims arising in the United States from the activities of foreign trainees from countries which have ratified the NATO-SOFA, see AR 27-20. For information concerning claims which arise incident to the activities of foreign trainees in oversea areas, see pertinent command claims directives. If an inquiry is made concerning a claim involving non-NATO personnel, the claimant should be advised to seek redress from the trainee or his government.

e. Reports. All legal questions concerning foreign trainees should be referred to the local staff judge advocate. In addition, any incident involving foreign trainees which might lead to the exercise of criminal jurisdiction by State or Federal authorities, or which has already led to the exercise of some form of jurisdiction by such authorities, such as arrest, indictment, or the issuance of a summons or charges, should be reported immediately as provided in paragraph 2-11, with information copy to The Judge Advocate General, ATTN: International Affairs Division, Department of the Army, Washington, DC 20310.

2-23. Leave and holidays. *a. Leave.* The total of MAP supported leave and constructive travel time between the training duty station and the POE will not exceed 15 days.

(1) Fifteen days leave in CONUS (includes Hawaii and Alaska) including constructive travel time between the training duty station and the POE is authorized MAP trainees upon completion of a course of instruction or observer/OJT (as defined in chap 3) prior to return to home country. Leave will be approved prior to student's departure from home country and authority for leave and payment of living allowance included in student's basic ITO. Living allowances prescribed in paragraph 6-27 apply.

(2) Requests for leave, other than that indicated above, should be submitted by the student directly to the student's military attaché in Washington, D.C. Upon receipt of approval from his military attaché, the student will inform the commandant of the school concerned of the amount of leave he is authorized so that an appropriate entry may be made in an administrative

indorsement to the ITO which authorized the attendance at the school. A copy of the leave approval (in English) will be furnished the student by his military attaché as evidence of the student's leave status while he is traveling in the United States.

(3) Matters pertaining to leave taken in CONUS by non-MAP and FMS students will be resolved by the student concerned with appropriate representatives of his country.

(4) Leave outside CONUS is governed as follows:

(a) Homeward travel from the U.S. will be by the most direct route feasible. When a trainee is permitted by his government to deviate from that route for the purpose of visiting other countries, MAP sponsorship will terminate at the point and time of such deviation, except that he may later commence onward travel from that point to homeland under MAP sponsorship. Further, if a trainee elects to remain at a point en route to his homeland beyond the time normally required to make travel connections, MAP funding of allowances during that excess time is not authorized. Travel arrangements for leave outside CONUS, to include necessary documentation, are the responsibility of the individual trainee. The ITO will be indorsed by the training installation to indicate the foregoing provisos, as appropriate.

(b) Trips by MAP students on an individual basis outside CONUS are limited to trips to Canada and Mexico for students located in those commands near U.S. borders. The duration of such trips by MAP students for which living allowance is authorized is 72 hours. Foreign students will be responsible for fulfilling all immigration requirements.

(5) In the event additional training is scheduled for foreign students at another installation immediately following the Christmas holiday period, the losing installation will be responsible for such students during the holiday period.

(6) Foreign students and trainees desiring to travel outside CONUS in excess of 72 hours must obtain prior approval from appropriate country respresentatives in Washington, DC. Students will comply with all immigration regulations. Living allowance in excess of 72 hours is not authorized MAP students.

b. Holidays.

(1) CONUS school and installation commanders are authorized to grant leave (not deductible from leave authorized in *a* above) during the following periods:

(a) Christmas holiday period when training activities have been curtailed.

(b) Other authorized holidays observed by the U.S. Army.

(c) Major national and religious holidays (not Christian) of the student's country not to exceed 1 academic day for each holiday authorized. Academic progress will be the deciding factor in each case.

(2) MAP trainees are authorized the same living allowances for leave granted during these periods as that authorized under *a* above. Table 6-1 applies.

c. Delay en route.

(1) Between consecutive courses or training duty stations, the commandant of a service school or the commanding officer of an installation at which training is given may authorize leave or delay en route not to exceed 7 days when OJT is not feasible.

(2) Upon termination of training in CONUS, a maximum of 7 days delay en route may be authorized between the training duty station and the POE in the event the trainee's port call is delayed through no fault of his own and the trainee has not been granted leave in accordance with *a* above.

d. Travel on space availability basis. Travel on a space availability basis in military aircraft while a student is on leave is not authorized.

2-24. Mail. *a.* Students attending military schools in oversea commands may use military postal facilities for purchase of stamps and receipt and dispatch of mail to and from their home country only, when consistent with any agreement between the U.S. and the government of the nation concerned. When corresponding with persons in their home country, students will address mail in the same manner as they would if they were in their home country. The use of a U.S. military postal address in the home country is not authorized.

b. CONUS schools and installations will not address mail direct to a foreign student incountry through APO facilities. Material should be addressed to the MAAG with instructions for delivery on the inner wrapper.

2-25. Medical service. *a.* Medical services furnished foreign trainees will be in accordance with AR 40-3. Medical services furnished MAP trainees will be reimbursed under the provisions of paragraph 25, AR 40-3.

b. The procedure for the administration of MAP, MASF, FMS, and non-MAP students who require hospitalization follows:

(1) Students who have been selected for training are examined as to physical condition and freedom from communicable diseases as a prerequisite to selection. If, upon arrival, it is discovered that the student cannot qualify for training by reason of physical or mental disability and will require more than nominal treatment before entering training, he will be returned to his home country immediately or as soon thereafter as his condition permits travel.

(2) When a foreign student requires hospitalization or is disabled after beginning a course of instruction, he will be returned to his home country as soon as practicable when, in the opinion of the U.S. medical authorities the hospitalization or the disability will prevent continuation of the training for a period in excess of 90 days.

(3) When action indicated by (1) and (2) above is required, the installation commander will notify the Deputy Chief of Staff for Military Operations,

ATTN: OPS-IA-MA, Department of the Army, by message with information to ACSI, ATTN: FLO, the Unified Commander and the MAAG. Authority for disposition of the student will be furnished by the Deputy Chief of Staff for Military Operations.

(4) In the event the student's hospitalization or disability will prevent the continuation of training in excess of 90 days, DCSOPS ATTN: IA MA, will be advised at the termination of the 90-day period and each 30-day period thereafter of the trainee's condition and the estimated date the student will be able to return to the country or to resume training, if practicable. The appropriate Unified Commander and MAAG will be furnished an information copy of all communications to DCSOPS.

c. In the event student cross-trains, the student's medical records will be forwarded to the next training activity.

2-26. Off-duty employment. Foreign trainees are not permitted to accept off-duty employment.

2-27. Officers in enlisted classes. Foreign officers are permitted to attend enlisted classes upon request of the MAAG. Such officers will, prior to departure from their home country, be thoroughly briefed that they are to attend enlisted classes and informed that their officer status does not entitle them to special academic privileges while attending these classes. Officers will be accorded officer privileges when not participating in training.

2-28. Public information. *a.* Public information activities will be conducted under the provision of AR 360-5.

b. Hometown-type releases of stories and pictures of foreign students and visitors will be provided. CONUS installation and oversea Army commanders will dispatch hometown releases directly to the MAAG by airmail, if appropriate. All releases require coordination by the MAAG concerned with Ambassadors and/or Public Affairs Officers of the United States Information Service (USIS).

c. Information concerning the number of students by nationality who are actually in attendance at any given time and a general description of the training being conducted may be released. No cumulative figures will be released except through the Chief of Information, Department of the Army.

d. No news releases will be made in CONUS or home country when it is in violation of applicable agreements between the Governments of the United States and the foreign country involved.

e. Stories for foreign release on students from countries without diplomatic relations with the United States will be submitted to the Chief of Information, Department of the Army.

f. No press coverage will be initiated for MAP orientation tour participants without their prior consent.

2-29. Purchase and use of power-driven vehicles. *a.* Students may purchase a power driven vehicle following arrival at a training installation unless specifically prohibited by their ITO.

b. Students must comply with post and State regulations for registration and operation of such vehicles. The students will be required to purchase public liability and property damage insurance in the amount required by law in the State in which the vehicle is registered, or in the amount required by the military post on which the vehicle is registered, whichever is higher.

2-30. Quarters. Where bachelor quarters are available they are authorized to be furnished to officer and enlisted personnel on a scale equivalent to that authorized for U.S. Army officer and enlisted personnel without dependents. Foreign students should be housed in the same barracks or quarters as U.S. students rather than in separate quarters by language groups. Where quarters are not available, the individuals will make their own arrangements at no cost to the U.S. Government. Installation commanders are authorized to assist.

2-31. Release of instructional material. Release of instructional material to foreign students, observers, and orientation tour visitors is authorized as outlined below:

a. Unclassified material. Commanders of U.S. Army installations and activities, and school commandants concerned may authorize release of unclassified Army publications used in courses of instruction.

b. Classified material. Classified material will not be furnished directly to foreign students, orientation tour participants, or observers for retention except as indicated in *c* below. Requests for any such material will be submitted through the MAAG to the installation concerned by the individual after he returns to his homeland.

c. Advanced weapons material. Foreign military students participating in missile training will be issued classified and unclassified publications used as texts and schematics for retention. Classified publications and student notes will be collected at the end of the course and shipped to the appropriate MAAG through U.S. postal channels.

d. Maps. Release of maps and related material will be in accordance with AR 115-11.

2-32. Content of instructional material. Content of material presented or released to foreign students will conform to approved doctrine of the U.S. Army.

2-33. Training with Army National Guard of the United States and U.S. Army Reserve units. The procedure and the conditions for providing training with Army National Guard and Army Reserve units for those active and reserve military personnel of foreign nations who are currently resident in the United States in other than a military training capacity are as prescribed below:

a. Application for training will be made by the individual to his military attaché who will submit the application to the Assistant Chief of Staff for Intelligence, ATTN: FLO, Department of the Army, Washington DC 20310.

b. The applicant must be in an active or reserve status with the military forces of a nation currently entitled to military training in the United States by intergovernmental agreement.

c. Participation will be voluntary on the part of the individual and without cost or other liability to, or claim against, the U.S. Government.

d. Training will be on an unclassified basis.

e. Application for training with military intelligence, counterintelligence, or Army security units will not be considered.

f. None of the services authorized MAP trainees will be provided to individuals participating in this training or to their dependents.

g. During training the foreign personnel will wear the uniform prescribed and furnished by their country. In the event the uniform so furnished is not suitable for the training being presented (e.g., field work or maintenance and repair training) foreign personnel (or their governments) are authorized to purchase and wear U.S. Army fatigue uniforms in accordance with paragraph 2-14 provided U.S. insignia are not worn.

2-34. Training cancellation. HQDA will be notified of cancellation of programmed training a minimum of 45 days in advance of reporting date to permit reallocation of training to other agencies.

2-35. Warrant officers. For administrative purposes all foreign warrant officers will be considered officers unless otherwise indicated in student's ITO. They are, therefore, entitled to living allowance rates applicable to officers and officer quarters while in training in U.S. Army Service schools and installations.

Section III. Invitational Travel Orders

2-36. Authorization. Invitational travel orders will reference the Department of the Army message or letter which approves the training. Appendixes A through D indicate formats to be followed in issuance of ITO for MAP, MASF, FMS and non-MAP students. In all cases, ITO will be issued for round trip travel and when applicable will provide for excess baggage, for on-the-job training (OJT) following formal instruction, and for leave authorized at MAP expense. Subject to approval of the students' government, ITO will include authorization to participate in flights in military aircraft during training and to participate in parachute jumps during training and for proficiency. Modifications of appendixes A and B (omission of last sentence of para 8 and entire para 9) will be made when appropriate, e.g., ITO for general officers. The subject of the orders will include student status (MAP, MASF, FMS, non-MAP) except where this would necessitate classification of orders.

2-37. Appropriation Citation. The invitational travel orders for MAP students will cite the appropriation to which the travel, living allowance, and other authorized expenses are chargeable (para 6-1*b*). To insure proper payment, at least one copy of each invitational travel order with all pertinent amendments and indorsements will be given to the MAP student and will be produced upon request. This copy will become the master copy and will be returned to the authority which originally issued the orders upon completion of the travel.

2-38. Indorsement and amendment of invitational travel orders. *a.* Upon arrival at the port of entry in CONUS or Hawaii, port authorities will indorse the master copy of the travel orders and a minimum of five other copies indicating the date and time of arrival at the port and the mode of transportation from the port to the next installation (commercial or military carrier).

b. The authorities at each school or installation visited will indorse the master copy and a minimum of five other copies of the travel orders showing dates and times of arrival and departure and the modes of transportation involved.

c. Each payment made to a MAP student will be shown by indorsement on the master copy and five other copies of the travel orders, indicating the amount, purpose, and date of the payment.

d. Upon arrival at the port of departure from CONUS or Hawaii, port authorities will indorse the master copy and five other copies of the travel orders showing the date and time of arrival at the port and mode of transportation and date of departure from CONUS or Hawaii.

e. Commanding officers of CONUS Army installations to which MAP students are attached will obtain approval of Headquarters, Department of the Army, prior to amending invitational travel orders, for periods in excess of 2 weeks.

2-39. Distribution. *a.* Invitational travel orders and amendments, or indorsements thereto will be distributed to reach the following addressees 30 days in advance of the student's reporting date for all CONUS training: (Distribution by activity will be listed at the end of each ITO.)

(1) Deputy Chief of Staff for Military Operations, ATTN: OPS-IA-MA, Department of the Army, Washington, DC 21310–1 copy.

(2) Appropriate unified command–1 copy.

(3) Major oversea Army command, when required–1 copy.

(4) Each U.S. Army service school or installation at which the nominee will be training–10 copies.

(5) Each student—50 copies.

(6) CONUS Army commander in whose area the port of embarkation is located—1 copy.

(7) Port of debarkation—5 copies.

(8) CONUS Army commander in whose area the training is to be conducted—1 copy.

(9) CGUSAMC, ATTN: AMCIL-MR, Washington, DC 20135—1 copy (FMS training only).

(10) Assistant Chief of Staff for Intelligence, ATTN: FLO, Washington, DC 20310—1 copy (orientation tours only).

(11) CG U.S. Army Intelligence Command, ATTN: ICDS, Fort Holabird, MD 21219—1 copy.

(12) Other addressees as considered appropriate by the issuing authority.

b. In addition to distribution above, MAAG will be provided 2 copies of amendments or indorsements prepared by other agencies.

c. Issuing authorities may distribute copies of ITO to the government of the country concerned when requested.

d. Distribution of ITO for oversea training will be prescribed by the appropriate component Army commander.

2-40. Security screening. *a.* Prior to students departure from his home country the provisions of paragraph 28*b*, AR 380-10, will be accomplished. In addition, the appropriate country security clearance will be obtained from the host government if training has been authorized on a classified basis. Compliance with security requirements will be indicated in the body of the original ITO by one of the following statements, as appropriate, dependent on classification of training authorized.

"(1) U.S. security requirements have been complied with.

"(2) U.S. security requirements have been complied with. The Government of__________ has granted the individual(s) concerned a security clearance of__________."

b. Statement in *a*(2) above will be used only if training is authorized by HQDA on a classified basis.

c. Orders will not be classified on the basis of the statements in *a* above.

d. U.S. training installations and/or activities will not enter students into training until above security requirements are met. If the appropriate statement is not included in the trainee's invitational travel orders, the training installation will contact the appropriate MAAG for compliance. The statement of country security clearance as stated in the ITO is not in itself authority to disclose U.S. classified information to foreign nationals.

e. The statement indicated in *a* above, insofar as pertains to U.S. security requirements, is not required to be included in ITO for trainee from

Australia, Canada, and Great Britain. Country security clearance will be included if appropriate.

Section IV. The Informational Program for Foreign Military Trainees and Visitors in the United States

2-41. Responsibility. The commanding officer or the officer in charge of the activity or installation conducting training for foreign military personnel is responsible for establishing procedures to implement the Informational Program.

2-42. Foreign training officer appointment. Foreign training officers (FTO) should be appointed at each installation to monitor and coordinate activities concerned with all phases of the foreign student's training including the implementation of the Informational Program. Personnel should be assigned as required based on the number and type of the foreign trainees present. To insure continuity of implementation, FTO should be assigned for a period of 2 years, if possible.

2-43. Priority of emphasis and scope of program. The Informational Program essentially applies to foreign officer personnel. The inclusion of enlisted personnel in Informational Program activities is encouraged when considered appropriate by the installation commander. Adaptation of this program to foreign students attending special advanced weapons courses is at the discretion of the installation commander. The program should commence upon arrival of the student at the CONUS installation and should continue throughout the training period.

2-44. Program content. *a. Sponsors.* Every effort should be made to insure that foreign trainees are properly received within the military and civilian communities. To achieve this the primary vehicle is through the use of military and civilian sponsors as indicated below.

(1) *Military sponsor.* A student and/or faculty member sponsor should be provided for the foreign trainee before he arrives in the United States. The sponsor will be of comparable rank whenever possible.

(2) *Civilian sponsor.* Through liaison with the civilian community, an individual or group who will act as a civilian sponsor for the student should be secured.

b. Precourse orientation. Prior to the commencement of formal training, a precourse program will be arranged for all foreign military trainees attending a course of instruction over 8 weeks. The length of the precourse program will be determined by the major commander. The presentation of specific informational topics, indicated in appendix J, should commence during this period of training and continue throughout the course of instruction.

c. Informational topics. The following guidance will assist in presenting the informational topics:

(1) *Lectures, roundtable discussions, and film showings.* Lectures, roundtable discussions, and film showings will be based on the topics outlined in appendix J and upon guidance from higher headquarters.

(2) *Local trips.* Visits suggested in appendix K should complement the activities listed in (1) above. Trips should be limited to 1000 miles round trip. In isolated areas, an exception to this policy will be considered by HQDA on a case-by-case basis. Provisions of AR 380-25 for visits to military installations apply. School commandants are encouraged to include the student's dependents in local visits where appropriate and where this can be done at no additional cost to the U.S. Government.

(3) *Tours to the Washington DC area.* Foreign officers who are in CONUS for a period of 120 days or more undergoing training in senior, advanced, and other selected courses, to include officer and cadet special training courses, when appropriate, will at some time during this training period be scheduled for a tour to Washington DC. This tour is an integral part of the Informational Program, and it should normally be programed for students near the end of their CONUS training period. No trainee should participate in more than one Washington tour.

(4) *Community relations.* Community participation in the Informational Program is essential. Chambers of Commerce and other civic groups make a worthwhile contribution in the introduction of foreign students to civilian communities. Members of these civic groups should be briefed thoroughly on the basic objectives of the Informational Program.

(5) *Civilian organizations.* Civilian organizations established for the purpose of welcoming foreign visitors to the United States exist within a 200-mile radius of most training installations. These organizations should be utilized to the maximum extent.

(6) *Civilian Aides to the Secretary of the Army.* Civilian aides represent the Secretary of the Army in their respective states or designated areas. They should be briefed on the Informational Program and, whenever possible, requested to participate in the planning of offpost Informational Program activities. Civilian aides also may be requested to act as sponsors for local events, to obtain entreé to civilian homes, and to arrange tours of State capitals and other major cities.

2-45. Followup. School commandants are encouraged to maintain contact with graduates of career and similar top level courses after the student's return to their home country. Such programs may include–

a. Forwarding letters from the commandant and annual school newsletters or similar school publications; encouraging students to enroll in nonresident extension courses; and encouraging informal correspondence between classmates.

b. Providing professional publications for students enrolled in CONUS staff and career courses. Each subscription must be appropriate to the course taken by the student and will be initiated prior to the departure of the officer from CONUS. The subscription will be for a maximum of 1 year and will be funded under the Informational Program.

c. Providing and maintaining current school publications to in-country counterpart military schools.

2-46. Information officer responsibility. The information officer should assist in promoting the foreign student training program by–

a. Portraying the activities of foreign students and the U.S. Army's role in training assistance through the news media.

b. Devoting a Command Information Hour to the foreign military training program.

c. Furnishing advice concerning the various aspects of community relations within his area of interest.

d. Notifying news media in localities to be visited by touring foreign students.

2-47. Identification booklets and pocket emblems. Foreign military trainees will be provided a suitable identification booklet in English upon arrival at a U.S. Army training installation. Upon graduation the student will be presented with a special emblem with authorization certificate. The emblem to be presented will incorporate the distinctive insignia for each school on a common background as developed by the Institute of Heraldry.

2-48. Orientation. *a.* Conferences of persons charged with training, administration, and orientation of foreign military trainees will be conducted annually by major commanders.

b. Shortly after his appointment, the Foreign Training Officer at each installation may be sent TDY to CONRAC and to HQDA for briefings and consultation in order to provide a full understanding of the MAP, its objectives and administration.

c. All U.S. personnel on duty where foreign students are training should understand the broad objectives of the MAP and be fully conversant with the purposes of the Informational Program and its relationship to the MAP.

d. Professors of Military Science (PMS) at civilian institutions of higher learning should know the purposes of the Informational Program and should encourage U.S. military personnel on civilian campuses, on a voluntary basis, to seek out foreign students, both military and civilian, to assist them to acquire a genuine understanding of the United States. Efforts of this kind include accompanying them to sports events and cultural and social activities. Home hospitality is appropriate.

2-49. Reporting requirements. In order to give the foreign trainee recognition for his contribution to the program, a brief report of the foreign student's participation in extracurricular and community affairs should be provided on DA Form 3288-R (fig 2-1).

ACADEMIC REPORT
FOREIGN STUDENTS ATTENDING SERVICE SCHOOLS

For use of this form, see AR 550-50; the proponent agency is the Office of the Deputy Chief of Staff for Military Operations.

TO: *(Include ZIP Code)*				FROM: *(Include ZIP Code)*		
1. LAST NAME - FIRST NAME - MIDDLE INITIAL		2. GRADE		3. COUNTRY	4. SSAN	5. BRANCH
6. NAME OR TITLE OF COURSE				7. DURATION OF COURSE FROM:		TO:
8. DID STUDENT SUCCESSFULLY COMPLETE COURSE ☐ YES ☐ NO *(Explain in item 15)*				9. STUDENT WAS AWARDED ☐ DIPLOMA ☐ CERTIFICATE OF COMPLETION		

10. ENGLISH LANGUAGE PROFICIENCY		EXCELLENT	GOOD	FAIR	POOR
	SPEECH				
	READING				
	WRITING				
	COMPREHENSION				

11. MILITARY SPONSOR *(Name and Address, Including ZIP Code)*

12. CIVILIAN SPONSOR AND OTHER CLOSE CONTACTS *(Names and Addresses, Including ZIP Codes)*

13. BRIEF REPORT ON STUDENT'S PARTICIPATION IN EXTRA-CURRICULAR AFFAIRS *(On Post)*

(Continue in item 15, if necessary.)

14. BRIEF REPORT ON STUDENT'S PARTICIPATION IN COMMUNITY AFFAIRS *(Off Post)*

(Continue in item 15, if necessary.)

15. REMARKS

(Continue on reverse, if necessary.)

DATE	TYPED NAME AND TITLE	SIGNATURE

DA FORM 3288-R, 1 Oct 70 EDITION OF 1 NOV 67 IS OBSOLETE. *(Paper size, 8" x 10½"; image size, 7-4/10" x 9-5/6")*

Figure 2–1.

Chapter 4—Orientation Tours

4-1. General. Orientation tours by specially selected key foreign officer personnel to the Continental United States or to United States commands overseas are for the purpose of acquanting those officers with the doctrine and training methods of the U.S. Army and for the purpose of acquainting them with the basic institutions and ideals of the United States. Normally orientation tours will be programed for a duration of 21 calendar days and will include a minimum of three individuals. When fully justified, orientation tours of a shorter duration may be programed. Personnel who have participated in training in the United States will be scheduled for an orientation tour only when fully justified by the MAAG and the unified commander, and approved by HQDA.

4-2. Procedure. *a.* Orientation tours will be programed in fiscal year military assistance programs in the same manner as other formal training.

b. Orientation tours in CONUS will be developed from the objectives to be accomplished and the areas of interest provided by the MAAG in the fiscal year military assistance program and indorsed by the unified commander. The nonmilitary aspects which should be emphasized and the contribution each can make to the orientation tour should be included. One itinerary for each group will be planned. The objectives and areas of interest for those orientation tours which have been recommended by a unified commander and are within the capability of the DA to conduct but have not been included in the fiscal year military assistance program will be submitted to HQDA a minimum of 120 days in advance of the proposed arrival date.

c. The biographical data shown in appendix M will be submitted in duplicate to the Deputy Chief of Staff for Military Operations, ATTN: OPS-IA-MA, Washington, DC 20310, a minimum of 60 days prior to the scheduled arrival date of each tour in CONUS. Defense Attaché channels will not be used to forward biographical information.

d. Information as to the mode of travel to CONUS and return, including the confirmed flight schedules and ports for arrival in and departure from the United States will be furnished a minimum of 60 days prior to arrival date to the following:

(1) DCSOPS, ATTN: OPS-IA-MA, DA, Washington, DC 20310.

(2) ACSI, ATTN: FLO, DA, Washington, DC 20310.

(3) CG USAINTC, ATTN: ICDS, Fort Holabird, MD 21219.

(4) CONUS POE and POD and the Commanding General of the U.S. Army area in which the ports are located.

(5) Each oversea headquarters through which the orientation tour is routed.

4-3. Baggage. A baggage allowance of 100 pounds (inbound and outbound) (not to exceed 2 bags) is authorized each visitor when costs are borne by the

Military Assistance Program and reflected in the ITO. The baggage will accompany the visitor. No additional allowance for instructional material is authorized for orientation tour participants. Dispositon of unauthorized baggage will be in accordance with paragraph 6-23*b*.

4-4. Invitational travel orders (ITO). *a.* ITO for orientation tour visitors will be prepared in accordance with appendix C, and distributed as prescribed in paragraph 2-39, in sufficient lead time to reach the appropriate installations 30 days in advance of the scheduled arrival date of the orientation tour in CONUS.

b. All ITO will include CIPAP (AR 310-50) to permit emergency changes as required by the agencies within CONUS.

c. ITO will include a statement concerning security screening in accordance with paragraph 2-40.

4-5. Leave. Living allowance is not authorized for leave taken in CONUS. Leave authorized upon completion of an orientation tour and/or on the return route will be included in the ITO with a notation that no MAP living allowance is authorized.

4-6. Allowances. *a.* Orientation tour personnel are considered to be senior officers and in a travel status during the entire period while in CONUS, Alaska, or Hawaii. They are entitled to the living allowance prescribed in table 6-1 for orientation tour participants. Orientation tour participants will defray all costs for meals, quarters, and incidental expenses. To meet expenses prior to the payment of living allowance at the conclusion of the orientation tour, the visitors will be provided sufficient funds in American dollar instruments (at the maximum rate prescribed in table 6-1) by their government prior to departure from home country.

b. MAP funds are authorized for extraordinary expenses for MAP orientation tours in accordance with paragraph 6-19.

4-7. In-country briefing. Orientation tour participants will be thoroughly briefed by the MAAG prior to their departure for CONUS. Briefings will cover the appropriate items outlined in appendix F and the following additional items:

a. Itinerary. MAAG should emphasize that no changes will be made in the scheduled itinerary after arrival of the orientation tour group in CONUS.

b. Civilian clothing. Participants should bring a minimum of one civilian suit and casual clothes, as appropriate.

c. Dependents. Dependents will not be scheduled to accompany orientation tour visitors during tours.

d. Instructional material. Information on the release of instructional material and procedures for transmitting the material to the home country will be given (para 2-31).

e. Organization of the Department of the Army. "The Department of the Army" manual should be used as a guide for the briefing.

f. Medical services. Only emergency medical services will be provided. Physical examinations cannot be authorized.

4-8. Rank. *a.* MAP orientation tours are provided on an austere basis with a minimum of official entertainment and are not appropriate for officers who, due to their positions, receive privileged treatment. Participation in MAP sponsored orientation tours is limited to personnel of the armed services of foreign countries who occupy positions below the U.S. Army Chief of Staff level.

b. In those cases in which an individual holds both a military rank and a responsible position in the governmental structure of the country, his eligibility for a MAP sponsored orientation tour will be determined by the position, rank, or title considered highest in his country.

c. Requests for orientation tours for personnel whose positions are equivalent to positions higher than that of the U.S. Army Chief of Staff should be submitted to the Office of the Secretary of Defense, ATTN: ISA, for consideration and, if approved, designation of the appropriate service as executive agency.

d. Requests for VIP tours for foreign personnel who hold positions equivalent to the U.S. Army Chief of Staff should be submitted through the unified commander to the Chief of Staff, U.S. Army.

4-9. Transportation. Transportation provided for orientation tour groups will be in accordance with paragraph 6-25.

4-10. Escort officers. *a.* The Department of the Army will furnish escort officers from CONUS sources. Language proficient officers will be furnished when such service is required and available. The escort officer will accompany the orientation tour group from the time of its arrival in CONUS until the group departs CONUS from its home country, except during leave periods.

b. The use of indigenous escort officers is not authorized. U.S. Army personnel from the host country may act as the escort officer when recommended by the unified commander and approved by HQDA on a case-by-case basis. To act in this capacity the United States officer must be language proficient and prepared to assume all administrative and interpreter duties.

4-11. Tour reports. Within 10 days after completion of each orientation tour and CONUS schools tour, a tour report will be prepared by the escort officer and furnished to the Assistant Chief of Staff for Intelligence, ATTN: FLO, Department of the Army, Washington, DC 20310, with information copy to DCSOPS, IA-MA.

4-12. Orientation tour limitations. *a.* Orientation tours to installations conducting primarily classified functions must be fully justified and will be approved on a case-by-case basis.

b. Orientation tours to the United States Military Academy and to joint or other senior colleges must be fully justified on a "need-to-know" basis and held to a minimum. Orientation tours to the United States Military Academy will not be arranged during examination and graduation periods (approximately 15 May to 15 July) or to Fort Benning, Ga, during the period of the Joint Civilian Orientation Conference (normally the second week of May).

c. Orientation tours to the U.S. Army War College will not be arranged.

4-13. Foreign Military Sales orientation tours. Orientation tours on a sales basis will be programed in fiscal year sales programs and executed under the same procedures as Military Assistance Program orientation tours. Funding procedures will be in accordance with the chapter 7.

4-14. Informational Program activities. Escort officers will be thoroughly conversant with the purposes of the Informational Program and will be instructed on the nonmilitary areas to be stressed on each orientation tour prior to commencement of the orientation tour. Local commanders will arrange itineraries to include, within time available, meetings with prominent citizens and visits to local points of interest. The advice and assistance of the Civilian Aides to the Secretary of the Army should be obtained in planning nonmilitary activities in accordance with paragraph 2-44. Those Army installations visited or those nearby will provide escort officers with Government transportation for the above purposes, to include transportation for weekends. Use of community service organizations (COSERV) should be used whenever possible. No more than 20 percent of the total training time available for any tour will be devoted to Informational Program activities.

4-15. Orientation tours to oversea units and installations. These tours will be scheduled in accordance with the policies established by the commander of the unified commander concerned.

4-16. Visits. Visits similar to Military Assistance Program orientation tours as defined in paragraph 4-1 may be arranged through diplomatic channels if the country concerned assumes all costs for transportation and living allowance. The military attaché or diplomatic representative in Washington, DC will submit a detailed visit request to the Assistant Chief of Staff for Intelligence, ATTN: FLO, Department of the Army, Washington, DC 20310, in accordance with AR 380-25.

Appendix F–Orientation in Home Country

F-1. The informational topics set forth in appendix J will guide the orientation given a foreign visitor or student before his departure for the United States and also subsequent contacts with him after his return. In conjunction with the

Country Team, the Chief MAAG will plan predeparture orientation, debriefing upon return, and followup contacts so as to make full use of available resources, including USIS materials and facilities, to support the Informational Program.

F-2. Informational materials for individual study to include USIS material on the United States, information about the school courses or special trips planned should be provided.

F-3. Personal interviews with U.S. officers who can explain the purpose of the trip and of any courses to be taken and who can explain or answer questions on the study material furnished should be arranged.

F-4. After a foreign visitor or trainee returns to his home country from the United States, a U.S. official should discuss fully with him the results and impressions of his trip and determine any improvements which can be made for subsequent similar trips.

F-5. Prior to his departure, each trainee will be thoroughly briefed on the following specific items, and other items considered appropriate based on country requirements as known by the MAAG:

a. Basic organization of the U.S. Army.

b. Mode of travel to the CONUS and within CONUS.

c. Reporting and processing at ports of debarkation and embarkation and at CONUS schools.

d. Reporting to the foreign liaison officer upon arrival at CONUS Army service schools and installations.

e. Ownership, operation, and insurance of privately owned automobiles.

f. Monetary allowances authorized under MAP.

g. Baggage allowances.

h. Uniform requirements and restrictions on the procurement of certain U.S. uniforms. To enable the trainee to be readily recognized as a foreign visitor and properly met by U.S. personnel the requirement should be emphasized for the trainee to wear his country uniform while in transit from home country to the initial training installation and return and when reporting to any installation.

i. Type of living accommodations at the service schools.

j. American civil and military customs.

k. Leave policy.

l. Discipline and applicability of U.S. civil and military law.

m. Claims policy.

n. Procedures for obtaining training other than that authorized by invitational travel orders.

o. The requirement for 30 days notice to MAAG of dependent travel will be stressed.

p. Released of instructional material.
q. Currency required during travel status and in the United States.
r. Use of international mail.
s. Security limitations applicable to training.

Appendix J–Informational Program for Foreign Military Trainees and Visitors in the United States

The following specific topics, or themes, are established. Each topic bears on a significant facet of life in the United States, and understanding of which contributes to a sound grasp of our institutions and ideals. In developing an informational program, each commander responsible for foreign trainees or visitors is expected to supplement or modify the topics where necessary to fit the character and background of the foreigners involved and the time and resources available for such purposes. Exposure of the foreign trainee to the non-military aspects of life in the United States is considered to be of importance to the Military Assistance Program second only to the strictly military training of that Program.

For a better understanding of life in the United States, foreign military trainees should be acquainted directly or indirectly with–

J-1. U.S. Government institutions. local, State and national governments; relationships between them; principle of checks and balances and its effect upon executive initiative.

J-2. The Judicial System: the Federal and State judicial systems and doctrine of judicial review; constitutional and legal status of the Armed Forces, with emphasis on their nonpolitical character.

J-3. Political Parties: American political parties and electoral procedures; the role of the opposition in a two-party system.

J-4. Press: role of free press and other communications media.

J-5. The diversity of life in the United States: the geographic, ethnic, religious, and social diversity of American life; how recent technological changes and urbanization processes are affecting this historic trait.

J-6. Minorities: minority groups in the United States, with reference to recent progress in applying American ideals to the Negro and the current steps underway to improve his opportunities.

J-7. Agriculture: the factors underlying agricultural productiveness; the changing life and role of the farmer today.

J-8. Economy: the national economy, diversity of industrial and business enterprises; role of government; role of private and commercial credit.

J-9. Labor and labor-management relations.

J-10. Education: Purpose and range of secondary and higher educational institutions; relationship between education and a responsible citzenry.

J-11. Public and social welfare: The care of the indigent, particularly the sick and aged; assistance to the underprivileged; unemployment benefits; the Social Security system.

Appendix K–Suggested Trips and Visits for Foreign Military Trainees and Visitors in the United States

K-1. Listed below are typical trips or visits which may appropriately be scheduled for foreign military trainees to acquaint them with various aspects of life in the United States. Local commanders will use this list as a guide, programing actual visits after taking into account their own staffs and other assets, local conditions, and the other means employed to meet the same objectives.

K-2. The USAFI guidance material provided to commanders should prove useful as preparatory material for many of these trips and visits. For full value, commanders must plan trips carefully, provide prior briefings, where appropriate, with knowledgeable and well-briefed escorts, and insure that persons who speak to each group are made aware of its level of English language comprehension.

a. Local governments: Commanders should bring foreign trainees in contact with agencies and principal personnel of local government at the city, township or country level at the earliest opportunity. This may best be accomplished when trainees are formally presented to local officials. One purpose of such an introduction is to make a point many, if not most, foreigners misunderstand, that is, local government officials are locally elected and responsible, within broad limits, to local people rather than to the central authorities.

b. State government: At some time during their stay here as many foreign trainees as possible should be taken to the State capital to be presented to the governor, or other officials, and to have an opportunity to observe selected operations of the State government. One purpose of this visit, like those outlined above, is to stress the autonomy of State governments and the independence of governors and legislatures. Where possible, the state supreme court should be included in such visits.

c. The opposition: On the occasion of the above mentioned visits, if convenient, or at other times, commanders should arrange for foreign military trainees to meet and talk with leaders of opposition parties, preferably office holders rather than party workers. Such a visit should be designed to show trainees the nature of the "loyal" opposition in this country, that its leaders perform official duties and have official status and that the parties in power and opposition are in fact more united than divided on most of the basic problems facing American society.

d. The political party system: An understanding of the "grass roots" character of American party organization is best gained by bringing visiting trainees in touch with representatives of one or the other party–sometimes with both–to give them some idea of the problems of local party organization, of means by which candidates are chosen, of the use of publicity and other means to convince voters, and of the relationships between local and national party organizations.

e. The press, radio, and TV: As a free press is one of the American institutions some foreigners find most difficult to grasp, visits to newspaper editorial offices should be arranged in order to underline how a free press works, and the ways in which editors and publishers define for themselves their responsibility to the public. Radio and TV stations and the printing plants of newspapers are interesting from a technical point of view but do not make this point quite as firmly as will acquaintance with the editorial and newsgathering functions of a newspaper.

f. Minority groups: Foreign trainees who interest themselves in the affairs of American minority groups should be put in touch with responsible leaders of minorities in order to give the trainees an idea of the goals of minority groups, as well as of their programs and procedures.

g. Farms: Commanders should arrange tours to modern "scientific" farms to show foreign trainees the character of American agriculture. On such trips, it may be advisable to match the interest and regional background of trainees with certain specialized types of farming operations in the vicinity. Especially worth underlining in such visits are marketing procedures, farmer credit facilities, and the question of the kinds of aid farmers receive from Federal, State, and other agricultural services in combatting pests and diseases, controlling breeding stock, and introducing improved varieties of crop plants.

h. Agricultural experiment stations: Such trips will permit foreign military trainees to view development of new and hybrid plants, animal and fish stock, experiments in controlling local soil conditions, pests, and diseases. The financing of the station and the means it uses to make information available to farmers are worthy of emphasis.

i. Business: The following four kinds of trips are designed to suggest the scope and diversity of American business enterprise:

(1) Visits to industrial enterprises should be designed to give foreign trainees an idea of the range of different kinds of industrial enterprises in the American "mixed economy," including Government-operated dams and hydroelectric institutions, local affiliates of large national corporations and smaller locally owned industries. Among other matters which company officials should be encouraged to discuss are—the relations between ownership and management of the company, management-union relationships, decisionmaking procedures in the field of product research and development, production scheduling, marketing and cost controls, and character and effect of governmental controls over operations.

(2) Visits to banks, savings and loan associations, Federal Housing Administration offices, and agricultural cooperative credit facilities will underline the range and ease of credit facilities available to the average American.

(3) Visits to local brokerage houses and discussions with brokers will emphasize the principles on which American financial investment is based and the procedures through which it is undertaken.

(4) Visits to large transportation centers for rail, air, water, truck, or pipeline will give foreign military trainees an opportunity to discuss the problems of management, maintenance, scheduling, and interconnection with transport officials.

j. Labor unions: In addition to putting interested foreign trainees in touch with local union officials where appropriate, tours to regional and national union headquarters will serve the useful purpose of emphasizing the scope of such organizations, the objectives of their leadership and their political and financial independence. In addition, trainees should be introduced to plant union officials during visits to industrial plants where that seems indicated.

k. Schools and colleges: Visits to nearby schools and colleges should be undertaken to show foreign trainees classes, laboratories, research facilities, extension course programs, agricultural experiment stations and cultural activities such as symphony performances and drama workshops. Area study programs, where they exist, will be of special interest to foreigners. Such visits should seek to underline the role of our schools and universities—to teach and learn, not to function as political instruments—and to show the diversity of our educational institutions, including privately endowed colleges, State or city colleges, land-grant universities, and church-affiliated institutions. Visits to high schools also may be useful.

l. Housing development: Visits to model homes, apartments, and publicly supported housing developments designed for low and middle income groups will be of particular interest to foreign military trainees.

m. Historic sites and national or State parks: In such trips emphasis should be laid on battlefield parks of our various wars to underline the care we have taken to preserve and commemorate the military side of American history.

n. Sporting events: Visits to baseball and football games, golf matches and other sporting events such as rodeos, regattas, horse and automobile races will show the trainee the multiplicity of American athletic interests.

o. Public and private welfare agencies: Visits to public health agencies, clinics, welfare agencies, national and State employment services, a local Social Security Office, Red Cross, or community chest will give the trainee an overall picture of the welfare facilities available in this country.

p. Churches and religious institutions. Trainees should be given a balanced picture of the role that religion and churches have played in American history and attitudes, and they should have an opportunity as appropriate, to visit churches of various religious denominations.

Appendix L–Biographic Data Requirements for Foreign Trainees

1. Country:
2. Date of information:
3. Full name (in natural order; use all capitals for surname or name/names by which the individual prefers to be addressed):
4. Present position:
5. Date assumed position:
6. Present rank/title:
7. Date of rank:
8. Service:
9. Branch of service:
10. Date and place of birth:
11. Present address:
12. Nationality:
13. Race:
14. Citizenship (indicate dual citizenship where applicable):
15. Religion:
16. Wife's name:
17. Number of children:
18. Military service (include military schooling, foreign service, units served with, awards and decorations):
19. Language proficiency:
20. Preferences (personal preferences in food, drink, tobacco, entertainment, sports, hobbies, dietary restrictions):
21. U.S. acquaintances/relatives (name, position or occupation, domicile, nature and duration of relationship):

The following additional information concerning the officer's family will be provided when the officer's wife will accompany or join him in the U.S.:

22. Wife's date and place of birth:

23. Nationality:
24. Race:
25. Religion:
26. Background (education, languages, preferences in food and drink, hobbies, preferences in entertainment, special interests, professional societies):
27. Children (names, sex, ages, marital status, other items of interest such as schools, health, military service):

Appendix M–Format for Biographical Data for Orientation Tour

Date ______________

SUBJECT: Biographical Data for Orientation Tour, ________, __________
(Country) Number
____
(and fiscal year)

1. Full name: (Use all capitals for surname or the name which individual prefers used when addressed.)
2. Phonetic spelling of name in all capitals:
3. Present military rank or title:
4. Date of rank:
5. Branch of service:
6. Present position:
7. U.S. Army equivalent position:
8. Date and place of birth:
9. Nationality:
10. Wife: (Name)
11. Children: (Number):
12. Education:
13. Religion:
14. Dietary restrictions:
15. English language proficiency:
16. Proficiency in languages other than English:
17. Previous assignments:
18. Publicity release: (State preferences and/or objections)

DISTRIBUTION: 1 copy– Department of the Army
Deputy Chief of Staff for Military Operations
ATTN: OPS-IA-MA
Washington, DC 20310

Notes

Notes

Chapter 1
Introduction

1. U.S., Congress, House, *Foreign Assistance and Related Programs Appropriation Bill, 1970, Report*, no. 91-708 together with Supplemental and Additional Views to Accompany H.R. 15149, House of Representatives, 91st Cong., 1st sess., 1969, p. 38. From the Supplemental Views.

2. Edgar S. Furniss, Jr., *Some Perspectives on American Military Assistance* (Princeton, N.J.: Center of International Studies, Princeton University, Memo no. 13, June 18, 1957), pp. 1-2, 37.

3. "Editorial," *The Nation*, July 19, 1971, p. 36.

4. Norman A. Graebner, "Global Containment: The Truman Years," *Current History*, 57 (August 1969), 77-83ff. Neal D. Houghton, ed., *Struggle Against History* (New York: Simon & Schuster, 1968), pp. 42-43, 76, 272-74. Cf.: David Horowitz, *From Yalta to Vietnam* (England: Penguin Books, 1967), pp. 68-75, 98: David Calleo, *The Atlantic Fantasy: The U.S., NATO, and Europe* (Baltimore, Md.: Johns Hopkins Press, 1970). Athan G. Theokaris, *The Yalta Myths: An Issue in U.S. Politics* (Columbia, Mo.: University of Missouri Press, 1970).

5. Walter LaFeber, *The New Empire* (Ithaca, N.Y.: Cornell University Press, 1963); William Appleman Williams, *The Tragedy of American Diplomacy* (New York: Dell, 1962); Lloyd C. Gardner, *Economic Aspects of New Deal Diplomacy* (Madison: University of Wisconsin Press, 1964); Gabriel Kolko, *The Politics of War* (New York: Random House, 1969).

6. Quoted in Kolko, *Politics of War*, p. 254.

7. Quoted by Andre Gunder Frank, "The Underdevelopment Policy of the United Nations in Latin America," *NACLA Newsletter*, 3 (December 1969), 1.

8. Gardner, *Economic Aspects*, p. 50. The following are instructive on the first decades of this century: Williams, *Tragedy of American Diplomacy*: Alonso Aguilar, *Pan-Americanism: From Monroe to the Present* (New York: Monthly Review Press, 1968); Ronald Radosh, *American Labor and United States Foreign Policy* (New York: Random House, 1969), pp. 3-303; William J. Pomeroy, *American Neo-Colonialism: Its Emergence in the Philippines and Asia* (New York: International Publishers, 1970); Scott Nearing and Joseph Freeman, *Dollar Diplomacy* (New York: Monthly Review Press, 1925 & 1966); David Green, *The Containment of Latin America* (Chicago: Quadrangle Books, 1971); and the work cited in the note which follows.

9. Gabriel Kolko, *The Roots of American Foreign Policy* (Boston: Beacon Press, 1969), pp. 17-22. G. William Domhoff, "Who Made American Foreign

Policy, 1945-1963?" in *Corporations and the Cold War*, David Horowitz, ed. (New York: Monthly Review Press, 1969), pp. 25-69, Cf., *Economic Issues in Military Assistance, Hearings*, Joint Economic Committee, pp. 46, 167.

10. Documented in Kolko's *Politics of War*.

11. Ibid., pp. 247-50, 275-76, 477-78, 589-90; and the same author's *Roots of American Foreign Policy*, pp. 88-96. John Gimbel, *The American Occupation of Germany: Politics and the Military, 1945-1949* (Stanford, Calif.: Stanford University Press, 1969), pp. 117-19.

12. Thus a recent Pentagon director of military assistance testified that "we are available to assist if needed, but we would rather they carry the ball." U.S., Congress, House, Committee on Foreign Affairs, *Foreign Assistance Act of 1963, Hearings*, before the Committee on Foreign Affairs of the House of Representatives, 88th Cong., 1st sess., p. 684. Cf., pp. 803-04. Cf., U.S. Congress, House, Committee on Foreign Affairs, *Military Assistance Training, Hearings*, before the Subcommittee on National Security Policy and Scientific Developments of the Committee on Foreign Affairs, House of Representatives, 91st Cong., 2d sess., October-December 1970, pp. 112, 116, 123-28.

13. Great Britain, House of Commons, *Parliamentary Debates*, 5th ser., vol. 694 (1963-1964), pp. 107-110. Cf., Barnet, *Intervention and Revolution*, pp. 237-43.

14. Barnet, *Intervention and Revolution*; Horowitz, *From Yalta to Vietnam*. Cf., David Wise and Thomas B. Ross, *The Invisible Government* (New York: Random House, 1964). Within a decade, the CIA-Special Forces role in overthrowing Cambodia's neutralist government will become apparent although currently available evidence is inconclusive. With regard to the latter, see the *New York Times*, January 28, 1970, where the CIA-Special Forces link to the Khmer Serei movement is reported, and: *I.F. Stone's Bi-Weekly*, May 18, 1970; T.D. Allman, *Far Eastern Economic Review*, April 23, 1970, p. 32; Cambodia, *Bulletin Mensuel de Documentation* (Peking: Office of the Chief of State, January 1971), p. 7. Military aid was resumed just eight months prior to the coup against Sihanouk, who years earlier had thrown "out our military mission because he said it had been trying to turn his armed forces against him, and gave up economic aid, too, rather than have it used as a cover for U.S. agents trying to overthrow him. This was not a figment of Sihanouk's imagination. As far back as 1958, in a police raid on the villa of one of his Generals, Sihanouk found a letter from President Eisenhower pledging full support to a projected coup and to a reversal of Cambodian neutrality. This was part of a 'Bangkok plan'. . . . When Sihanouk resumed relations last August, in his desperate see-saw between the two sides, his condition was that the U.S. mission be kept small. He didn't want too many CIA agents roaming around." *I.F. Stone's Bi-Weekly*, May 18, 1970, pp. 1-2. The letter itself could, of course, have been a forgery. Cf., N. Sihanouk, "The Story of My Overthrow and Resistance," *Ramparts*, July 1972, pp. 19-23, 42-47.

15. U.S., Congress, Senate, Committee on Foreign Relations, *Foreign Assistance, 1966*, Hearings, before the Committee on Foreign Relations, United States Senate, 89th Cong., 2d sess., p. 693. Cf., U.S., Congress, House, Committee on Appropriations, *Foreign Assistance and Related Agencies Appropriations for 1970, Hearings* before the Subcommittee on Foreign Operations and Related Agencies, House of Representatives, 91st Cong., 1st sess., p. 727.

16. For several examples of such scholarship, see: Kenneth F. Johnson, "Causal Factors in Latin American Political Stability," *Western Political Quarterly*, 17 (Summer 1964), 432-46; Harry Eckstein, "On the Etiology of Internal War," *History and Theory*, 4 (1965), 133-63; D. P. Bwy, "Political Instability in Latin America: The Cross-Cultural Test of a Causal Model," *Latin American Research Review*, 3 (Spring 1968), 17-66; J.A. Brill, "The Military and Modernization in the Middle East," *Comparative Politics*, 2 (October 1969), 41-62; Merle Kling, "Violence and Politics in Latin America," in *Latin American Sociological Studies*, by Paul Halmos, ed. (Keele, U.K.: University of Keele, 1967), pp. 119-32.

17. Thus, of the seventy-three studies annotated, no more than five held that external military training exerted a significant influence upon internal political roles. Sixty-five abstracts failed to even mention such intervention. Peter B. Riddleberger, *Military Roles in Developing Countries: An Inventory of Past Research and Analysis* (Washington: Special Operations Research Office, American University AD 463188, 1965). Cf.: Eric A. Nordlinger, "Soldiers in Mufti: The Impact of Military Rule upon Economic and Social Change in the Non-Western States," *American Political Science Review*, 64 (December 1970), 1131-48; D. Nelkin, "The Economic and Social Setting of Military Takeovers in Africa," *Journal of Asian and African Studies*, 2 (July-October 1967), pp. 230-44; Amos Perlmutter, "From Obscurity to Rule: The Syrian Army and the Baath Party," *Western Political Quarterly*, 22 (December 1969), 827-45.

18. For a brief summary of their role in one underdeveloped country, see the author's "Chile's Left: Structural Factors Inhibiting an Electoral Victory in 1970," *Journal of Developing Areas*, 3 (January 1969), 215-29.

19. An instructive analysis of a number of these factors is Susanne Bodenheimer's "Dependency and Imperialism: The Roots of Latin American Underdevelopment," *Politics & Society*, 1 (May 1971), 327-57. Cf.: C. Gordon, "Has Foreign Aid Been Overstated?" *Inter-American Economic Affairs*, 21 (Spring 1968), 3-18; Robert I. Rhodes, ed., *Imperialism and Underdevelopment* (New York: Monthly Review Press, 1970).

20. "The average military mind is conservative in its relation to the revolutionary processes of history from the Protestant Reformation to our own times." Alfred Vagts, *A History of Militarism: Civilian and Military* (New York: Free Press, 1967), pp. 29-30. W. Eckhardt and A.G. Newcombe, "Militarism, Personality and Other Social Attitudes," *Journal of Conflict Resolution*, 13 (June 1969), pp. 210-19. Riddleberger, *Military Roles*, pp. 6-7. Martin C.

Needler, "The Latin American Military: Predatory Reactionaries or Modernizing Patriots?" *Journal of Inter-American Studies*, 11 (April 1969), 239-42. U.S., Department of the Army, *Report of the Thirteenth Annual U.S. Army Human Factors Research and Development Conference, October 25-27, 1967* (Fort Monmouth, N.J.: U.S. Army Signal Center and School, 1967), p. 180. Edwin Lieuwen, *Generals vs. Presidents* (New York: Praeger, 1964), pp. 130-35. Eliezer Be'eri, *Army Officers in Arab Politics and Society* (New York: Praeger, 1970), pp. 464-72. Harold A. Hovey, *United States Military Assistance: A Study of Policies and Practices* (New York: Praeger, 1965), pp. 70-71. James Clotfelter, "The African Military Forces," *Military Review*, 48 (May 1968), 23-31. Stephen S. Kaplan, "United States Military Aid to Brazil and the Dominican Republic: Its Nature, Objectives and Impact," (paper presented at the Symposium on Arms Control, Military Aid and Military Rule in Latin America, University of Chicago, May 26-27, 1972), pp. 40-51. On secular trends in officers' social origins, see Bengt Abrahamsson and Steven Kaplan's "Social Recruitment of the Military: World-Wide Research Inventory," (paper prepared under the auspices of the Research Group on Armed Forces and Society, International Sociological Association, Morris Janowitz, University of Chicago, Chairman, for the Seventh World Congress, Varna, September 13-19, 1970).

21. Nordlinger, "Soldiers in Mufti," pp. 1131-48. Cf.: Robert M. Price, "A Theoretical Approach to Military Rule in New States: Reference-Group Theory and the Ghanaian Case," *World Politics*, 23 (April 1971), 399-430; Jerry Weaver, "Assessing the Impact of Military Rule: Alternative Approaches," (paper presented at the Symposium on Arms Control, Military Aid and Military Rules in Latin America, University of Chicago, May 26-27, 1972).

Chapter 2
Foreign Officers as Instruments of American Policy

1. "Democracy and the Military Base of Power," *Middle East Journal*, 22 (Winter 1968), 42.

2. *United States Military Assistance: A Study of Policies and Practices* (New York: Praeger, 1965), p. 70.

3. "Defense of the Western Hemisphere: A Need for Reexamination of United States Policy," *Midwest Journal of Political Science*, 3 (November 1959), 397.

4. U.S., Congress, House, Committee on Foreign Affairs, *Reports of the Special Study Mission to Latin America on Military Assistance Training and Developmental Television*, submitted by Hon. Clement J. Zablocki, Hon. James G. Fulton and Hon. Paul Findley to the Subcommittee on National Security Policy and Scientific Developments of the Committee on Foreign Affairs, House of Representatives, 91st Cong., May 7, 1970, p. 31.

5. *Military Assistance Training, Hearings*, House, p. 145. Cf., *Economic Issues in Military Assistance, Hearings*, Joint Economic Committee, p. 176.

6. U.S., Congress, Senate, Committee on Foreign Relations, *United States Security Agreements and Commitments Abroad: Morocco and Libya, Hearings*, before the Subcommittee on United States Security Agreements and Commitments Abroad, Committee on Foreign Relations, United States Senate, 91st Cong., 2d sess., pp. 1991-2004. Cf., *Military Assistance Training, Hearings*, House, pp. 113-15, 125.

7. U.S., Congress, Senate, Committee on Foreign Relations, *United States Security Agreements and Commitments Abroad: The Republic of the Philippines, Hearings*, before the Subcommittee on United States Security Agreements and Commitments Abroad, Committee on Foreign Relations, United States Senate, 91st Cong., 1st sess., p. 228. Cf.: pp. 208-10, 221-29, 231; U.S., Congress, Senate, Committee on Foreign Relations, *United States Military Policies and Programs in Latin America, Hearings*, before the Subcommittee on Western Hemisphere Affairs of the Committee on Foreign Relations, United States Senate, 91st Cong., 1st sess., June 24th and July 8, 1969, pp. 46-51, 65, 80-81; *Foreign Assistance Act of 1969, Hearings*, House, pp. 554-57.

8. Lt. Gen. Andrew P. O'Meara, USA, "Opportunities to the South of US," *Army*, 13 (November 1962), 41-43. U.S., Congress, House, Committee on Foreign Affairs, *Foreign Assistance Act of 1965, Hearings*, before the Committee on Foreign Affairs, House of Representatives, 89th Cong., 1st sess., pp. 634, 809. Congress, House, *Foreign Assistance Act of 1966, Hearings*, before the Committee on Foreign Affairs, House of Representatives, 89th Cong., 2d sess., p. 263. Capt. Robert F. Farrington, USN, "Military Assistance at the Crossroads," *U.S. Naval Institute Proceedings*, 88 (June 1966), 68-76. Lt. Col. David R. Hughes, USA, "The Myth of Military Coups and Military Assistance," *Military Review*, 47 (December 1967), 7-8. U.S., Congress, Senate, Committee on Foreign Relations, *Foreign Assistance Act of 1967, Hearings*, before the committee on Foreign Relations, 90th Cong., 1st sess., p. 316. U.S., Congress, House, *Foreign Assistance Act of 1968, Hearings*, before the Committee on Foreign Affairs, House of Representatives, 90th Cong., 2d sess., pp. 822, 824-26, 836. U.S., Congress, House, Committee on Foreign Affairs, *Foreign Assistance Act of 1969, Hearings*, before the Committee on Foreign Affairs, House of Representatives, 91st Cong., 1st sess., pp. 583, 637, 650, 696. *Foreign Assistance Act of 1972, Hearings*, House, pp. 104, 107.

9. Gibert, "Soviet-American Military Aid Competition," p. 1122.

10. *Foreign Assistance Appropriations for 1966, Hearings*, House, p. 246. " 'Follow Me' United States Army Infantry School," *Military Review*, 39 (July 1959), 39. For similar affirmations, see: U.S., Congress, House, Committee on Foreign Affairs, *Foreign Assistance Act of 1964, Hearings*, before the Committee on Foreign Affairs, House of Representatives, 88th Cong., 2d sess., p. 94; *Foreign Assistance Act of 1965, Hearings*, House, pp. 367, 647-48, 732, 740.

Cf., Col. William R. Kintner, USA, "The Role of Military Assistance," *U.S. Naval Institute Proceedings*, 87 (March 1961), 78, 82.

11. *Foreign Assistance Appropriations for 1966, Hearings*, House, pp. 322-23, 401. *Foreign Assistance Act of 1968, Hearings*, House, pp. 705-06, 812. Cf., Banning Garrett, "The Vietnamization of Laos," *Ramparts*, June 1970, pp. 37-45.

12. Gibert, "Soviet-American Military Aid Competition," p. 1128.

13. John M. Dunn, "Military Aid and Military Elites: The Political Potential of American Training and Technical Assistance Programs," (unpublished Ph.D. dissertation, Princeton University, 1961), pp. 295-96. Edgar S. Furniss, Jr., *Some Perspectives on American Military Assistance* (Princeton, N.J.: Center of International Studies, Princeton University, Memo no. 13, June 18, 1957), pp. 19, 38-39. M.J.V. Bell, *Military Assistance to Independent African States* (London: Institute for Strategic Studies, December 1964), p. 3. Charles D. Windle and Theodore R. Vallance, "Optimizing Military Assistance Training," *World Politics*, 15 (October 1962), 92-94. Victor, "Military Aid and Comfort," pp. 43-47. *Economic Issues in Military Assistance, Hearings*, Joint Economic Committee, pp. 83-4, 89, 285. Cf.: Stephen S. Kaplan, "United States Military Aid to Brazil and the Dominican Republic: Its Nature, Objectives and Impact," (paper presented at the Symposium on Arms Control, Military Aid and Military Rule in Latin America at the University of Chicago Arms Control and Foreign Policy Seminar, May 26-27, 1972), pp. 14, 18; Lenny Siegel, "The Future of Military Aid," *Pacific Research & World Empire Telegram*, 3 (November-December 1971), 2.

14. Corporate expansion as a military aid goal is referred to in the following: Furniss, *Perspectives*, pp. 17, 29-32; John L. Holcombe and Alan D. Berg, *MAP for Security: Military Assistance Programs of the United States* (Columbia: University of South Carolina, Bureau of Business and Economic Research, April 1957), p. 17; Charles Wolf, Jr., *Military Assistance Programs* (Santa Monica, Calif.: Rand Corporation, P-3240, October 1965), p. 22; Theodore R. Vallance and Charles D. Windle, "Cultural Engineering," *Military Review* 42 (December 1962), 62; Simon G. Hanson, "Third Year: Military," *Inter-American Economic Affairs*, 18 (Spring 1965), 20-37; Eduardo Galeano, "The De-Nationalization of Brazilian Industry," *Monthly Review* 21 (December 1969), 16-17. Cf., Irving Louis Horowitz, "The Military Elites," in *Elites in Latin America*, Seymour M. Lipset and Aldo Solari, eds. (London: Oxford University Press, 1967), pp. 175-77.

15. Hilton P. Goss, *Africa Present and Potential: A Strategic Survey of the Role of the African Continent in United States Defense Planning* (Santa Barbara, Calif.: General Electric Company Technical Military Planning Operation, RM57TMP-8, December 31, 1958), Summary.

16. The relationship between corporate investment and access to underpriced raw materials which deny Third World countries vitally needed potential

development capital is treated in: Pierre Jalee, *The Third World in World Economy* (New York: Monthly Review Press, 1969); Robert I. Rhodes, ed., *Imperialism and Underdevelopment* (New York: Monthly Review Press, 1970); Magdoff, *Age of Imperialism*. Cf., Ved P. Nanda, "Expropriation: Imperialism's Achilles' Heel," *Nation*, June 19, 1972, pp. 775-78.

17. "Address by General Robert W. Porter, Jr., Commander in Chief, U.S. Southern Command, Presented to the Pan American Society of the United States, New York, N.Y., Tuesday, March 26, 1968," as reprinted in the *Foreign Assistance Act of 1968, Hearings*, House, pp. 1204-05.

18. *Reports of the Special Study Mission*, House, p. 11.

19. U.S., Congress, House, Committee on Foreign Affairs, *Cuba and the Caribbean, Hearings*, before the Subcommittee on Inter-American Affairs of the Committee on Foreign Affairs, House of Representatives, 91st Cong., 2d sess., July 8-August 3, 1970, p. 98.

20. U.S., Congress, Senate, Committee on Foreign Relations, *Rockefeller Report on Latin America, Hearings*, before the Subcommittee on Inter-American Affairs of the Committee on Foreign Relations, United States Senate, 91st Cong., November 20, 1969, pp. 83, 148.

21. For obvious reasons, and perhaps because some are not aware of the fact that corporations withdraw more than they invest, there are no specific references to the *process of capital drainage*. See Chapter Seven, note 40 and footnote o. Cf.: U.S., Department of the Army, *U.S. Army Handbook for Venezuela* (Washington: GPO, 1964), pp. 267, 463, 468; "Why U.S. Military Assistance is Given to the Argentine Dictatorship: The Explanation of the U.S. Department of Defense," *Inter-American Economic Affairs*, 21 (Autumn 1967), 81-89.

22. *United States Security Agreements, Morocco and Libya, Hearings*, Senate, p. 1995. Cf.: *Foreign Assistance Act of 1969, Hearings*, House, pp. 840, 844, 1048, 1064-69, 1075-76, 1110, 1155-56, 1170; *Rockefeller Report, Hearing*, Senate, pp. 85-93, 117-19.

23. *United States Security Agreements, Morocco and Libya, Hearings*, Senate, pp. 1962-87.

24. *United States Security Agreements, Philippines, Hearings*, Senate, p. 1035. For more typical euphemistic references to MAP as a safeguard for overseas American "economic interests," see the *Foreign Assistance Act of 1972, Hearings*, House, pp. 4, 47, 50, 167.

25. The term is a euphemism for corporate capitalism. U.S., Congress, Senate, *Foreign Assistance, 1965, Hearings*, before the Committee on Foreign Relations, United States Senate, 89th Cong., 1st sess., pp. 212-13. U.S. Congress, House, Committee on Appropriations, *Foreign Assistance and Related Agencies Appropriations for 1969, Hearings*, before a subcommittee of the Committee on Appropriations, House of Representatives, 90th Cong., 2nd sess., pp. 840-844, 1048, 1064-69, 1075-77, 1110, 1155-56, 1170. *Foreign Assistance Appropriations for 1970, Hearings*, House, pp. 851-52. *United States Military Policies,*

Hearings, Senate, pp. 46-48, 80-81. *United States Security Agreements, Morocco and Libya, Philippines, Hearings*, Senate, pp. 1035, 1976, 1978, 1984, 1993-95. Cf.: U.S. President, *The Foreign Assistance Program: Annual Report to the Congress for Fiscal Year 1970* (Washington: GPO, n.d.), pp. 17-25, 68-71; U.S., Congress, House, *Foreign Assistance and Related Programs Appropriations Bill, 1970, Report*, no. 91-708, together with Supplemental and Additional Views to Accompany H.R. 15149, House of Representatives, 91st Cong., 1st sess., pp. 2-3.

26. *Foreign Assistance, 1965, Hearings*, Senate, pp. 212-13. *Foreign Assistance Appropriations for 1966, Hearings*, House, p. 266. *Foreign Assistance Appropriations for 1970, Hearings*, House, p. 628. *Rockefeller Report, Hearing*, House, pp. 9-11. *United States Security Agreements, Morocco and Libya, Hearings*, Senate, pp. 1962-87, 1995-2002. *Cuba and the Caribbean, Hearings*, House, pp. 94, 97. *Reports of the Special Study Mission*, House, pp. 5, 18-20. U.S., Congress, House, Committee on Foreign Affairs, *Military Assistance Training in East and Southeast Asia, A Staff Report*, prepared for the use of the Subcommittee on National Security Policy and Scientific Developments of the Committee on Foreign Affairs, House of Representatives, 91st Cong., 2d sess., p. 6.

Frequently, if not always, the gravamen of liberal criticism is that in terms of long-run influence, it is dysfunctional for the United States to openly associate itself and aid dictatorial regimes. See, e.g., *Foreign Assistance Act of 1972, Hearings*, House, pp. 39-40. Cf., Lenny Siegel, "The Future of Military Aid," *Pacific Research & World Empire Telegram*, 3 (November-December 1971), p. 7.

27. Thayer, *The War Business*. Cf.: *Military Assistance Training, Hearings* House, pp. 165, 171; *Economic Issues in Military Assistance, Hearings*, Joint Economic Committee, pp. 246-47.

28. *Foreign Assistance Appropriations for 1966, Hearings*, House, pp. 425-56. Cf., *Military Assistance Training in East Asia, Report*, p. 24.

29. *Military Assistance Training, Hearings*, House, pp. 40, 45. Similarly, "Brazil was the only nation in the Organization of American States to join the United States in sending troops to the Dominican Republic in 1965." *Reports of the Special Study Mission*, House, p. 4.

30. Windle and Vallance, "Optimizing Military Assistance Training," p. 96. Col. G.A. Lincoln, "Factors Determining Arms Aid," *Academy of Political Science Proceedings*, 25 (1954), 264. Col. Lowell T. Bondshu, "Keeping Pace with the Future–Toward Mutual Security," *Military Review* 39 (June 1959), 58-60, 63-64, 73. U.S., President's Committee to Study the United States Military Assistance Program, *Composite Report*, 1 (Washington, D.C.: GPO, 1959), pp. 153-54. *Military Assistance Training, Hearings*, House, p. 170.

31. What among other things the secretary of defense had in mind may have been indicated two years earlier when he testified: "You cannot supply military

assistance for internal security purposes to any Latin American country and expect them not to use that equipment in the hour of a coup. Therefore, we have to take a long hard look, it seems to me, at the stability of a country, the possibility of a coup, and what kind of a coup it is likely to be. . . . " *Foreign Assistance Act of 1965, Hearings*, House, p. 632. U.S., Congress, Senate, Committee on Foreign Relations, *Foreign Assistance Act of 1963, Hearings*, before the Committee on Foreign Relations, United States Senate, 88th Cong., 1st sess., p. 208.

32. *Foreign Assistance Act of 1964, Hearings*, House, pp. 728-29. In recent testimony, a Brookings Institution expert also stressed the lead time factor: "It seems to me that military assistance is not a very good instrument for manipulating day-by-day events—a faucet to be turned on and off to show approval or disapproval of current policies in a friendly government. It is, rather, a long-range investment, and this is particularly true of training. Person-to-person contacts and the transfer of capability constitute a long-range investment, the results of which we cannot fully control, but in which we have some degree of faith." *Military Assistance Training, Hearings*, House, p. 52.

Chapter 3
Social Aspects of Training

1. *War and the Minds of Men* (New York: Harper & Brothers, 1950), p. 72, as quoted in John M. Dunn, "Military Aid and Military Elites: The Political Potential of American Training and Technical Assistance Programs," (unpublished Ph.D. dissertation, Princeton University, 1961), p. 306.

2. "The Nationbuilder: Soldier of the Sixties," *Military Review* 45 (January 1965), 64.

3. U.S., Congress, House, Committee on Foreign Affairs, *Military Assistance Training, Hearings*, before the Subcommittee on National Security Policy and Scientific Developments of the Committee on Foreign Affairs, House of Representatives, 91st Cong., 2d sess., October-December 1970, p. 112.

4. U.S., Congress, Senate, Committee on Foreign Relations, *United States Military Policies and Programs in Latin America, Hearings*, before the Subcommittee on Western Hemisphere Affairs of the Committee on Foreign Relations, United States Senate, 91st Cong., 1st sess., June 24 and July 8, 1969, p. 71.

5. U.S., Congress, House, Committee on Appropriations, *Foreign Assistance and Related Agencies Appropriations for 1967, Hearings*, before a subcommittee of the Committee on Appropriations, House of Representatives, 89th Cong., 2d sess., p. 634.

6. U.S., Congress, House, Committee on Foreign Affairs, *Reports of the Special Study Mission to Latin America on Military Assistance Training and Developmental Television*, Submitted by Hon. Clement J. Zablocki, Hon. James

G. Fulton and Hon. Paul Findley to the Subcommittee on National Security Policy and Scientific Developments, Committee on Foreign Affairs, House of Representatives, 91st Cong., May 7, 1970, p. 7.

7. Institute of International Education, *Military Assistance*, p. 15. The nonescorted groups or "package tours" may be multinational in composition. Dunn, "Military Aid," (dissertation), p. 119.

8. Dunn, "Military Aid," pp. 113-14.

9. Harold A. Hovey, *United States Military Assistance: A Study of Policies and Practices* (New York: Praeger, 1965), p. 174. Cf., *Reports of the Special Study Mission*, p. 7.

10. U.S., Congress, Senate, Committee on the Armed Services, *Military Cold War Education and Speech Review Policies, Hearings*, before the Subcommittee on Special Preparedness of the Committee on Armed Services, United States Senate, 87th Cong., 2d sess., pp. 2633, 3143.

11. Dunn, "Military Aid," pp. 115-16.

12. Hovey, *United States Military Assistance*, p. 174.

13. U.S., Congress, House, Committee on Foreign Affairs, *Foreign Assistance Act of 1963, Hearings*, before the Committee on Foreign Affairs, House of Representatives, 88th Cong., 1st sess., pp. 701, 715, 727-28. U.S., Congress, House, Committee on Appropriations, *Foreign Assistance and Related Agencies Appropriations for 1966, Hearings*, before a subcommittee of the Committee on Appropriations, House of Representatives, 89th Cong., 1st sess., pp. 279, 406-07.

14. *Foreign Assistance Appropriations for 1969, Hearings*, House, p. 540.

15. U.S., Congress, House, Committee on Appropriations, *Foreign Assistance and Related Agencies Appropriations for 1968, Hearings*, before a subcommittee of the Committee on Appropriations, House of Representatives, 90th Cong., 1st sess., pp. 567-69.

16. U.S., Congress, Senate, Committee on Foreign Relations, *Foreign Assistance Act, 1969, Hearings*, before the Committee on Foreign Relations, United States Senate, 91st Cong., 1st sess., p. 136.

17. *Reports of the Special Study Mission*, p. 7.

18. U.S., Congress, House, Committee on Foreign Affairs, *Winning the Cold War: The U.S. Ideological Offensive, Hearings,* before the Subcommittee on International Organizations and Movements of the Committee on Foreign Affairs, House of Representatives, 88th Cong., 2d sess., pp. 1031-32. The text of this highly important memo is reprinted in Appendix 3.

On the greatly increased priority assigned to ideological indoctrination during the 1962-65 period, see: U.S., Congress, Senate, Committee on Foreign Relations, *Foreign Assistance, 1965, Hearings*, before the Committee on the Foreign Relations, United States Senate, 89th Cong., 1st sess., p. 211.

19. United States Naval Mine Warfare School, *Information Booklet for Senior Foreign Naval Officers* (Yorktown, Virginia, 1958), cited in Dunn, "Military Aid," p. 109.

20. United States Naval War College, *Naval Command Course for Senior Officers: Directive for Washington Orientation Visit, Class of June 1961* (Newport, R.I., 1960), passim; Ibid., *Directive for New London-United Nations Visit, Class of June 1961* (Newport, R.I., 1960), passim; Ibid., *Directive for Norfolk, Cape Canaveral, Key West, Charleston Orientation Visit, Class of June 1961* (Newport, R.I., 1960), passim; Ibid., *The Curriculum 1960-61, Naval Command Course for Senior Foreign Officers* (Newport, R.I., 1960), pp. 27-28; all of the preceding cited in Dunn, "Military Aid," pp. 225, 376-82.

21. United States Naval War College, *Naval Command Course for Senior Foreign Officers: Directive for Final Orientation Visit, Class of June 1960* (Newport, R.I., 1960), passim, cited in Dunn, "Military Aid," pp. 223-24.

22. Ibid., pp. 355-57.

23. 1st Lt. James M. Tiffany, USA, "Training Foreign Students," *Ordnance* 50 (January-February 1966), 439. Cf., U.S. Army, *Foreign Nationals*, p. 2-18.

24. U.S., Department of the Navy, *Information Program for FMT (Foreign Military Trainees) and Visitors in the United States* (Chapter Six, OPNAVINIST 4950 ID CH-1, 14 December 1966), pp. 32e-f.

25. *Winning the Cold War, Hearings*, House, pp. 1036-38.

26. Donn R. Grand Pre, "A Window on America: The Department of Defense Informational Program," *International Educational and Cultural Exchange*, 6 (Fall 1970), 91.

27. With regard to the general credibility of Army allegations that a "balanced" picture is conveyed, see: J.W. Fulbright, *The Pentagon Propaganda Machine* (New York: Liveright, 1970), pp. 67-70, 88-91.

28. *Winning the Cold War, Hearings*, House, p. 1044.

29. U.S. Dept. of the Navy, *Informational Program for FMT*, p. 32d.

30. W.F. Gutteridge, *The Military in African Politics* (London: Methuen, 1969), p. 132.

31. Grand Pre, "Window on America," p. 90.

32. U.S., Dept. of the Navy, *Informational Program for FMT*, p. 32b. U.S., Armed Forces Institute, *Informational Program on American Life and Institutions for Foreign Military Trainees: Guidance to Instructors and Training Officers* (Madison, Wisconsin, 1965), pp. 2, 4. Cf., U.S. Army, *Foreign Nationals*, pp. 2-18-20.

33. Quoted by Drew Middleton, "Thousands of Foreign Military Men Studying in US," *New York Times*, November 1, 1970, p. 18.

34. Middleton, "Thousands," p. 18.

35. Charles Windle and T.R. Vallance, "Optimizing Military Assistance Training," *World Politics* 15 (October 1962), 107. Cf., *Winning the Cold War, Hearings*, House, p. 1038.

36. Col. Lowell T. Bondshu, "Keeping Pace with the Future–Toward Mutual Security," *Military Review* 39 (June 1959), 66.

37. U.S. Army Command and General Staff College, *U.S. Army English for Allied Students* (Fort Leavenworth, Kansas: Reference Book 20-8, April 1968), p. 15.

38. Dunn, "Military Aid," p. 198.

39. U.S. Army, *Foreign Nationals*, pp. 2-4-7.

40. Fulbright, *Pentagon Propaganda*, pp. 9, 25-34, 44-45, 60-61, 88-91, 95-101.

41. Middleton, "Thousands," p. 18. "By spending a few evenings with these [*sic*] families, students become familiar with the basic strength of our society–The American family." Tiffany, "Training," p. 439.

42. Dunn, "Military Aid," pp. 107-08, 110, 214-18, 305-06, 352-67. U.S., Dept. of the Navy, *Informational Program for FMT*, pp. 32-32c.

43. Tiffany, "Training," p. 439. Dunn, "Military Aid," pp. 205-06, 264. Hovey, *United States Military Assistance*, p. 174.

44. Dunn, "Military Aid," pp. 350-54.

45. Bondshu, "Keeping Pace," p. 64.

46. Conversation with Maj. Worthington at the Latin American Studies Association annual meeting, Washington, D.C., April 17, 1970.

Chapter 4
Ideological Stereotypes in Military Training

1. Col. Edward L. King, USA, Ret., "Making It in the U.S. Army," *New Republic*, May 30, 1970, p. 20.

2. Escuela de las Americas, Comite de Seguridad, Ejercito de los EE. UU., *Teoria, Practica e Instituciones de los Gobiernos Totalitarios* (Fuerte Gulick, Zona del Canal, MI 4606, Enero 1969), p. 3.

3. U.S., Congress, House, Committee on Foreign Affairs, *Foreign Assistance Act of 1965, Hearings*, before the Committee on Foreign Affairs, House of Representatives, 89th Cong., 1st sess., p. 783.

4. U.S., Congress, House, Committee on Foreign Affairs, *Military Assistance Training, Hearings*, before the Subcommittee on National Security Policy and Scientific Developments of the Committee on Foreign Affairs, House of Representatives, 91st Cong., 2d sess., October-December 1970, p. 6.

5. T.G. Harvey, "Political Socialization Research: Problems and Prospects," (paper prepared for presentation at the annual meeting of the Canadian Political Science Association, McGill University, Montreal, June 1972).

6. King, "Making It," pp. 19-20. J.W. Fulbright also speaks of "a near unanimity of outlook" and links it to professional or institutional constraints. *The Pentagon Propaganda Machine* (New York: Liveright, 1970), pp. 142-43.

7. With regard to Defense and State Department "security reviews" of such articles, see: U.S., Congress, Senate, Committee on the Armed Services, *Military Cold War Education and Speech Review Policies, Hearings*, before the Subcommittee on Special Preparedness of the Committee on Armed Services, United States Senate, 87th Cong., 2d sess., pp. 574-81, 591.

8. W. Eckhardt and A.G. Newcombe, "Militarism, Personality and other Social Attitudes," *Journal of Conflict Resolution*, 13 (June 1969), 212-16. Cf., R.R. Izzett, "Authoritarianism and Attitudes toward the Vietnam War as Reflected in Behavioral and Self-Report Measures," *Journal of Personality and Social Psychology* 17 (February 1971), 145-48.

9. Fulbright, *Pentagon Propaganda*, p. 138. Cf.: Capt. Robert J. Hanks, USN, "Against All Enemies," *U.S. Naval Institute Proceedings*, 96 (March 1970), 23-29; Col. Baylor Gibson, USMC, "The Universities: Trial by Fire," ibid. (April 1970), 33-39; Frank Donner, "The Theory and Practice of American Political Intelligence," *New York Review of Books*, April 22, 1971, pp. 27-30.

10. While the following works reveal certain biases, they nevertheless treat the social and welfare gains which have brought such regimes significant though not universal domestic public support. Edgar Snow, *The Other Side of the River: Red China Today* (New York: Random House, 1962). Dudley Seers, Andrew Bianchi, Richard Jolly and Max Nolff, *Cuba: The Economic and Social Revolution* (Chapel Hill: University of North Carolina Press, 1964). Maurice Zeitlin, *Revolutionary Politics and the Cuban Working Class* (Princeton, N.J.: Princeton University Press, 1967). William Mandel, *Russia Re-Examined* (New York: Hill & Wang, 1967). Harry G. Shaffer, *The Communist World: Marxist and Non-Marxist Views* (New York: Appleton-Century-Crofts, 1967).

11. *Military Cold War Education, Hearings*, Senate, pp. 1130, 1157, 1172, 1333, 1353-54, 1358, 1576, 1701, 2190, 2527, 2633.

12. U.S., Department of the Army, *Alert: Heritage of Freedom, Facts for the Armed Forces* (Washington, D.C.: Pamphlet No. 355-140, February 1, 1963), p. 9.

13. Lt. Col. Harry F. Walterhouse, USA, "Civic Action: A Counter and Cure for Insurgency," *Military Review* 52 (August 1968), 47-54.

14. U.S., Congress, House, Committee on Appropriations, *Foreign Operations Appropriations for 1965, Hearings*, before a subcommittee of the Committee on Appropriations, House of Representatives, 88th Cong., 2d sess., p. 530.

15. U.S., Congress, Senate, Committee on Foreign Relations, *Foreign Assistance 1964, Hearings*, before the Committee on Foreign Relations, United States Senate, 88th Cong., 2d sess., p. 518. Cf., Neville Maxwell, *India's China War* (New York: Pantheon, 1970).

16. U.S., Congress, House, Committee on Foreign Affairs, *Foreign Assistance Act of 1969, Hearings*, before the Committee on Foreign Affairs, House of Representatives, 91st Cong., 1st sess., p. 606. Cf., U.S., Congress, House, Committee on Foreign Affairs, *Foreign Assistance Act of 1972, Hearings*, before the Committee on Foreign Affairs, House of Representatives, 92d Cong., 2d sess., March 1972, pp. 52, 176-79, 236.

17. U.S., Department of Defense, *Military Assistance Facts: May 1969* (Washington, D.C.: Office of the Assistant Secretary of Defense for International Security Affairs, 1969), pp. 1, 7. Cf., Fulbright, *Pentagon Propaganda*, pp. 129-33.

18. Lt. Col. David R. Hughes, USA, "The Myth of Military Coups and Military Assistance," *Military Review* 57 (December 1967), 5-6.

19. Fulbright, *Pentagon Propaganda*, pp. 134-35.

20. George W. Grayson, Jr., "The Era of the Good Neighbor," and Lester D. Langley, "Military Commitments in Latin America: 1960-1968," *Current History* 56 (June 1969), 329, 349-50.

21. U.S., Congress, House, Committee on Foreign Affairs, *Foreign Assistance Act of 1966, Hearings*, before the Committee on Foreign Affairs, House of Representatives, 89th Cong., 2d sess., pp. 236, 243.

22. J.C. Hurewitz, *Middle East Politics: The Military Dimension* (New York: Praeger, 1969), p. 71.

23. Gene M. Lyons and Louis Morton, *Schools for Strategy: Education and Research in National Security Affairs* (New York: Praeger, 1965), pp. 121-22. Cf.: Waldemar A. Nielsen, "Huge, Hidden Impact of the Pentagon," *New York Times Magazine*, June 25, 1961, p. 36; *Military Cold War Education, Hearings*, Senate, pp. 1990-91, 1156-61.

24. Col. Richard F. Rosser, "The Air Force Academy and the Development of Area Experts," *Air University Review* 20 (November-December 1968), 28. Berkely Rice, "The Cold-War College Think Tanks," *Washington Monthly*, June 1969, pp. 22-34. Lewis H. Lapham, "Military Theology: The Ethos of the Officer Class," *Harper's Magazine*, July 1971, pp. 73-85.

25. Typical articles include the following: Charles S. Stevenson, "The Far East MAAGs: Good Investment in Security," *Army* 11 (November 1960), 60-63; Lt. Col. A.C. Smith, Jr., "United Kingdom Joint Services Staff College," *Marine Corps Gazette* 52 (July 1968), 10; Cmdr. E.H. Van Rees, R. Neth. N., "Reflections of a Foreign Student," *U.S. Naval Institute Proceedings* 8 (February 1960), 40; Col. William R. Kintner, USA, "The Role of Military Assistance," ibid., 87 (March 1961), 76-77; Capt. Robert F. Farrington, USN, "Military Assistance at the Crossroads," ibid., 92 (June 1966), 69; Col. A.H. Victor, Jr., USA, "Military Aid and Comfort to Dictatorships," ibid., 95 (March 1969), 43-47; Frank R. Barnett, "A Proposal for Political Warfare," *Military Review* 51 (March 1961), 7; Carl M. Guelzo, "The Higher Level Staff Advisor," ibid., 47 (February 1967), 97; Gen. Porter, "Look South," pp. 85-90. Cf., note 9 above.

26. Rosser, "The Air Force Academy," p. 28. Col. Anthony L. Wermuth, USA, "Bittersweet Roots of Policy," *Military Review* 42 (August 1962), 29-38. Cf., *Military Cold War Education, Hearings*, Senate, p. 2047; Lawrence L. Whetten, "Strategic Parity in the Middle East," *Military Review* 50 (September 1970), 24-31.

27. *Military Cold War Education, Hearings*, Senate, p. 1392.

28. U.S., Department of Defense, *Know Your Communist Enemy: Communism in Red China* (Washington, D.C.: Office of Armed Forces Information and Education, no. 5, 1955). Excerpts from this tract appear in appendix 5.

29. U.S., Department of the Army, *Officers' Call: The Soviet Armed Forces* (Washington, D.C.: Pamphlet no. 355-22, January 1956), p. 2.

30. Even the Committee for a Sane Nuclear Policy was considered to be "helping the Communist cause." *Military Cold War Education, Hearings*, Senate, pp. 1352-53, 2232.

31. Ibid., p. 1680. Lyons and Morton, *Schools for Strategy*, p. 114.

32. *Military Cold War Education, Hearings*, Senate, p. 1373.

33. Ibid., pp. 1391-93, 2127, 2527-28.

34. Ibid., pp. 2169-70.

35. U.S., Department of Defense, *Facts About the United States* (Washington, D.C.: Armed Forces Information and Education, Dept. of the Army Pamphlet 355-123, 1961), p. 56. U.S., Department of the Army, *Africa Today: Army Information Support Series* (Washington, D.C.: Pamphlet 355-200-11, February 14, 1961), pp. 8, 20. U.S., Department of Defense, *Republic of South Africa: Capsule Facts for the Armed Forces* (Washington, D.C.: Armed Forces Information Service, Dept. of the Army Pamphlet 360-764, January 1969), 4pp. U.S., Department of the Army, *Field Manual 19-15: Civil Disturbances and Disasters* (Washington, D.C.: GPO, 1968), Ch. 19.

36. U.S., Department of Defense, *Background: United States Foreign Policy in a New Age* (Washington, D.C.: Armed Forces Information and Education, Dept. of the Army Pamphlet 355-108, July 24, 1961), p. 14.

37. U.S., Department of Defense, *ALERT: The Truth About Our Economic System, Facts for the Armed Forces* (Washington, D.C.: Armed Forces Information and Education, Dept. of the Army Pamphlet 355-136, May 29, 1962–not superceded as of March 15, 1969), p. 3. Cf., U.S., Department of Defense, *Facts About the United States* (Washington, D.C.: Armed Forces Information and Education, Dept. of the Army Pamphlet 360-516, 1964), pp. 3, 11.

38. U.S., Department of Defense, *Ideas in Conflict: Liberty and Communism* (Washington, D.C.: Armed Forces Information and Education, Dept. of the Army Pamphlet 360-523, June 1967–not superseded as of March 15, 1969), p. 113.

39. The catalog cited below lists eight Officers' Conference Films. All incorporate political themes stressing such topics as patriotism, liberty, cold war roles, etc. Two emphasize anti-Communism in terms of alternatives allegedly represented by the "Free World" and the United States. With regard to the remaining films, about thirteen are largely non-political, while fifty-five present patriotic/ideological themes. Of these, thirty-seven focus upon anti-Communism. U.S., Department of Defense, *Catalog of Information Materials: Publications and Motion Pictures* (Washington, D.C.: Dept. of the Army, Pamphlet 355-50, August 1961), 35pp. Cf.: *Military Cold War Education, Hearings*, Senate, pp. 1021, 1058-78, 1377, 1576, 2650; Nielsen, "Huge, Hidden Impact," pp. 9, 31, 34; U.S. Department of the Army, *Army Information Basic Guidance* (Washington, D.C.: Pamphlet 360-4, June 1964), pp. 29-32; Fulbright, *Pentagon Propaganda*, pp. 62, 70-71, 90-91, 105-10.

40. This is not to imply that the works selected do not differ with regard to the most efficacious means for countering social revolution. See, e.g.: "Ex-

Libris," *Marine Corps Gazette* 51 (June 1967), 14-15; U.S., Department of the Army, *U.S. National Security and the Communist Challenge: The Spectrum of the East-West Conflict* (Washington, D.C.: Pamphlet 20-60, August 11, 1961), 93pp.

41. Alvin Whitely, "U.S. Army–The Free World's Biggest School," *Army Digest* 20 (April 1965), 54.

42. William T.R. Fox, "Military Representation Abroad," in *The Representation of the United States Abroad*, ed. The American Assembly (New York: Graduate School of Business, Columbia University, 1956), 145.

43. John W. Masland and Laurence I. Radway, *Soldiers and Scholars: Military Education and National Policy* (Princeton, N.J.: Princeton University Press, 1957), pp. 323-63, 375-83, 390-91, 398, 402-04. Cf., Fulbright, *Pentagon Propaganda*, pp. 38-39.

44. *Military Cold War Education, Hearings*, Senate, pp. 1103, 3143.

45. U.S., Industrial College of the Armed Forces, *National Security Seminar, 1968-1969: Presentation Outlines and Reading List* (Washington, D.C.: Fort Lesley J. McNair, December 1968), pp. 49-59. Cf., Fulbright, *Pentagon Propaganda*, pp. 40-41.

46. U.S., Department of the Army, *U.S. Army Area Handbook for Venezuela* (Washington, D.C.: GPO, 1964), pp. 267, 463-68, 523-33.

47. Masland and Radway, *Soldiers and Scholars*, p. 443.

48. S.H. Hays, "Military Training in the U.S. Today," *Current History*, 55 (July 1968), 10.

49. William E. Simons, *Liberal Education in the Service Academies* (New York: Teachers College of Columbia University, 1965), pp. 150-59.

50. Richard de Neufville, "Education at the Academies," *Military Review* 47 (May 1967), 6-7.

51. *United States Air Force Academy: Catalog 1967-1968* (Washington, D.C.: GPO, 1967), pp. 101-03.

52. And between 1946-56, approximately 3,100 Army officers enrolled on a full-time basis in civilian institutions of higher education. Many, of course, were in technical fields. Masland and Radway, *Soldiers and Scholars*, pp. 304-305.

53. Ibid., pp. 170, 172, 200, 238.

54. *Military Cold War Education, Hearings*, Senate, pp. 1102, 1768-69.

55. Ibid., pp. 1700-1; H.O. Werner, "Breakthrough at the U.S. Naval Academy," *U.S. Naval Institute Proceedings* 89 (October 1963), 60.

56. *United States Naval Academy Catalog: 1967-1968* (Washington, D.C.: GPO, 1967), pp. 127-28.

57. *United States Military Academy: 1968-1969 Catalog*, West Point, New York (Washington, D.C.: GPO, 1968), pp. 62-65.

58. Masland and Radway, *Soldiers and Scholars*, p. 195.

59. Ibid., p. 250.

60. John M. Dunn, "Military Aid and Military Elites: The Political Potential

of American Training and Technical Assistance Programs," (unpublished Ph.D. dissertation, Princeton University, 1961), pp. 193-204. U.S. Congress, House, Committee on Foreign Affairs, *Winning the Cold War: The U.S. Ideological Offensive, Hearings*, before the Subcommittee on International Organizations and Movements of the Committee on Foreign Affairs, House of Representatives, 88th Cong., 2d sess., pp. 1024-34.

61. Dunn, "Military Aid," pp. 201-04.

62. Ibid., p. 252. Cf., *Military Cold War Education, Hearings*, Senate, pp. 2260-61.

63. Donn R. Grand Pre, "A Window on America: The Department of Defense Informational Program," *International Educational and Cultural Exchange* 6 (Fall 1970), p. 90.

64. *Military Assistance Training, Hearings*, House, p. 4.

65. Grand Pre, "Window on America," p. 90.

66. Lt. Col. Frank H. Robertson, "Military Assistance Training Program," *Air University Review* 19 (September-October 1968), 35. Cf., Institute of International Education, *Military Assistance Training Programs of the U.S. Government* (New York: I.I.E., Committee on Educational Interchange Policy Statement no. 18, July 1964), pp. 16-17.

67. Col. Lowell T. Bondshu, "Keeping Pace with the Future–Toward Mutual Security," *Military Review* 39 (June 1959), 64-65. U.S. Army Command and General Staff College, *U.S. Army English for Allied Students* (Fort Leavenworth, Kansas: Reference Book 20-8, April 1968–supersedes 1967 edition), pp. 6-80.

68. U.S. Army Command and General Staff College, *Communism in Selected Countries: Bibliography* (Ft. Leavenworth: R2809/0, 1969). USACGSC, *The Communist Powers: USSR–Lesson 1* (Ft. Leavenworth: R2811-1/0). USACGSC, *The Communist Powers: Communist China–Lesson 2* (Ft. Leavenworth: R2811-2/0). USACGSC, *Nature of Communism: Advance Sheet* (Ft. Leavenworth: R1808). USACGSC, *Nature of Communism: Command and General Staff Officer Course, School Year 1968-1969 Lesson Plan* (Ft. Leavenworth: R1808/9). Cf., Lt. Col. Selwyn P. Rogers, Jr., USA, "Unshuttered Vision," *Military Review* 54 (September 1964), 17-21.

69. U.S. Army Command and General Staff College, *The Communist Powers: Communist China, School Year 1969-1970 Lesson Plan* (Ft. Leavenworth, R2811-2/0, 1969), LP 2-5-9, LP 2-74, et passim.

70. Escuela de las Americas, Ejercito de los EE. UU., *Plan de Conquista del Comunismo: Seccion I* (Fuerte Gulick, Z.C.: MI 4541, Marzo 1964), 3pp. EAEEU, *El Atractivo del Comunismo* (Fuerte Gulick, Z.C.: MI 4514, Abril 1964), 5pp. EAEEU, *Expansion en Europa: Seccion III* (Fuerte Gulick, Z.C.: HA 4525, Mayo 1964), 9pp. EAEEU, *La Política Exterior de la Union Soviética: Seccion I* (Fuerte Gulick, Z.C.: MI 4535, Junio 1964), 40pp. EAEEU, Seccion de Inteligencia, *Lectura Supplementaria: Principios y Practicas del Gobierno*

Totalitario (Fuerte Gulick, Z.C.: HA 4600, Febrero 1965), 25 pp. EAEEU, *Expansión del Comunismo en América Latina* (Fuerte Gulick, Z.C.: MI 4531A, Mayo 1968), 71pp. EAEEU, *Ideologia Comunista II* (Fuerte Gulick, Z.C.: MI 4501, Junio 1965), 40 pp. EAEEU, Comite de Seguridad, *Expansión Comunista en Europa Oriental* (Fuerte Gulick, Z.C.: MI 4525, Agosto 1968) EAEEU, *El Desarrollo de Tacticas Revolucionarias–Primera Parte* (Fuerte Gulick, Z.C.: HA 4526, Septiembre 1965) EAEEU, *El Desarrollo de las Tacticas Revolucionarias I* (Fuerte Gulick, Z.C.: MI 4526, Octubre 1965), 47pp. EAEEU, *Partido Comunista de la Union Sovietica* (Fuerte Gulick, Z.C.: MI 4516, Octubre 1965), 56pp. EAEEU, *Táctica y Técnica del Comunismo Internacional* (Fuerte Gulick, Z.C.: MI 4530, Marzo 1964), 3 pp. EAEEU, *La Teória del Comunismo* (Fuerte Gulick, Z.C.: HA 4500-2, Abril 1966), 23pp.; EAEEU, Comité de Seguridad, *Teória, Practica e Instituciones de los Gobiernos Totalitários* (Fuerte Gulick, Z.C.: MI 4606, Enero 1969), 26pp. In addition to the foregoing instructor manuals, the following pamphlets were distributed to students: Secretary of War, Army Information Service, *Así el Comunismo; Los Agentes Enemígos y Usted: Temas para Tropas*, 25pp.; J. Edgar Hoover, *La Respuesta de una Nación al Comunismo* (Enero 1961); Ibid., *Illusión Comunista ye Realidad Democratica: Hoja Avanzada* (Fuerte Gulick, Z.C.: Escuela de las Americas, Marzo 1966), 17pp.; *La Democrácia contra el Comunismo* (Folleto no. 1); 17pp.; *Qué es el Comunismo*? (Folleto no. 2), 20pp.; *Que hacen los Comunistas de la Libertad*? (Folleto no. 3), 35pp.; *Como controla el Comunismo las Ideas del Pueblo* (Folleto no. 4), 24pp.; *Como Funciona el Partido Comunista* (Folleto no. 5), 24pp.; *Somo logran y Retienen el Poder los Comunistas* (Folleto no. 6), 22pp.; *El Dominio del Partido Comunista en Rusia* (Folleto no. 7), 14pp.; *Como viven los Obreros y Campesinos bajo el Comunismo* (Folleto no. 9), 14pp.; *Conquista y Colonización Comunista* (Folleto no. 10), 24pp. Although the folletos are undated, textual citations suggest preparation during the 1960-1963 period. For excerpts from Folletos nos. 1 and 2, see Appendix 6.

71. U.S., Department of Defense, *Informational Program on American Life and Institutions for Foreign Military Trainees: Vol. VI, Economic Aspects of American Life* (Madison, Wisconsin: United States Armed Forces Institute, 1965), pp. 5-6 17, 25, 31-36. U.S. Air Force University, *Allied Officer Familiarization Course Lecture Outline* (Maxwell Air Force Base: May 1966), A0-409. Escuela de las Americas, Ejercito de los EE. UU., *Ideologia Comunista I* (Fuerte Gulick, Zona del Canal: MI 4500, Octubre 1965), pp. 40-43. EAEEU, *Desarrollo de las Tacticas Revolucionarias, II Parte, Seccion I* (Fuerte Gulick, Zona del Canal: MI 4527, Noviembre 1965), pp. 9-14. *Como controla el comunismo la Vida Economica del Pueblo* (Folleto no. 8), pp. 14-15. U.S. Army Command and General Staff College, *Nature of Communism: Command and General Staff Officer Course, School Year 1968-1969 Lesson Plan* (Ft. Leavenworth: R1808/9), LP 67-71. U.S., Department of the Army, *U.S. Army Area Handbook for Colombia* (Washington, D.C.: DA Pamphlet, no. 550-26, 2d ed., June 1964), p. 578.

72. With regard to Latin America, see: *Military Cold War Education, Hearings*, Senate, p. 2514.

73. Ibid., pp. 2501, 2516; Harry F. Walterhouse, *A Time to Build: Military Civic Action as a Medium for Economic Development and Social Reform* (Columbia: University of South Carolina Press, 1964), pp. 33, 122.

74. Defined by General Conway as a "low-pitched, low-key approach." *Foreign Assistance Act of 1969, Hearings*, House, pp. 601-602.

Chapter 5
Major Foreign Officer Training Installations

1. Harry F. Walterhouse, *A Time to Build: Military Civic Action as a Medium for Economic Development and Social Reform* (Columbia: University of South Carolina Press, 1964), p. 33.

2. U.S., Congress, House, Committee on Foreign Affairs, *Reports of the Special Study Mission to Latin America on Military Assistance Training and Developmental Television*, submitted by Hon. Clement J. Zablocki, Hon. James G. Fulton and Hon. Paul Findley to the Subcommittee on National Security and Scientific Developments of the Committee on Foreign Affairs, House of Representatives, 91st Cong., May 7, 1970, p. 28.

3. Ibid., p. 21.

4. In FY 1970, at least 4000 officers and men were being trained abroad under MAP. U.S., Congress, House, Committee on Foreign Affairs, *Military Assistance Training, Hearings*, before the Subcommittee on National Security Policy and Scientific Developments of the Committee on Foreign Affairs, House of Representatives, 91st Cong., 2d sess., October-December 1970, p. 3.

5. U.S., Congress, House, Committee on Appropriations, *Foreign Assistance and Related Agencies Appropriations for 1970, Hearings*, before the Subcommittee on Foreign Operations and Related Agencies, Committee on Appropriations, House of Representatives, 91st Cong., 1st sess., p. 624.

6. Ralph E. Lapp, *The Weapons Culture* (New York: Norton, 1968), p. 180.

7. U.S., Congress, Senate, Committee on Foreign Relations, *Foreign Assistance Act, 1969, Hearings*, before the Committee on Foreign Relations, United States Senate, 91st Cong., 1st sess., p. 140.

8. Col. Amos A. Jordan, *Foreign Aid and the Defense of Southeast Asia* (New York: Praeger, 1962), p. 56.

9. "The Spanish-Speaking Man in the Green Beret," *Army Digest* 21 (March 1966), 60. Cuba, Haiti, and possibly Mexico did not receive such units.

10. Waldemar A. Nielsen, "Huge Hidden Impact of the Pentagon," *New York Times Magazine*, June 25, 1961, p. 31.

11. Mike Klare, "U.S. Military Operations," *NACLA Newsletter* 2 (October 1968), 6. Cf., U.S. Army, *Foreign Countries and Nationals: Training of Foreign Personnel by the U.S. Army* (Washington, D.C.: Headquarters, Dept. of the

Army, Army Regulation AR 550-50, October 1970, Effective 1 December 1970), p. 2-5.

12. Drew Middleton, "Thousands of Foreign Military Men Studying in U.S.," *New York Times*, November 1, 1970, p. 18.

13. Disclosed in the course of MAP hearings by the Joint Economic Committee of Congress. *I.F. Stone's Bi-Weekly*, January 25, 1971, p. 1.

14. *Military Assistance Training, Hearings*, House, p. 103. In addition to the grant aid program's 2,800, it is likely that another 2,000 foreign officers received ideological indoctrination. Those undergoing unit training are excluded although some of them were also at least marginally exposed.

15. *Foreign Assistance Act of 1969, Hearings*, Senate, p. 139. Another source suggests that between 100 and 150 officers were being brought to the United States for this purpose during the early 1960s. Institute of International Education, *Military Assistance Training Programs of the U.S. Government* (New York: I.I.E., Committee on Educational Interchange Policy Statement no. 18, July 1964), p. 12.

16. U.S., Department of the Army, *Industrial College of the Armed Forces: 1968-1969 Catalog* (Fort McNair, Washington, D.C.), pp. 21-22, 52, 55. U.S., Department of State, *A Guide to U.S. Government Agencies Involved in International Educational and Cultural Activities*, Bureau of Educational and Cultural Affairs, International information and Cultural Series 97, Dept. of State Publication 8405 (Washington, D.C.: GPO, September, 1968), p. 49. Cf., U.S., Bureau of Naval Personnel, *List of Training Manuals and Correspondence Courses*, NAVPERS 10061-AC (Washington, D.C.: GPO, 1969).

17. U.S., *Guide to Government Agencies*, p. 49.

18. Institute of International Education, *Military Assistance*, p. 12. In FY 1970, FMTs were enrolled at about 125 CONUS installations. *Military Assistance Training, Hearings*, House, p. 3.

19. Harold A. Hovey, *United States Military Assistance: A Study of Policies and Practices* (New York: Praeger, 1965), p. 171. Yet a third list released in 1964 contains only 75 training centers, although it includes the number processed at each one. See our Appendix IV.

20. *Foreign Assistance Appropriations for 1970, Hearings*, House, p. 829.

21. Institute of International Education, *Military Assistance*, p. 20. Fort Belvoir, Va. may be typical of the larger installations in terms of foreign trainees as a percentage of the total: "Today the Engineer School graduates approximately 2,500 enlisted men a year from courses conducted on the college-like campus at Fort Belvoir. In the 1961-62 school year approximately 500 foreign officers and enlisted men from 32 Allied countries also graduated from the school." Brig. Gen. George H. Walker, USA, "The Engineer School Builds for the Future," *Army Digest* 17 (November 1962), 14. Cf., "U.S. Army Air Defense School," *Military* Review 39 (April 1959), 25.

22. On these matters, see the following: *Military Assistance Training,*

Hearings, House, pp. 60-61, 65, 79-83, 165, 172; U.S., Congress, House, Committee on Foreign Affairs, *Military Assistance Training in East and Southeast Asia, A Staff Report,* prepared for the use of the Subcommittee on National Security Policy and Scientific Developments of the Committee on Foreign Affairs, House of Representatives, 91st Cong., 2d sess., p. 25.

23. *Military Assistance Training, Hearings,* House, p. 130.

24. U.S., Congress, House, Committee on Foreign Affairs, *Foreign Assistance Act of 1965, Hearings*, before the Committee on Foreign Affairs, House of Representatives, 89th Cong., 1st sess., p. 796. For numbers of officers trained, see *New York Times*, June 25, 1969, p. 3. *Reports of the Special Study Mission*, House, p. 27.

25. *Foreign Assistance Act of 1965, Hearings*, House, p. 782. *Foreign Assistance Act of 1969, Hearings*, House, p. 640.

26. Philip B. Taylor, Jr., "Hemispheric Defense in World War II," *Current History* 56 (June 1959), 33-39.

27. James C. Hahr, "Military Assistance to Latin America," *Military Review* 49 (May 1969), 13-20.

28. U.S., Congress, House, Committee on Foreign Affairs, *Castro-Communist Subversion in the Western Hemisphere, Hearings*, before the Subcommittee on Inter-American Affairs of the Committee on Foreign Affairs, House of Representatives, 88th Cong., 1st sess., pp. 18-19.

29. Lt. Gen. Andrew P. O'Meara, USA, "Opportunities to the South of Us," *Army* 13 (November 1962), 41-43. For an analogous association of civic action training with the prevention of revolution, see: Lt. Col. Harry F. Walterhouse, USA, Ret., "Good Neighbors in Uniform," *Military Review* 45 (February 1965), 16-18.

30. "Spanish-Speaking Man," p. 60.

31. *Foreign Assistance Act of 1965, Hearings*, House, p. 785. Since 1960 there have also been nine "conferences of American Armies" for high level officers. Collective security, subversion and communism have served as the primary themes for these gatherings. For a summary of a recent meeting see Lt. Col. Donald C. Shuffstall, USA, "Ninth Conference of American Armies," *Military Review* 50 (April 1970), 82-7.

32. *Foreign Assistance Act of 1969, Hearings*, House, pp. 116, 642-63.

33. John M. Goshko, "Latins Blame the United States for Military Coups," *Washington Post*, February 5, 1968, reprinted in U.S., Congress, Senate, Committee on Foreign Relations, *Amendments to the OAS Charter, Hearings*, before the Committee on Foreign Relations, United States Senate, 90th Cong., 2d sess., p. 13.

34. "The School of the Americas Shows How Armies can be Builders," *Army Digest* 20 (February 1965), 18. Capt. John R. Deats, "Bridge of the Americas," *Army Digest* 23 (September 1968), 14. "U.S. Army School of the Americas," *Military Review* 50 (April 1970), 92.

The "teacher-pupil ratio is a low 2.2 students per instructor. From 20 to 25 percent of the faculty are guest instructors from Latin nations." *Reports of the Special Study Mission*, House, p. 27.

35. "School of the Americas," pp. 16-18. Deats, "Bridge," p. 13. "U.S. Army School," p. 89. David Wood, *Armed Forces in Central and South America* (London: Institute for Strategic Studies, 1967), pp. 4, 6.

According to a recent investigation, the school "has a professional emphasis in its course offerings. As a result, a majority of its students are officers or cadets." In FY 1970, there were 217 officers, 116 cadets and 103 enlisted enrolled. The four departments into which the school is now organized "are the department of command, which provides command and general staff training courses; the combat operations department which specializes in civic action, counterinsurgency, and jungle warfare; the support operations department which offers courses in military intelligence, military police, medical aid and supply; and the technical operations department which trains in aspects of mechanics." *Reports of the Special Study Mission*, House, p. 27.

36. Maj. Charles J. Bauer, USA, "USACARIB's Biggest Little School," *Army Digest* 17 (October 1962), 25.

37. U.S., Department of the Army, *The United States School of the Americas: 1969 Catalog* (Fort Gulick, Canal Zone, n.d.).

38. Ibid.

39. Maj. Gen. Theodore F. Bogart, USA, "The Army's Role in Latin America," *Army* 12 (October 1961), 62.

40. Bauer, "USACARIB's," pp. 27-28.

41. *Foreign Assistance Act of 1965, Hearings*, House, p. 348.

42. Maj. Gen. Theodore F. Bogart, USA, "The United States Army Caribbean Command and Operation Friendship," *Army Digest* 17 (October 1962), 23. "Alumni have risen to such key positions as Minister of Defense and Chief of Staff in Bolivia, Director of Mexico's War College, Minister of War and Chief of Staff in Columbia [*sic*], Chief of Staff for Intelligence in Argentina and Under-Secretary of War in Chile." Deats, "Bridge," p. 14.

43. *New York Times*, February 15, 1970, section 1, p. 29. Currently, about 80 Latin American officers are being trained each year. Sixty percent of the school's enrollment is civilian. *Reports of the Special Study Mission*, House, p. 25.

44. Col. Richard J. Stillman, USMC, "The Inter-American Defense College," *Marine Corps Gazette* 40 (September 1965), 25-26.

45. "Defense College of the Americas," *Army Digest* 19 (May 1964), 38.

46. *Generals vs. Presidents: Neo-Militarism in Latin America* (New York: Praeger, 1964), p. 148.

47. Mike Klare, "U.S. Military Operations in Latin America," *NACLA Newsletter* 2 (October 1968), 6; "Defense College of the Americas," pp. 37-39; Col. Stillman, "The Inter-American Defense College," *Military Review* 50 (April 1970), 24-6.

48. Lt. Col. Bernard L. Pujo, "Two Colleges, Two Styles," *Military Review* 45 (March 1965), 68.

49. Maj. Gen. John H. Caughey, USA, "The Army's Schools of Battle," *Army* 40 (November 1964), 99.

50. Middleton, "Thousands," p. 18.

51. See for example the following: U.S. Army Command and General Staff College, *Nature of Communism*, Advance Sheet, R1808; U.S. Army Command and General Staff College, *Nature of Communism*, Command and General Staff Officer Course, Lesson Plan R1808/9, School Year 1968-69, Ft. Leavenworth; U.S. Army Command and General Staff College, *Communism in Selected Countries: Bibliography*, R2809/0, Ft. Leavenworth, 1969; U.S. Army Command and General Staff College, *Strategic Subjects Handbook: RB 100-1*, Fort Leavenworth, Kansas, 1 August 1969; U.S. Army Command and General Staff College, *The Communist Powers* (USSR-Lesson 1), R2811-1/0; U.S. Army Command and General Staff College, *The Communist Powers: Communist China*, School Year 1969-70, Lesson Plan Subject R2811-2/0, Ft. Leavenworth, 1969.

52. Col. Lowell T. Bondshu, USA, "Toward Mutual Security," *Military Review* 39 (June 1959), 59-60.

53. Ibid., U.S. Army Command and General Staff College, *U.S. Army English for Allied Students*, Ref. Book: RB20-8, Fort Leavenworth, Kansas, 1968, p. 15. *New York Times*, November 1, 1970, I, p. 18. At the end of the 1960s, the annual cost per foreign officer trainee was $2,810. U.S., Congress, House, Committee on Foreign Affairs, *To Amend the Foreign Military Sales Act, Hearings*, before the Committee on Foreign Affairs, House of Representatives, 91st Cong., 2d sess., February 1970, p. 64. See also Middleton, "Thousands," p. 18.

54. U.S. Congress, House, Committee on Foreign Affairs, *Foreign Assistance Act of 1963, Hearings*, before the Committee on Foreign Affairs, House of Representatives, 88th Cong., 1st sess., pp. 727-28.

55. U.S., Congress, House, Committee on Foreign Affairs, *Military Assistance Training, Report*, of the Subcommittee on National Security Policy and Scientific Developments to the Committee on Foreign Affairs, House of Representatives, April 2, 1971, p. 12.

56. David W. Chang, "Military Forces and Nation Building," *Military Review* (September 1970), 78-88.

57. "What it Takes: Essentials of Special Operations," *Military Review* 41 (August 1961), 12-16.

58. U.S. Congress, Senate, Committee on Government Operations, *State Department Advisors to Military Training Institutions: Analysis and Assessment*, Submitted by the Subcommittee on National Security and International Operations of the Committee on Government Operations, United States Senate, 91st Cong., 1st sess., p. 14.

59. "Military Notes: Cadet Exchange Program," *Military Review* 47 (April 1967), 100.

60. " 'Follow Me' United States Army Infantry School," *Military Review* 39 (July 1959), 39.

61. Ibid., p. 40: *Foreign Assistance Act of 1963, Hearings*, House, pp. 701, 727-28.

62. " 'Follow Me,' " p. 40.

63. *Foreign Assistance Act of 1963, Hearings*, House, p. 685.

64. Maj. Gen. Orwin C. Talbott, "Where Nobody Ever Rests," *Army* March 1970, p. 40.

65. *Military Cold War Education, Hearings*, Senate, pp. 2037, 2039, 2047.

66. Ibid., pp. 2040, 2087-88.

67. *Military Cold War Education, Hearings*, Senate, p. 1681; U.S., Congress, House, Committee on Foreign Affairs, *Winning the Cold War: The U.S. Ideological Offensive, Hearings*, before the Subcommittee on International Organizations and Movements of the Committee on Foreign Affairs, House of Representatives, 88th Cong., 2d sess., pt 8, pp. 1027, 1043-44; Capt. Jones, USA, "The Nationbuilder," pp. 63-65.

68. U.S., Department of the Army, *Counterinsurgency Operations (Spanish Version): A Handbook for the Suppression of Communist Guerrilla Terrorist Operations* (Ft. Bragg, North Carolina: U.S. Army Special Warfare School, n.d.).

69. U.S., Department of the Army, *Stability Operations–U.S. Army Doctrine*, Field Manual 31-23, December 1967, p. 110.

70. U.S., *State Department Advisers*, Senate, p. 14. Charles Windle and T.R. Vallance, "Optimizing Military Assistance Training," *World Politics* 15 (October 1962), 96.

71. *New York Times*, August 15, 1969, p. 10. Elsewhere it has been noted that "ideological factors are of fundamental importance in training and operational missions of U.S. Army special forces." *Winning the Cold War, Hearings*, House, p. 1026.

72. Salvatore Alfred Arcilesi, "The Development of United States Foreign Policy in Iran: 1949-1960" (unpublished Ph.D. dissertation, University of Virginia, 1965), pp. 80-84. David Wise and Thomas B. Ross, *The Invisible Government* (New York: Random House, 1964), pp. 110-14. For tentative evidence of similar working relationships in neutral Laos during and since the late 1950s, see: Banning Garrett, "The Vietnamization of Laos," *Ramparts*, June 1970, pp. 34-45; Hugh Toye, *Laos* (London: Oxford University Press, 1968), pp. 182-84; U.S., Congress, House, Committee on Appropriations, *Foreign Assistance and Related Agencies Appropriations for 1966, Hearings*, before a subcommittee of the Committee on Appropriations, House of Representatives, 89th Cong., 1st sess., pp. 322-23, 375.

73. *Foreign Assistance Act of 1965, Hearings*, House, p. 80.

74. *Foreign Assistance Act of 1965, Hearings*, House, p. 809; Col. Margaret Brewster, USA, "Foreign Military Training Bolsters Free World," *Army Digest*, 20

(May 1965), 18. Institute of International Education, *Military Assistance*, p. 20.

75. U.S., Congress, House, Committee on Appropriations, *Foreign Assistance and Related Agencies Appropriations for 1967, Hearings*, before a subcommittee of the Committee on Appropriations, House of Representatives, 89th Cong., 2d sess., pp. 520, 654-55. Cf., Hovey, *Military Assistance*, pp. 107-8.

76. *Military Assistance Training, Hearings*, House, pp. 136-37. *Military Assistance Training in East Asia, Report*, House, pp. 11-13.

77. Lt. Col. Frank H. Robertson, "Military Assistance Training Program," *Air University Review* 19 (September-October 1968), 30, 36.

78. Col. Frank L. Gailer, Jr., USAF, "Air Force Missions in Latin America," *Air University Quarterly Review* 13 (Fall 1961), 54. Dr. A. Glenn Morton, "The Inter-American Air Forces Academy," *Air University Review* 18 (November-December 1966), 16-17, 19.

79. Maj. Matthew T. Dunn and Maj. James B. Jones, USAF, "Preventive Medicine Civic Action Training Program," *Air University Review* 18 (November-December 1966), 21-25.

80. Morton, "Inter-American Air," pp. 17-19.

81. Maj. Gen. Kenneth O. Sanborn, USAF, "United States Air Force's Southern Command," *Air Force and Space Digest* 51 (September 1968), 140.

82. Ibid. Morton, "Inter-American Air," pp. 19-20.

83. Ibid. Wood, *Armed Forces*, p. 4. *Reports of the Special Study Mission*, House, pp. 25-26.

84. Morton, "Inter-American Air," pp. 13-14.

85. John W. Masland and Laurence I. Radway, *Soldiers and Scholars: Military Education and National Policy* (Princeton: Princeton University Press, 1957), p. 279. *Military Cold War Education, Hearings*, Senate, p. 2180.

86. Ibid., pp. 2196-97.

87. Masland and Radway, *Soldiers and Scholars*, pp. 288-94.

88. *Military Cold War Education, Hearings*, Senate, p. 2180.

89. Masland and Radway, *Soldiers and Scholars*, p. 505; *Military Cold War Education, Hearings*, Senate, p. 2180.

90. Dunn, "Military Aid," pp. 355-57. Masland and Radway, *Soldiers and Scholars*, p. 295.

91. Masland and Radway, *Soldiers and Scholars*, pp. 300, 302.

92. Cmdr. E.H. Van Rees, R. Neth. N., "Reflections of a Foreign Student," *U.S. Naval Institute Proceedings* 86 (February 1960), 40. Cf.: Masland and Radway, *Soldiers and Scholars*, p. 299; Dunn, "Military Aid," pp. 264-66.

93. Cmdr. Van Rees, "Reflections," pp. 39-40.

94. Dunn, "Military Aid," pp. 231-77.

95. Brig. Gen. F.J. Karch, USMC, "Marine Corps Command and Staff College," *Marine Corps Gazette* 51 (June 1967), 27-30.

96. *New York Times*, June 22, 1969, p. 4.

Chapter 6
American Military Representation Abroad

1. George T. McKahin, Guy J. Pauker and Lucien W. Pye, "Comparative Politics of Non-Western Countries," *American Political Science Review* 49 (December 1955), 1023, cited in John M. Dunn, "Military Aid and Military Elites: The Political Potential of American Training and Technical Assistance Programs," (unpublished Ph.D. dissertation, Princeton University, 1961), pp. 305-06.

2. U.S., Department of the Navy, *Information Program for FMT (Foreign Military Trainees) and Visitors in the United States* (Chapter 6, OPNAVINIST 4950 ID CH-1, 14 December 1966), p. 32a.

3. U.S., Congress, House, Committee on Foreign Affairs, *Military Assistance Training, Hearings*, before the Subcommittee on National Security Policy and Scientific Developments of the Committee on Foreign Affairs, House of Representatives, 91st Cong., 2d sess., October-December 1970, p. 17.

4. Robert I. Rhodes, "The Disguised Conservativism in Evolutionary Development Theory," *Science & Society* 32 (1968), 383-412. James Petras, "Latin American Studies in the United States: A Critical Assessment," in his *Politics and Social Structure in Latin America* (New York: Monthly Review, 1970), pp. 327-44. Andre Gunder Frank, *Latin America: Underdevelopment or Revolution?* (New York: Monthly Review, 1969), pp. 21-145.

5. United States Naval War College, Naval Command Course for Senior Foreign Officers, *Annual New Year's Letter, Class of June 1959* (Newport, R.I.: 1960), p. 10. Ibid., p. 25. *Letter, Class of 1958*, op. cit., p. 16. United States Department of Defense, Office of the Assistant Secretary of Defense for International Security Affairs, *Collateral Benefits Derived from Military Assistance and Other DOD Training Programs* (unpublished memorandum, classified by paragraph, all paragraphs quoted here are unclassified, Washington, 1960), p. 3. *Letters, Classes of 1957, 1958, and 1959*, op. cit., Foreword: Ibid., *Letter, Class of 1959*, op. cit., p. 17. Ibid., p. 30. Ibid., p. 39, all of the preceding cited in Dunn "Military Aid," pp. 266-69.

6. U.S., Department of State, Bureau of Educational and Cultural Affairs, *A Guide to U.S. Government Agencies Involved in International Educational and Cultural Activities*, Dept. of State Publication 8405 (Washington, D.C.: GPO, International Information and Cultural Series 97, September 1968), p. 49.

7. U.S., Congress, Senate, Committee on the Armed Services, *Military Cold War Education and Speech Review Policies, Hearings*, before the Subcommittee on Special Preparedness of the Committee on Armed Services, United States Senate, 87th Cong., 2d sess., pp. 2918-19.

8. Maj. Gen. Theodore F. Bogart, USA, "The Army's Role in Latin America," *Army* 12 (October 1961), 64.

9. Lt. Col. Ralph F. Mendenhall, "Sports Ambassadors," *Parks and Recreation* 2 (February 1967), 20, 39-40.

10. With regard to Indonesia, see: U.S., Congress, House, Committee on Foreign Affairs, *Military Assistance Training in East and Southeast Asia, A Staff Report*, prepared for the use of the Subcommittee on National Security Policy and Scientific Developments of the Committee on Foreign Affairs, House of Representatives, 91st Cong., 2d sess., pp. 18-20. Similarly, in the case of King Hassan's oligarchic regime in Morocco, the low profile MAAG has been retitled the "Morocco-United States Liaison Office." U.S., Congress, Senate, Committee on Foreign Relations, *United States Security Agreements and Commitments Abroad: Morocco and Libya, Hearings*, before the Subcommittee on United States Security Agreements and Commitments Abroad, Committee on Foreign Relations, United States Senate, 91st Cong., 2d sess., pp. 1991-2004.

11. Lt. Col. David R. Hughes, USA, "The Myth of Military Coups and Military Assistance," *Military Review* 47 (December 1967), 5.

12. *Military Assistance Training, Hearings*, House, p. 92.

13. J.M. Lee, *African Armies and Civil Order* (New York: Praeger, 1969), p. 118.

14. Dunn, "Military Aid," (dissertation), pp. 206-08. A clearly implied intelligence role for the MAAG formerly stationed in Pakistan appears in: U.S., Congress, House, Committee on Appropriations, *Foreign Assistance and Related Agencies Appropriations for 1970, Hearings*, before the Subcommittee on Foreign Operations and Related Agencies, Committee on Appropriations, House of Representatives, 91st, Cong., 1st sess., p. 681. Cf., William T.R. Fox, "Military Representation Abroad," in *The Representation of the United States Abroad*, ed. by The American Assembly (New York: Columbia University Graduate School of Business, 1956), pp. 132-33.

15. Dunn, "Military Aid," pp. 198-200. Cf. U.S., Congress, House, Committee on Appropriations, *Foreign Assistance and Related Agencies Appropriations for 1969, Hearings*, before a subcommittee of the Committee on Appropriations, House of Representatives, 90th Cong., 2d. sess., p. 649.

16. *Foreign Assistance Appropriations for 1970, Hearings*, House, pp. 727-29. Institute of International Education, *Military Assistance Training Programs of the U.S. Government* (New York: IIE, Committee on Educational Interchange Policy Statement no. 18, July 1964), pp. 16-17. Dunn, "Military Aid," pp. 138-42, 193-98.

17. Lt. Gen. Samuel T. Williams, USA, Ret., "The Practical Demands of MAAG," *Military Review* 51 (July 1961), 11.

18. These were normally submitted through the unified commands to the directorate of military assistance and deputy assistant secretary of defense for military assistance and sales. US., Congress, House, Committee on Foreign Affairs, *Winning the Cold War: The U.S. Ideological Offensive, Hearings*, before the Subcommittee on International Organizations and Movements of the Committee on Foreign Affairs, House of Representatives, 88th Cong., 2d. sess., pp. 1028-29. John E. Lynch, "Analysis for Military Assistance," *Military Review* 48 (May 1968), 43. *Foreign Assistance Appropriations for 1970, Hearings*, House, pp. 830-31, 856.

19. *Military Assistance Training, Hearings*, House, p. 151. Cf., pp. 59, 67.

20. And MAAG officers submit classified reports on their "effectiveness" abroad. Dunn, "Military Aid," pp. 159, 176.

21. William A. Bell, "The Culture of Bureaucracy," *Washington Monthly*, July 1969, p. 28.

22. *Intelligence and Foreign Policy*, p. 17.

23. U.S., Congress, House, Committee on Appropriations, *Foreign Assistance and Related Agencies Appropriations for 1968, Hearings*, before a subcommittee of the Committee on Appropriations, House of Representatives, 90th Cong., 1st sess., pp. 762-63.

24. Maj. Matthew T. Dunn and Maj. James B. Jones, USAF, "Preventive Medicine Civic Action Training Program," *Air University Review*, 18 (November-December 1966), 24.

25. U.S., Congress, Senate, Committee on Foreign Relations, *United States Military Policies and Programs in Latin America, Hearings*, before the Subcommittee on Western Hemisphere Affairs, Committee on Foreign Relations, United States Senate, 91st Cong., 1st sess., June 24 and July 8, 1969, p. 75. Cf., *Economic Issues in Military Assistance, Hearings*, Joint Economic Committee, pp. 89, 103, 239, 296.

26. *United States Security Agreements, Hearings*, Senate, pp. 805-06. *Economic Issues in Military Assistance, Hearings*, Joint Economic Committee, pp. 133, 135.

27. *U.S. Military Policies in Latin America, Hearings*, Senate, pp. 36-7.

28. Ibid., p. 83. Charles Wolf, Jr., *Military Assistance Programs* (Santa Monica, Calif.: Rand Corporation, P-3240, October 1965), pp. 1-2, 7, 8. *Foreign Assistance Appropriations for 1970, Hearings*, House, pp. 830-31.

29. *Military Assistance Training, Hearings*, House, p. 11.

30. U.S., Congress, House, Committee on Foreign Affairs, *Cuba and the Caribbean, Hearings*, before the Subcommittee on Inter-American Affairs of the Committee on Foreign Affairs, House of Representatives, 91st Cong., 2d sess., pp. 91-92.

31. *Economic Issues in Military Assistance, Hearings*, Joint Economic Committee, p. 176.

32. *Military Assistance Training, Hearings*, House, p. 75.

33. On MAP's concern with *domestic policies* in recipient countries, ibid., p. 50. Here there is considerable jurisdictional overlap with the CIA. See note k, above.

34. Lt. Col. Donald E. Sampson, "Military Missionaries," *Army Digest* 22 (April 1967), 46.

35. Ibid., pp. 45-46. Support for this view was provided by Secretary of Defense Melvin Laird when he testified that the Nixon administration's request for $21.4 million in grant aid for Latin America was required "to maintain our military training and advisory roles in this area," which results in "United States

influence on these governments." *Foreign Assistance Act of 1969, Hearings*, House, p. 553.

36. *Military Assistance Training, Hearings*, House, p. 169.

37. U.S., Congress, House, Committee on Foreign Affairs, *Foreign Assistance Act of 1968, Hearings*, before the Committee on Foreign Affairs, House of Representatives, 90th Cong., 2d sess., p. 827. *Foreign Assistance Act of 1963, Hearings*, House, p. 783. Maj. Gary Wilder, USMC, "M.A.P. Orientation," *Marine Corps Gazette*, (February 1965), 40. Col. Stilson H. Smith, "Assignment to MAAG: How to Win Friends and Influence People," *Army* 12 (July 1962), 57-9. Dunn, "Military Aid," pp. 150-53, 166-67. *Military Assistance Training, Hearings*, House, p. 101.

38. *Military Assistance Training, Hearings*, House, p. 41. Cf.: *Reports of the Special Study Mission*, House, p. 6; U.S. Army, *Foreign Nationals*, p. 2-1.

39. John M. Goshko, "Latins Blame the United States for Military Coups–Aid is Suspect," *Washington Post*, February 5, 1968, inserted by Senator Wayne Morse in: U.S., Congress, Senate, Committee on Foreign Relations, *Amendments to the OAS Charter, Hearings*, before the Committee on Foreign Relations, United States Senate, 90th Cong., 2d sess., p. 12.

This may explain former Ambassador to Chile Ralph Dungan's complaint that "there are platoons of military brass gingering up the military in these countries and only squads of civilians gingering up the civilian sector." Quoted in George Thayer, *The War Business: The International Trade in Armaments* (New York: Simon & Schuster, 1969), p. 199. Cf., *Foreign Assistance Act of 1968, Hearings*, House, p. 701.

40. *U.S. Military Policies in Latin America, Hearings*, Senate, pp. 37, 63, 82. Other sources on intimate relationships between MAAG/Attache officers and foreign officers are: U.S., Congress, Senate, Committee on Foreign Relations, *Foreign Assistance Act of 1968, Hearings*, before the Committee on Foreign Relations, United States Senate, 90th Cong., 2d sess., p. 437. *Foreign Assistance Act of 1969, Hearings*, House, p. 639. Charles S. Stevenson, "The Far East MAAGs: Good Investment in Security," *Army* 11 (November 1960), 60-61. Thayer, *The War Business*, p. 199. David Wood, *Armed Forces in Central and South America*, (London: Institute for Strategic Studies, 1967), p. 5. *Reports of the Special Study Mission* House, pp. 5, 12-13, 15.

41. *Reports of the Special Study Mission*, House, p. 5.

42. The American counterinsurgency goal was to be achieved by strengthening the military leadership until they could forestall or suppress insurgencies. Rebellions were exclusively attributed to Soviet inspired "massive aggression." This unqualified counterrevolutionary posture has been adhered to despite: (1) an acknowledgment that the "roots of insurgency . . . are deep" in the "socioeconomic and political structures;" and (2) the admission by Secretary of Defense McNamara that "of 164 instances of internationally significant violence in the eight years between 1958 and 1966 . . . only 38 percent of the 149

insurgencies involved communism in any degree." Clearly, then, the open door goals discussed in Chapter 2 appear to be the operative ones. See: Maj. Gen. W.R. Peers, Asst. Deputy Chief of Staff for Special Operations, "Meeting the Challenge of Subversion," *Army* 15 (November 1964), 95-7, 158. Hughes, "The Myth of Military Coups," p. 5. Gen. Harold K. Johnson, Chief of Staff, USA, "The Army's Role in Nation Building and Preserving Stability," *Army Digest*, 20 (November 1965), 7, 10-12. U.S., Department of the Army, *Stability Operations–U.S. Army Doctrine*, Field Manual 31-23, December 1967, pp. 13, 56. Cf., *Military Assistance Training, Hearings*, House, p. 51.

43. *Military Assistance Training, Hearings*, House,; p. 128.

44. *Military Cold War Education, Hearings*, Senate, pp. 2499-2501, 2522, 2115-16. Cf.: Maj. Gen. Bogart, "The Army's Role," p. 62. Col. Frank L. Gailer, Jr., "Air Force Missions in Latin America," *Air University Quarterly Review*, 12 (Fall 1961), 45. Maj. Trevor W. Swett, Jr., "Assignment: Latin America; Duty: Army Advisor; Mission: Be Ever Helpful," *Army*, 14 (October 1961), 62. Dunn, "Military Aid," pp. 150-53, 305-06.

45. *Military Assistance Training, Hearings*, House, pp. 124-25.

46. *Military Assistance Training in East Asia, Report*,, House, p. 18.

47. Sidney Mintz, Review of *Haiti: The Politics of Squalor*, by Robert I. Rotberg (Boston: Houghton Mifflin, 1971), in the *New York Times Book Review*, March 28, 1971, p. 7.

48. *United States Security Agreements, Hearings*, Senate, p. 508. *Military Assistance Training, Hearings*, House, p. 113.

49. *Reports of the Special Study Mission*, House, p. 16.

50. Harry F. Walterhouse, *A Time to Build: Military Civic Action as a Medium for Economic Development and Social Reform* (Columbia: University of South Carolina Press, 1964), pp. 33-35. Gen. W.B. Palmer, "Military Assistance Program: A Progress Report," *Army Information Digest* 17 (April 1962), 43.

51. *Foreign Assistance Appropriations for 1970, Hearings*, House, p. 713. *Foreign Assistance Act of 1972, Hearings*, House, p. 166.

52. Only Cuba lacks U.S. Military Representation. MAAGs or MILGRPs act in this capacity elsewhere except for Haiti where military attachés are assigned. A Joint Defense Commission is stationed in Mexico, although military advisors may have been added recently. *New York Times*, June 25, 1969, p. 3. *U.S. Military Policies in Latin America, Hearings*, Senate, p. 77. *Foreign Assistance Appropriations for 1970, Hearings*, House, p. 635. *New York Times*, November 1, 1970, I, p. 18. For the much lower strengths of the early 1960s, see: Bogart, "The Army's Role," p. 61.

53. *Reports of the Special Study Mission*, House, p. 16.

54. U.S., Congress, Senate, Committee on Foreign Relations, *Foreign Assistance Act of 1969, Report*, no. 91-603, To Accompany H.R. 14580, 91st Cong., 1st sess., p. 27. *Foreign Assistance Appropriations for 1970, Hearings*, House, p. 877.

55. *U.S. Military Policies in Latin America, Hearings*, Senate, p. 75.

56. *Foreign Assistance Appropriations for 1970, Hearings*, House, p. 807.

57. Official worldwide entertainment allowances for *both* MAAGs and MILGRPs were $165,997.00 in FY 1968, $263,000.00 in FY 1969 and $230,823 proposed for FY 1970. Ibid., p. 787. Cf., U.S., Congress, Senate, Committee on Foreign Relations, *Foreign Assistance Act, 1969, Hearings*, before the Committee on Foreign Relations, United States Senate, 91st Cong., 1st sess., p. 142. There are both MAAGs and MILGRPs in Latin America.

58. *Foreign Assistance Act, 1969, Hearings*, p. 136. And in addition to this possibility, $1,800,000.00 was requested from Congress to reimburse the Dept. of State for "training support" in: Greece, Libya, Morocco, Spain, Tunisia, Turkey, Dominican Republic, Congo Ethiopia, Iran, Liberia, Saudi Arabia, China, Philippines. Among other items, this includes publications (USIA propaganda?) and language training. *Foreign Assistance Appropriations for 1970, Hearings*, House, p. 787.

59. For an illustration of such mutuality, see *Foreign Assistance Appropriations for 1970, Hearings*, House, pp. 830-32. On the otherhand, "AID and MAP are constantly at war over exactly what constitutes economic and military aid." Thayer, *The War Business*, p. 201. For the view that often they are genuinely inseparable, see *Global Defense* (Washington, D.C.: Congressional Quarterly Service, 1969), p. 39.

60. Robert E. Elder, *The Information Machine: The United States Information Agency and American Foreign Policy* (Syracuse, N.Y.: Syracuse University Press, 1968), p. 310.

61. Lynch, "Analysis," p. 41.

62. U.S., Department of the Army, *Report of the Thirteenth Annual U.S. Army Human Factors Research and Development Conference* (Fort Monmouth, N.J.: U.S. Army Signal Center and School, October 25-27, 1967), p. 179.

63. Taylor Branch, "The Ugly American and Flexible Response," *Washington Monthly*, April 1972, p. 60.

64. Dunn, "Military Aid," pp. 150-53.

65. Ibid., p. 162.

66. Carl M. Guelzo, "The Higher Level Staff Advisor," *Military Review* 47 (February 1967), 96.

67. *Economic Issues in Military Assistance, Hearings* Joint Economic Committee, p. 146. Cf.: *Military Assistance Training, Hearings*, House, p. 17; Capt. Richard A. Jones, USA, "The Nationbuilder: Soldier of the Sixties," *Military Review* 45 (January 1965), 63-65. Harlan Cleveland, Gerard J. Mangone, and John Clark Adams, *The Overseas Americans* (New York: McGraw Hill, 1960), p. 280.

68. Dunn, "Military Aid," pp. 175ff. *Military Assistance Training, Hearings*, House, p. 104.

69. *Foreign Assistance Act of 1963, Hearings*, House, p. 724.

70. Cleveland et al., *The Overseas Americans*, p. 280. Cf., Dunn, "Military Aid," p. 68.

71. James C. Shelburne and Kenneth J. Groves, *Education in the Armed Forces* (New York: Center for Applied Research in Education, Inc., 1965), p. 81. Cf., Institute of International Education, *Military Assistance*, p. 11.

72. *Military Cold War Education, Hearings*, Senate, pp. 2240-60. Dunn, "Military Aid," pp. 154-58, 169-72, 178-79. Cf., U.S., Department of Defense, *Facts About the United States* (Washington, D.C.: Armed Forces Information and Education, DA Pamphlet 360-516, 1964), 36pp.

73. *Military Cold War Education, Hearings* Senate, p. 2656. Joint Colleges: National War College, Industrial College of the Armed Forces, Armed Forces Staff College, *Information Data Sheet No. 49 to the Army Information Program*, (Reference DA Circular 11-1, 29 October 1958). U.S., Department of the Army, *U.S. Army Handbook for Colombia* (Washington, D.C.: DA Pamphlet No. 550-26, 2d ed., June 1964), pp. 199, 356-59, 376-77, 389, 394, 425-33, 578.

74. Center for Research in Social Systems, *CRESS Work Program: Fiscal Year 1968* (Washington, D.C.: CRESS, American University, n.d.), pp. 1, 13, 21. Cf.: Arthur J. Hoehn, *The Design of Cross-Cultural Training for Military Advisors* (Alexandria, Va.: Human Resources Research Office, George Washington University, Operating under contract with the Department of the Army, December 1966), pp. 2-5. Mike Klare, "U.S. Military Operations," *NACLA Newsletter* 2 (October 1968), 7. Edward C. Stewart, "American Advisors Overseas," *Military Review* 45 (February 1965), 3-9. Lewis D. Overstreet, "Strategic Implications of Developing Areas," *Military Review* 46 (October 1966), 70-77.

75. *Foreign Assistance Act, 1969, Hearings*, Senate, p. 140. *Foreign Assistance Appropriations for 1970, Hearings*, House, p. 627. Following the 1959 Draper Committee Report, MAP responsibility had also been decentralized to the unified commands (currently: CINCARIB-Latin America; CINPAC-Pacific; NESA-Near East and South Asia; CINCEUR-Europe, CONUS-United States). *Military Cold War Education, Hearings*, Senate, p. 22. *Military Assistance Training, Hearings*, House, p. 105.

76. *Military Assistance Training, Hearings*, House, p. 104.

Chapter 7
Effectiveness of the Program

1. Annotated by Peter B. Riddleberger, *Military Roles in Developing Countries: An Inventory of Past Research and Analysis* (Washington, D.C.: American University, Special Operations Research Office, AD-463188, March 1965), p. 50.

2. U.S., Congress, House, *Foreign Assistance Act of 1966, Hearings*, before the Committee on Foreign Affairs, 89th Cong., 2d sess., p. 242.

3. U.S. Congress, Joint Economic Committee, *Economic Issues in Military*

Assistance, Hearings, before the Subcommittee on Economy in Government of the Joint Economic Committee, Congress of the United States, 92d Cong., 1st sess., January-February, 1971, p. 37.

4. John J. Johnson, *The Military and Society in Latin America* (Palo Alto, Calif.: Stanford University Press, 1964), pp. 143-44, as quoted by Irving Louis Horowitz, "The Military Elites," in *Elites in Latin America*, Seymour M. Lipset and Aldo Solari, eds. (London: Oxford University Press, 1967), p. 178. Johnson's view would seem to be strongly supported by the first post-World War II scholarly investigation of these relationships. See: Edwin Lieuwen, *Arms and Politics in Latin America* (New York: Praeger, 1960).

5. See Appendix 10.

6. Eric A. Nordlinger, "Soldiers in Mufti: The Impact of Military Rule upon Economic and Social Change in the Non-Western States," *American Political Science Review* 64 (December 1970), 1131-48.

7. Donn R. Grand Pre, "A Window on America–The Department of Defense Informational Program," *International Educational and Cultural Exchange* 6 (Fall 1970), 87.

Upon another occasion, Gen. Warren stated that "over the past 20 years, military assistance has been one of the most important instruments in the attainment of American security objectives abroad." U.S., Congress, House, Committee on Appropriations. *Foreign Assistance and Related Agencies Appropriations for 1970, Hearings*, before the Subcommittee on Foreign Operations and Related Agencies, Committee on Appropriations, House of Representatives, 91st Cong., 1st sess., p. 832. Cf.: Gen. Robert W. Porter, USA, "Look South to Latin America," *Military Review* 48 (June 1968), 88. Frank Robbins Pancake, "Military Assistance as an Element of U.S. Foreign Policy in Latin America, 1950-1968," (unpublished Ph.D. dissertation, University of Virginia, 1969). U.S., Congress, Senate, Committee on the Armed Services, *Military Cold War Education and Speech Review Policies, Hearings*, before the Subcommittee on Special Preparedness of the Committee on Armed Services, United States Senate, 87th Cong., 2d sess., p. 2124. U.S., Congress, House, Committee on Foreign Affairs, *Foreign Assistance Act of 1963, Hearings,* before the Committee on Foreign Affairs, House of Representatives, 88th Cong., 1st sess., p. 207.

8. U.S., Congress, House, Committee on Foreign Affairs, *Military Assistance Training, Hearings*, before the Subcommittee on National Security Policy and Scientific Developments of the Committee on Foreign Affairs, House of Representatives, 91st Cong., 2d sess., October-December 1970, p. 152.

9. Ibid., p. 151.

10. General Fondren in: U.S., Congress, House, Committee on Foreign Affairs, *Foreign Assistance Act of 1965, Hearings*, before the Committee on Foreign Affairs, House of Representatives, 89th Cong., 1st sess., p. 800.

More recently, Col. Amos A. Jordan, a military scholar, held that despite Pakistan's acceptance of Chinese aid, training has been "reasonably good" in furthering American military and political objectives throughout the world. Thus a year earlier it was claimed that MAP was responsible for the maintenance of diplomatic relations with the United States by seven of fourteen Arab Governments after the 1967 Israeli blitzkrieg. *Military Assistance Training, Hearings*, House, p. 99. U.S., Congress, House, Committee on Foreign Affairs, *Foreign Assistance Act of 1969, Hearings*, before the Committee on Foreign Affairs, House of Representatives, 91st Cong., 1st sess., p. 599. Cf., *Foreign Assistance Appropriations for 1970, Hearings*, House, p. 844.

11. Lt. Col. David R. Hughes, USA, "The Myth of Military Coups and Military Assistance," *Military Review* 47 (December 1967), 7. James Clotfelter, "The African Military Forces," ibid., 48 (May 1968), 29-30. *New York Times*, November 1, 1970, I, p. 18.

12. See Chapter 2, note 31.

13. U.S., Congress, House, Committee on Foreign Affairs, *Military Assistance Training in East and Southeast Asia, A Staff Report*, prepared for the use of the Subcommittee on National Security Policy and Scientific Developments of the Committee on Foreign Affairs, House of Representatives, 91st Cong., 2d sess., February 16, 1971, pp. 18-22.

14. *New York Times*, November 1, 1970, I, p. 18. Cf., U.S., Congress, House, Committee on Foreign Affairs, *Reports of the Special Study Mission to Latin America on Military Assistance Training and Developmental Television*, submitted by Hon. Clement J. Zablocki, Hon. James G. Fulton and Hon. Paul Findley to the Subcommittee on National Security and Scientific Developments of the Committee on Foreign Affairs, House of Representatives, 91st Cong., May 7, 1970, p. 3.

15. U.S., Congress, House, Committee on Appropriations, *Foreign Assistance and Related Agencies Appropriations for 1967, Hearings*, before a subcommittee of the Committee on Appropriations, House of Representatives, 89th Cong., 2d sess., pp. 654-55.

Obviously other factors may also contribute to failure. "African territories," for example, "have had to face the threat of a *coup d'etat* launched from within. . . . The failure of all but one of these attempts (Zanzibar) was due not only to prompt intervention by British troops, but also to the sheer ineptitude of the leaders. . . ." CES, "The Coup d'Etat," *The Army Quarterly* (U.K.), reprinted in *Military Review* 47 (April 1967), 14.

16. Lt. Cmdr. D.A. Van Horssen, "The Royal Thai Navy," *U.S. Naval Institute Proceedings*, 92 (June 1966), 81-2. In an earlier chapter, we referred to the Americanization of curricula in several Latin American countries.

17. *Military Assistance Training in East Asia, Report*, House, p. 26.

18. Ibid., p. 24.

19. U.S., Department of Defense, *Military Assistance: Data for Debate* (Washington, D.C.: Office of the Assistant Secretary of Defense for International Security Affairs, August 1966), p. 1.

20. Charles Wolf, Jr., *Military Assistance Programs* (Santa Monica, Calif.: Rand Corporation, P-3240, October 1965), pp. 10-11. U.S., President's Committee to Study the United States Military Assistance Program, *Supplement to the Composite Report*, vol. 2 (Washington, D.C.: GPO, 1959), pp. 146-48, 158. Cf., U.S., Department of Defense, *Military Assistance: Data for Debate*, p. 15.

21. Institute of International Education, *Twenty Years of United States Government Programs in Cultural Relations* (New York: IIE, Committee on Educational Interchange Policy, 1959), pp. 22-27.

22. *Foreign Assistance Act of 1963, Hearings*, House, pp. 785-86.

23. Institute of International Education, *Military Assistance Training Programs of the U.S. Government* (New York: IIE, Committee on Educational Interchange Policy, Statement no. 18, July 1964), p. 21.

24. U.S. Congress, Senate, Committee on Foreign Relations, *Foreign Assistance, 1966, Hearings* before the Committee on Foreign Relations, United States Senate, 89th Cong. 2d. Sess., p. 216.

25. See the Hearings referred to in notes 7-19, above and especially: *Foreign Assistance Act of 1969, Hearings*, House, pp. 498-99. Cf., Harold A. Hovey, *United States Military Assistance: A Study of Policies and Practices* (New York: Praeger, 1965), pp. 66-69.

26. U.S. Congress, House, Committee on Foreign Affairs, *Military Assistance Training, Report*, of the Subcommittee on National Security Policy and Scientific Developments to the Committee on Foreign Affairs, House of Representatives, April 2, 1971, pp. 12-13. Cf., Anibal Quijano, "Nationalism and Capitalism in Peru: A Study in Neo-Imperialism," *Monthly Review* 23 (July-August 1971), 1-119.

27. *Reports of the Special Study Mission, House*, pp. 10-12. Yet, for some members of Congress, the existence of the nationalist regime, there means that the "U.S. somehow failed." *Military Assistance Training, Hearings*, House, p. 78. A commitment to "national independence" seems to have been the primary goal. Richard Lee Clinton, "The Modernizing Military: The Case of Peru," *Inter-American Economic Affairs* 24 (Spring 1971), 51-64.

28. For the record of those very few senators who have articulated verbal disconsolation over this tendency, see the Hearings referred to in note 24 above, and: U.S., Congress, Senate, Committee on Foreign Relations, *United States Military Policies and Programs in Latin America, Hearings*, before the Subcommittee on Western Hemisphere Affairs of the Committee on Foreign Relations, United States Senate, 91st Cong., 1st sess., June 24 and July 8, 1969, passim.

29. For an exemplary manifestation of concern by a congressman, see the colloquy between Hon. Donald Fraser and Assistant Secretary of Defense Nutter which is quoted in Chapter 2.

30. Hughes, "The Myth of Military Coups," p. 7.

31. See Rogers testimony in Chapter 2, footnote q.

32. Charles Wolf, "The Political Effects of Military Programs: Some indica-

tions from Latin America," *Orbis* 8 (Winter 1965), 87, et passim. Edward S. Mason, *Foreign Aid and Foreign Policy* (New York: Harper & Row, 1964), p. 66.

33. Edwin Lieuwen, *Generals vs. Presidents* (New York: Praeger, 1964), pp. 130, 135, 148, and his "The Changing Role of the Armed Forces: An Analysis," in *Latin American Politics: Twenty-Four Studies of the Contemporary Scene*, by Robert D. Tomasek, ed. (Garden City, N.Y.: Doubleday-Anchor, 1966), pp. 63-65, 77. Cf., U.S., Congress, Senate, Committee on Foreign Relations, *Survey of the Alliance for Progress–The Latin American Military: A Study*, for the Subcommittee on American Republics, Committee on Foreign Relations, United States Senate, 90th Cong., 1st sess., October 9, 1967, pp. 11, 13-14, 29-30.

34. "The Military Elites," in Lipset and Solari, eds., pp. 176, 179-80. Cf., John Duncan Powell, "Military Assistance and Militarism in Latin America," *Western Political Quarterly* 18 (June 1965), 382-92.

35. Jose Nun, *Latin America: The Hegemonic Crisis and the Military Coup* (Berkeley: University of California, Institute of International Studies, Politics of Modernization Series, no. 7, 1969), pp. 54-55.

36. John M. Dunn, "Military Aid and Military Elites: The Political Potential of American Training and Technical Assistance Programs," (unpublished Ph.D. dissertation, Princeton University, 1961), pp. 304-06, 331-36.

37. Edgar S. Furniss, Jr., *Some Perspectives on American Military Assistance* (Princeton, N.J.: Princeton University, Center of International Studies, Memo no. 13, June 18, 1957), pp. 29-32.

38. *Military Assistance Training, Hearings*, House, pp. 59-68, 151, 165, 171.

39. *Military Assistance Training, Hearings*, House, p. 5.

40. Ibid., p. 113. For similar testimony by Morton H. Halperin of the Brookings Institution, see *Economic Issues in Military Assistance, Hearings*, Joint Economic Committee, p. 156.

41. *Military Assistance Training, Hearings*, House, pp. 117-21.

42. Ibid., pp. 35, 97, 103.

43. *Army Officers in Arab Politics and Society* (New York: Praeger, 1970), p. 271. Referring to the incitement of coups, a journalist with access to Pentagon sources has reported "this type of meddling is common among U.S. military officers stationed abroad." George Thayer, *The War Business: The International Trade in Armaments* (New York: Simon and Schuster, 1969), p. 200.

44. With regard to the Dominican Republic–for which our evidence is weakest–see: Theodore Draper, *The Dominican Revolt: A Case Study in American Policy* (New York: Commentary, 1968), pp. 5-19. Richard J. Barnet, *Intervention and Revolution: America's Confrontation with Insurgent Movements Around the World* (New York: World, 1968), p. 166. Jerome Levinson and Juan de Onis, *The Alliance That Lost Its Way* (Chicago: Quadrangle, 1970), pp. 85-6. Lester D. Langley, "Military Commitments in Latin America: 1960-1968," *Current History* 56 (June 1969), 349-50.

Concerning Brazil, see: Thayer, *The War Business*, p. 200. New York Times, November 1, 1970, I, p. 18; *Foreign Assistance Act of 1965, Hearings*, House, p. 783. *Military Assistance Training, Hearings*, House, p. 128. Cf., footnote m, and note 14, above.

On Iran, the following are of great value: Salvatore Alfred Arcilesi, "The Development of United States Foreign Policy in Iran: 1949-1960," (unpublished Ph.D. dissertation, University of Virginia, 1965, pp. 82-84. David Wise and Thomas B. Ross, *The Invisible Government* (New York: Random House, 1964), pp. 110-14; Barnet, *Intervention and Revolution*, pp. 226-28. Bahman Nirumand, Iran: *The New Imperialism in Action* (New York: Monthly Review Press, 1969). Cf., *Military Assistance Training, Hearings*, House, pp. 40, 45.

The Indonesian takeover by Suharto is dealt with in: *Military Assistance Training in East Asia, Report*, House, pp. 18-22. U.S., Congress, House, *Foreign Assistance Act of 1968, Hearings*, before the Committee on Foreign Affairs, House of Representatives, 90th Cong., 2d. sess., pp. 705-06, 812. U.S., Congress, House, Committee on Appropriations, *Foreign Assistance and Related Agencies Appropriations for 1966, Hearings*, before a subcommittee of the Committee on Appropriations, House of Representatives, 89th Cong., 1st sess., p. 401. U.S., Congress, Senate, Committee on Foreign Relations, *Foreign Assistance, 1966 Hearings*, before the Committee on Foreign Relations, United States Senate, 89th Cong., 2d. sess., p. 693. Cf.: *Foreign Assistance Appropriations for 1970, Hearings*, House, p. 727. David Ranson, "The Berkeley Mafia and the Indonesian Massacre," *Ramparts*, October 1970, pp. 27-29, 40-49.

45. The order of the first two passages has been reversed. They appear in: Simon G. Hanson, "Third Year: Military," *Inter-American Economic Affairs* 18 (Spring 1965), 14, 36. The remainder is cited from Eduardo Galeano, "The De-Nationalization of Brazilian Industry," *Monthly Review* 21 (December, 1969), 16-17. Cf., E.D.B., "The Hanna Industrial Complex," *NACLA Newsletter*, 2 (May-June 1968), 1-5. Jerry L. Weaver, "Assessing the Impact of Military Rule: Alternative Approaches," (paper prepared for the Arms Control and Foreign Policy Seminar, Center for Policy Study, University of Chicago, May 26-27, 1972), pp. 17-22.

46. Military Assistance Training, Hearings, House, p. 107. Cf. footnote m and note 44, above.

47. Fred Goff and Michael Locker, "The Violence of Domination: U.S. Power and the Dominican Republic," in *Latin American Radicalism: A Documentary Report on Left and Nationalist Movements*, Irving Louis Horowitz, Josue de Castro, and John Gerassi, eds. (New York: Random-Vintage, 1969), pp. 264-67.

48. Lieuwen, *Generals vs. Presidents*, p. 127, as quoted in Goff and Locker, "The Violence of Domination," p. 269.

49. Goff and Locker, "The Violence of Domination," pp. 273-74.

50. Walter LaFeber, *America, Russia, and the Cold War: 1945-1966* (New York: Wiley, 1967), pp. 155-57. Cf., Nirumand, *Iran*.

51. Robert Coats, "Indonesian Timber," *Pacific Research and World Empire Telegram* 2 (May-June 1971), 9.

Chapter 8
Conclusions

1. A.M. Scott, "Nonintervation and Conditional Intervention," *Journal of International Affairs* 22 (1968), 208-16, from the abstract appearing in *International Political Science Abstracts*, 19 (1969), 138.

2. U.S., Congress, House, Committee on Foreign Affairs, *Military Assistance Training in East and Southeast Asia, A Staff Report*, prepared for the use of the Subcommittee on National Security Policy and Scientific Developments of the Committee on Foreign Affairs, House of Representatives, 91st Cong., 2d sess., p. 31.

3. Lenny Siegel, "The Future of Military Aid," *Pacific Research & World Empire Telegram* 3 (November-December 1971), 5.

4. Because this CONUS figure includes enlistees (about one-third of the total), it may be higher for officers. In any case, during the fifteen years, the per capita cost did not rise significantly. U.S., Congress, House, Committee on Foreign Affairs, *Military Assistance Training, Report*, of the Subcommittee on National Security Policy and Scientific Developments to the Committee on Foreign Affairs, House of Representatives, April 2, 1971, p. 6.

5. Thailand and Indochina (1967-70) are omitted from the calculation. Ibid., p. 7.

6. U.S., Congress, House, Committee on Foreign Affairs, *Military Assistance Training, Hearings*, before the Subcommittee on National Security Policy and Scientific Developments of the Committee on Foreign Affairs, House of Representatives, 91st Cong., 2d sess., October-December 1970, p. 172. On the increased "professional" as opposed to technical emphasis, see the *Military Assistance Training, Report*, House, p. 14.

7. "Increased Foreign Military Sales Inescapable under Nixon Doctrine," *Armed Forces Journal*, 2 November 1970, p. 22. With reference to the Nixon Doctrine's emphasis upon *increased* foreign officer training, see: U.S., Congress, House, Committee on Foreign Affairs, *Reports of the Special Study Mission to Latin America on Military Assistance Training and Developmental Television*, submitted by Hon. Clement J. Zablocki, Hon. James G. Fulton and Hon. Paul Findley to the Subcommittee on National Security and Scientific Developments of the Committee on Foreign Affairs, House of Representatives, 91st Cong., May 7, 1970, p. 29.

8. *Military Assistance Training, Hearings*, House, p. 92. *Foreign Assistance Act of 1972, Hearings*, House, p. 166.

9. U.S., Congress, Senate, Committee on Foreign Relations, *Rockefeller*

Report on Latin America, Hearing, before the Subcommittee on Inter-American Affairs of the Committee on Foreign Relations, United States Senate, 91st Cong., November 20, 1969, pp. 83, 148.

10. *Military Assistance Training, Report*, House, p. 21.

11. *Rockefeller Report, Hearing*, Senate, p. 121.

12. *Foreign Assistance Act of 1972, Hearings*, House, pp. 10, 48-50, 83, 154.

13. Ibid., p. 10.

14. *Intelligence and Foreign Policy* (Cambridge, Mass.: Africa Research Group, 1971), pp. 17-18.

15. *Military Assistance Training in East Asia, Report,* House, pp. 2-4. *Military Assistance Training, Report*, House, pp. 9, 14-15, 17. House, pp. 2-4. Cf., *Rockefeller Report, Hearing*, Senate, pp. 8-11, 35-6, 85, 93, 117-19.
Senate, pp. 8-11, 35-6, 85, 93, 117-19.

16. *Military Assistance Training in East Asia, Report*, House pp. 6-10, 15-16. Cf. U.S. Congress, Joint Economic Committee, *Economic Issues in Military Assistance, Hearings*, before the Subcommittee on Economy in Government of the Joint Economic Committee, Congress of the United States, 92d Cong., 1st sess., January-February, 1971, pp. 85-6, 363.

17. *Military Assistance Training, Report*, House, pp. 3-8, 45.

18. Ibid., p. 24. *Military Assistance Training, Hearings*, House, p. 150.

19. *Military Assistance Training, Report*, House, p. 23.

20. *Military Assistance Training, Hearings*, House, pp. 139-43.

21. Ibid., p, 138.

22. *Military Assistance Training, Report*, House, pp. 19-21. The "findings" of the Subcommittee—some of which are not in accord with our own—appear on pp. 9-18.

23. *Reports of the Special Study Mission*, House, p. 26.

24. U.S., Congress, Senate, Committee on Foreign Relations, *United States Security Agreements and Commitments Abroad, Hearings*, before the Subcommittee on United States Security Agreements and Commitments Abroad of the Committee on Foreign Relations, United States Senate, 91st Cong., 2d sess., p. 1088. Cf., John Rothchild, "Cooing Down the War: The Senate's Lame Doves," *Washington Monthly*, August 1971, pp. 6-19.

More recently Fulbright has declared that many underdeveloped countries need major socioeconomic changes, "and if the only way they can be brought about is by internal revolution, that is their business." *Economic Issues in Military Assistance, Hearings*, Joint Economic Committee, p. 48.

25. Jose Nun, *Latin America: The Hegemonic Crisis and the Military Coup* (Berkeley: University of California, Institute of International Studies, 1969), p. 9.

26. Thus civilians have not only established a tradition of systematic denigration of militarism, but such norms have also been internalized by officers because both professional rules and modern military technology were externally

imposed—by civilians. He stresses that this has not been the case in Latin America. Ibid., pp. 2-4. We would add that rapid economic growth reinforced this process of internalization by providing nonmilitary opportunities for upward social mobility. And officer openings were reserved for those from families with high social prestige.

27. Ibid., p. 55.

28. This view is developed from a counterrevolutionary normative perspective by Seymour Martin Lipset in his "Some Social Requisites of Democracy: Economic Development and Political Legitimacy," in *Empirical Democratic Theory*, Charles F. Cnudde and Deane E. Neubauer, eds. (Chicago: Markham, 1969). Cf., Martin C. Needler, "Political Development and Socioeconomic Development: The Case of Latin America," *American Political Science Review* 62 (September 1968), 889-98.

29. John Gerassi, "Latin America: The Left on the Move," *Ramparts*, September 1971, p. 25.

30. Richard Lee Clinton, "The Modernizing Military: The Case of Peru," *Inter-American Economic Affairs* 24 (Spring 1971), 49-50.

Appendix 10
Foreign Aid and Military Interventions

1. M.J.V. Bell, *Military Assistance to Independent African States* (London: Institute for Strategic Studies, December 1964), passim. William F. Gutteridge, *Armed Forces in the New States* (London: Oxford University Press, 1962), passim. David Ottaway and Marina Ottaway, *Algeria: The Politics of a Socialist Revolution* (Berkeley: University of California Press, 1970). U.S., Congress, House, Committee on Foreign Affairs, Foreign Assistance Act of 1969, Hearings, before the Committee on Foreign Affairs, House of Representatives, 91st Cong., 1st sess., p. 462. U.S. Congress, House, Committee on Appropriations, *Foreign Assistance and Related Agencies Appropriations for 1970, Hearings*, before the Subcommittee on Foreign Operations and Related Agencies, Committee on Appropriations, House of Representatives, 91st Cong., 1st sess., p. 763.

2. "General NeWin, coup leader, who will become the new Premier is a staunch anti-Communist. While in Washington for medical treatment last year he conferred with military and defense officials." *New York Times*, September 27, 1958, March 2, 1962. D.W. Chang, "The Military and Nation-Building in Korea, Burma and Pakistan," *Asian Survey* 9 (November 1969), 818-30. Gutteridge, *Armed Forces*; Peter B. Riddleberger, *Military Roles in Developing Countries: An Inventory of Past Research and Analysis* (Washington: Special Operations Research Office, AD 463188, American University, 1965), pp. 58-9, 74-8, 170. S.E. Finer, *The Man on Horseback: The Role of the Military in Politics* (New York: Praeger, 1962), p. 229. Harold A. Hovey, *United States Military*

Assistance: A Study of Policies and Practices (New York: Praeger, 1965), p. 36. John L. Sutton and Geoffrey Kemp, *Arms to Developing Countries* (London: Institute for Strategic Studies, 1966).

3. *New York Times*, March 20-August 19, 1970. U.S., Congress, House, Committee on Foreign Affairs, *Foreign Assistance Act of 1972, Hearings*, before the Committee on Foreign Affairs, House of Representatives, 92d Cong., 2d sess., March 1972, p. 8. And the sources cited in our text.

4. Bell, *Military Assistance*; Gutteridge, *Armed Forces*, p. 139. J.M. Lee, *African Armies and Civil Order* (New York: Praeger, 1969), p. 118. Dorothy Nelkin, "The Economic and Social Setting of Military Takeovers in Africa," *Journal of Asian and African Studies* 2 (January-April 1967), 230-44.

5. Bell, *Military Assistance*. C.E. Welch, "Soldier and State in Africa," *Journal of Modern African Studies* 5 (November 1967), 318. U.S., Congress, House, Committee on Foreign Affairs, *Report of Special Study Mission to West and Central Africa, March 29 to April 27, 1970*, by Hon. Charles C. Diggs, Jr., House of Representatives, 91st Cong., 2d sess., pp. 33-34.

6. During the early 1960s, a Kennedy administration spokesman testified: "We do sell them a certain amount of equipment and we do provide at their expense certain training." U.S., Congress House, Committee on Foreign Affairs, *Foreign Assistance Act of 1963, Hearings*, before the Committee on Foreign Affairs, House of Representatives, 88th Cong., 1st sess., p. 728. Foreign Assistance *Act of 1969, Hearings*, House, p. 462. U.S., Department of State, Bureau of Educational and Cultural Affairs, *A Guide to U.S. Government Agencies Involved in International Educational and Cultural Activities* (Washington, D.C.: GPO, International Information and Cultural Series 97, Dept. of State Publication 8405, September 1968), p. 45. George Thayer, *The War Business: The International Trade in Armaments* (New York: Simon & Shuster, 1969), p. 334. J.H. Hoagland, "Arms in the Developing World," *Orbis* 12 (Spring 1968), 169. Louis W. Koenig, "The Truman Doctrine and NATO," *Current History* 57 (July 1969), 23.

7. Bell, *Military Assistance*; Gutteridge, *Armed Forces*. Sutton and Kemp, *Arms to Developing Countries*, p. 5. For reports of CIA involvement in the 1966 coup, see the *New York Times*, February 22, 1966, p. 12. St. Clair Drake, "Nkruma: The Real Tragedy," *Nation*, June 5, 1972, pp. 721-24. Kenneth W. Grundy and Robert L. Farlow, "Internal Sources of External Behavior: Ghana's New Foreign Policy," *Journal of Asian and African Studies* 4 (July 1969), 161-71. Robert M. Price, "A Theoretical Approach to Military Rule in New States: Reference-Group Theory and the Ghanaian Case," *World Politics* 23 (April 1971), 399-430.

8. Constantine Tsoucalas, "Class Struggle and Dictatorship in Greece," *New Left Review*, no. 56 (July-August 1969), pp. 7-11. *New York Times*, April 21, 1967, p. 1.

9. In 1965 it was reported that "some 1,200 Indonesian officers have been

trained under the United States military aid program, including the Army and Air Force Chief of Staff and about two-fifths of the Army staff." Hovey, *Military Assistance*, p. 37. Sutton and Kemp, *Arms to Developing Countries*, pp. 5, 21, 25. Riddleberger, *Military Roles*, pp. 54-55, 70-1. Uri Ra'anan, *The USSR Arms the Third World: Case Studies in Soviet Foreign Policy* (Cambridge: MIT Press, 1969), pp. 210-11, 226, 233. L. Palmer, "The 30 Sept. Movement in Indonesia," *Modern Asian Studies* 5 (January 1971), 1-20. J.M. Van Der Kroef, "Interpretations of the 1965 Indonesian Coup: A Review of the Literature," *Pacific Affairs* 43 (Winter 1970-71), 557-77. Haldore Hanson, "A Gamble on Guided Democracy," *Reporter*, July 23, 1959, pp. 23-6. In the course of a 1965 exchange with Congressman Passman, Director of Military Assistance General Robert Wood intimated that the continuation of training was designed to strengthen a military pressure group that would ultimately force a change in Sukarno's neutralist policies. U.S., Congress, House, Committee on Appropriations, *Foreign Assistance and Related Agencies Appropriations for 1966, Hearings*, before a subcommittee of the Committee on Appropriations, House of Representatives, 89th Cong., 1st sess. pp. 235, 330, 401, 1161. Ibid., *Foreign Assistance and Related Agencies Appropriations for 1967, Hearings*, before a subcommittee of the Committee on Appropriations, House of Representatives, 89th Cong., 2d sess., pp. 647, 721. *Foreign Assistance Appropriations for 1970, Hearings*, p. 727. *Foreign Assistance Act of 1963, Hearings*, House, pp. 690-91, 759, 762, 773-74, 781, 797-800, 807. U.S., Congress, House, Committee on Foreign Affairs, *Foreign Assistance Act of 1965, Hearings*, before the Committee on Foreign Affairs, House of Representatives, 89th Cong., 1st sess., pp. 643-44.

10. Sutton and Kemp, *Arms to Developing Countries*. J.C. Hurewitz, *Middle East Politics: The Military Dimension* (New York: Praeger, 1969), pp. 72-3, testifying in 1965, Military Assistance Director General Robert Wood stated: "We believe that the strategic significance of Iran, its importance as an ally, its evolution toward democracy, its economic viability, and so on, are important to the United States." *Foreign Assistance Appropriations for 1966, Hearings*, House, p. 283. *Foreign Assistance Act of 1969, Hearings*, House, p. 681. Salvatore Alfred Arcilesi, "The Development of United States Foreign Policy in Iran; 1949-1960," (unpublished Ph.D. dissertation, University of Virginia, 1965), pp. 82-4.

11. Testifying after the February 1963 coup, a prominent Kennedy administration official reported "our dialog with the new Iraqi Government is much fuller than it ever was with the Quasim Government." *Foreign Assistance Act of 1963, Hearings*, House, p. 710. A Small U.S. training program for the Iraqi Army was launched in 1961, and then broadened to include the Air Force. By 1967, more than $4,000,000.00 was being budgeted by the U.S. to train Iraqi military personnel. This would normally allow training spaces for several hundred officers. *Foreign Assistance Appropriations for 1966, Hearings*, House, p. 276.

Foreign Assistance Appropriations for 1967, Hearings, House, pp. 637-39, 721. *Foreign Assistance Act of 1969, Hearings*, House, p. 595. Eleizer Be'eri, *Army Officers in Arab Politics and Society* (New York: Praeger, 1970), p. 193. Sutton and Kemp, *Arms to Developing Countries*, p. 5. Majid Khadduri, *Republican Iraq* (London: Oxford University Press, 1969), passim.

12. For the reputed role of U.S. attachés in Hussein's 1957 coup, see: *New York Times*, April 21, 1957, April 24, 1957, April 26, 1957, April 30, 1957. Gutteridge, *Armed Forces*. MAP training and equipment totaling $36.6 million from FY 1957 through FY 1965 was intended to ensure military loyalty to King Hussein, i.e. inhibit Arab nationalism or "Nasserism." *Foreign Assistance Act of 1965, Hearings*, House, pp. 731, 741. *Foreign Assistance Appropriations for 1966, Hearings*, House, p. 276

13. On U.S. policy of supporting the army as distinguished from the neutralist government, see the *Foreign Assistance Appropriations for 1966, Hearings*, House, pp. 322-23. *Foreign Assistance Act of 1969, Hearings*, House, p. 617.

14. "The objective of the military assistance program for Libya is to assist the Government in maintaining internal security, assure [censored] the training complex at Wheelus Air Base, and to promote the education of their military leaders, while precluding bloc influence and assistance." Military assistance from the United States and Britain prompted the expansion of the Libyan Army from 1,800 troops in 1957 to 7,000 in 1967-68. *Foreign Assistance Act of 1965, Hearings*, House, pp. 810-11. Robert C. Sellers, ed., *Reference Handbook of the Armed Forces of the World* (Garden City, N.Y.: Robert C. Sellers & Associates, 2nd ed., 1968). Gutteridge, *Armed Forces*. Bell, *Military Assistance*. *Guardian*, December 27, 1969. *New York Times*, September 2, 1969. *Christian Science Monitor*, November 1, 1969.

15. *Foreign Assistance Act of 1965, Hearings*, House, p. 812. *Foreign Assistance Appropriations for 1966, Hearings*, House, p. 277. J.N. Hazard, "Marxian Socialism in Africa," *Comparative Politics* 2 (October 1969), 1-15. According to a Defense Department spokesman, the 1968 coup "reduced Chinese Communist influence there." *Foreign Assistance Act of 1969, Hearings*, House, p. 442. *Report of Special Study Mission to West Africa*, House, pp. 18-19. Luis Joaquin Munoz, "Golpes de Estado en Africa: El Caso de Mali," *Revista de Politica Internacional*, no. 106 (Noviembre-Deciembre 1969), 164-65, 169-71.

16. For an indication that one of the purposes of military aid is to inhibit Pakistan from developing good relations with China, see the *Foreign Assistance Act of 1965, Hearings*, House, p. 680; MAAG representation in FY 1968 was programmed at 13 U.S. and 12 locals. U.S., Congress, House, Committee on Appropriations, *Foreign Assistance and Related Agencies Appropriations for 1968, Hearings*, before a subcommittee of the Committee on Appropriations, House of Representatives, 90th Cong., 1st sess., p. 762. *Foreign Assistance Act*

of 1969, Hearings, House, pp. 559, 576, 584. U.S., President, *The Foreign Assistance Program: Annual Report to the Congress for Fiscal Year 1969* (Washington, D.C.: GPO, 1970), p. 46. Gutteridge, *Armed Forces*. Hurewitz, *Middle East Politics*, pp. 494, 498. Riddleberger, *Military Roles*, pp. 72-4. Chang, "Military and Nation-Building."

17. *Foreign Assistance Act of 1969, Hearings*, House, p. 576; *New York Times*, October 22, 1969, p. 3; Thayer, *The War Business*; Gutteridge, *Armed Forces*, p. 133.

18. Fred Halliday, "Class Struggle in the Arab Gulf," *New Left Review*, no. 58 (November-December 1969), p. 36.

19. General Abboud was given a tour of China in 1964. Be'eri, *Army Officers*, p. 270. Gutteridge, *Armed Forces*. Lee, *African Armies*, p. 124. U.S., Congress, House, Committee on Foreign Affairs, *Foreign Assistance Act of 1968, Hearings*, before the Committee on Foreign Affairs, 90th Cong., 2d sess., p. 792. *Foreign Assistance Act of 1969, Hearings*, House, p. 576. *Guardian*, December 27, 1969. J. Howell and M.B. Hamid, "Sudan and the Outside World: 1964-1968," *African Affairs* 68 (October 1969), 299-315.

20. Eight U.S. diplomats were expelled on charges of attempting to organize a coup on 1957. Similar accusations were made in 1967 and 1968. *New York Times*, August 14, 1957, June 25, 1967, p. 16, July 28, 1968, p. 4, November 14-17, 1970. Training programmed for FY 1967 amounted to $81,000 or about 20 officer trainees. *Foreign Assistance Appropriations for 1967, Hearings*, House, p. 690. *Foreign Assistance for 1966, Hearings*, House, pp. 279, 405-06. *Foreign Assistance Appropriations for 1968, Hearings*, House, p. 691. *Foreign Assistance Act of 1963, Hearings*, House, pp. 727-28. *Foreign Assistance Act of 1968, Hearings*, House, pp. 647-48, 732. Riddleberger, *Military Roles*, pp. 132-34; Hurewitz, *Middle East Politics*, p. 88. Sutton and Kemp, *Arms to Developing Countries*, p. 23. Finer, *Man on Horseback*, pp. 165, 185-86.

21. Justifying continued military aid in 1965, an administration spokesman cited the danger to Thailand from Communist and Neutral states in the area. *Foreign Assistance Act of 1965, Hearings*, House, p. 367. U.S., Congress, Senate, Committee on Foreign Relations, *United States Security Agreements and Commitments Abroad of the Committee on Foreign Relations, Hearings*, United States Senate, 91st Cong., 2d sess., pp. 613, 751. Cf., pp. 758-59; Riddleberger, *Military Roles*, pp. 84-88, 90-91.

22. Bell, *Military Assistance*, p. 7. Thayer, *War Business*, p. 327.

23. *Foreign Assistance Act of 1963, Hearings*, House, pp. 695-96, 709, 727-28. *Foreign Assistance Act of 1965, Hearings*, House, p. 981. *Foreign Assistance Appropriations for 1966, Hearings*, House, pp. 276, 425. *Foreign Assistance Appropriations for 1970, Hearings*, House, p. 890. Hurewitz, *Middle East Politics*, pp. 88, 91. L. Dupree, "Democracy and the Military Base of Power," *Middle East Journal* 22 (Winter 1968), 41.

24. Bell, *Military Assistance*. Welch, "Soldier and State in Africa," p. 318.

25. In 1963, "Prince Sihanouk decided that the American missions in Cambodia were threatening his existence and asked that they leave." Hovey, *Military Assistance*, p. 36. Sutton and Kemp, *Arms to Developing Countries*, p. 5. U.S., Congress, Senate, Committee on Appropriations, *Foreign Assistance and Related Agencies Appropriations for 1964, Hearings*, before a subcommittee of the Committee on Appropriations, United States Senate, 88th Cong., 1st sess., p. 377. *Foreign Assistance Appropriations for 1967, Hearings*, House, p. 721. *Foreign Assistance Act of 1969, Hearings*, House, p. 617.

26. Bell, *Military Assistance*.

27. Ibid. James Clotfelter, "The African Military Forces," *Military Review*, 48 (May 1968), 29.

28. Gutteridge, *Armed Forces. Foreign Assistance Act of 1969, Hearings*, House, p. 617.

29. Bell, *Military Assistance*.

30. Thayer, *War Business*, p. 345.

31. Bell, *Military Assistance*. Nelkin, "Economic and Social Setting."

32. An abortive reformist or Nasserite coup in December 1960 was suppressed by Army and Air Force commanders who remained loyal to the Emperor. Ibid. C. Claphan, "The Ethiopian Coup d'Etat of December 1960," *Journal of Modern African Studies* 6 (December 1968), 495-507. U.S. aid to Ethiopia exceeds that to the rest of Africa combined. Gutteridge, *Armed Forces*, p. 133.

33. Bell, *Military Assistance*.

34. Ibid. Sutton and Kemp, *Arms to Developing Countries*, p. 5. Lee, *African Armies*, p. 116. U.S., Congress, Senate, Committee on Foreign Relations, *Foreign Assistance 1964, Hearings*, before the Committee on Foreign Relations, United States Senate, 88th Cong., 2d sess., p. 515. *Foreign Assistance Appropriations for 1966, Hearings*, House, p. 402.

35. Army pressure after defeat by Chinese forced Menon's resignation as minister of defense. *New York Times*, November 1, 1962. J.H. Hoagland, "Arms in the Developing World," *Orbis* 12 (Spring 1968), 176. Gutteridge, *Armed Forces*, Hovey, *Military Assistance*, p. 101. Sutton and Kemp, *Arms to Developing Countries*, pp. 5, 24, 25. *Foreign Assistance Act of 1965, Hearings*, House, p. 730. *Foreign Assistance Act of 1968, Hearings*, House, p. 799. *Foreign Assistance Act of 1969, Hearings*, House, pp. 582-84. *Foreign Assistance Program: Annual Report to the Congress for Fiscal Year 1969*, p. 46. The United States has supplied limited quantities of war material (and presumably training) since independence. After 1962, there was a major step-up in the level of aid. Despite the alleged termination of military assistance in 1965 after the war with Pakistan, 15 U.S. MAAG personnel were stationed in India three years hence. *Foreign Assistance Appropriations for 1968, Hearings*, House, p. 762. *Foreign Assistance Appropriations for 1970, Hearings*, House, p. 722.

36. Bell, *Military Assistance*.

37. With regard to U.S. political objectives in Japan and their furtherance through MAP training, see: *Foreign Assistance Act of 1965 Hearings*, House, 368.

38. Finer, *Man on Horseback*, pp. 153, 159. Riddleberger, *Military Roles*, p. 68.

39. Gutteridge, *Armed Forces.*

40. Most of the 7.5+ million dollars in equipment grant aid was delivered during the FY 1957-FY 1960 period. Since then, grants have been largely for training of which more than $1,150,000.00 was provided under attache supervision between FY 1957 and FY 1967. *Foreign Assistance Appropriations for 1967, Hearings*, House, p. 638.

41. Bell, *Military Assistance. Foreign Assistance Appropriations for 1966, Hearings*, House.

42. Bell, *Military Assistance.*

43. Ibid.

44. Gutteridge, *Armed Forces*. Sutton and Kemp, *Arms to Developing Countries*, p. 14. *Foreign Assistance Act of 1965, Hearings*, House, p. 368.

45. Bell, *Military Assistance.*

46. The following are typical executive justifications for military aid before Congress: "They have been very constructive in the arguments in the U.N. over the Congo and other events in Africa. Although Morocco has in the past obtained military assistance from the East, she has increasingly been moving closer to the United States [censored]." *Foreign Assistance Act of 1965, Hearings*, House, pp. 56, 1194, 1196. Bell, *Military Assistance*. Sutton and Kemp, *Arms to Developing Countries*, p. 5.

47. Gutteridge, *Armed Forces*. Michael Brecher, *The New States of Asia* (London: Oxford University Press, 1963), p. 155. *Foreign Assistance Appropriations for 1966, Hearings*, House, pp. 278, 402. *Foreign Assistance Appropriations for 1968, Hearings*, House, p. 366. *Foreign Assistance Act of 1969, Hearings*, House, p. 617.

48. Bell, *Military Assistance.*

49. Gutteridge, *Armed Forces*. Bell, *Military Assistance*. Nelkin, "Economic and Social Setting," pp. 236-37. Hoagland, "Arms in the Developing World," p. 175. *Foreign Assistance Act of 1969, Hearings*, House, p. 576.

50. In justifying continued military aid to the Philippines, Admiral Sharp (commander-in-chief, Pacific) referred to "the firmly pro-Western Philippine Armed Forces." *Foreign Assistance Act of 1965, Hearings*, House, p. 839.

51. Bell, *Military Assistance*. Welch, "Soldier and State in Africa," p. 318.

52. Ex-King Saud claimed that the regime which overthrew him was controlled by the CIA. *New York Times*, April 27, 1967, p. 1. *Foreign Assistance Appropriations for 1966, Hearings*, House, pp. 256-278.

53. Bell, *Military Assistance. Foreign Assistance Act of 1965, Hearings*, House, p. 812. *Foreign Assistance Appropriations for 1966, Hearings*, House, p. 278.

54. Gutteridge, *Armed Forces.*

55. Bell, *Military Assistance*, p. 7. *Foreign Assistance Act of 1969, Hearings*, House, p. 442.

56. "The training proposed will strengthen the favorable attitude of the Tunisian armed forces toward the free world. . . ." *Foreign Assistance Appropriations for 1966, Hearings*, House, p. 279.

57. Ten thousand U.S. officers and enlisted men were stationed in Turkey during FY 1969, but only 1,000 of them were assigned to NATO units. *Foreign Assistance Act of 1969, Hearings*, House, p. 731. *Foreign Assistance Appropriations for 1970, Hearings*, House, p. 612.

58. Lee, *African Armies*, pp. 105, 117.

59. *Foreign Assistance Appropriations for 1967, Hearings*, House, p. 637. Nelkin, "Economic and Social Setting," pp. 236-37.

60. Thayer, *War Business*, p. 327. Sutton and Kemp, *Arms to Developing Countries*, p. 5. On the political objectives of U.S. military training, see: *Foreign Assistance Act of 1965, Hearings*, House, p. 740; *Foreign Assistance Appropriations for 1966, Hearings*, House, pp. 279, 406-07. Forty-two thousand dollars or about ten to fifteen training spaces were programmed for FY 1967. *Foreign Assistance Appropriations for 1967, Hearings*, House, p. 690.

61. *Foreign Assistance Act of 1969, Hearings*, House, p. 442.

Bibliography

Books and Monographs

Arcilesi, Salvatore Alfred. "The Development of United States Foreign Policy in Iran: 1949-1960." Unpublished Ph.D. dissertation, University of Virginia, 1965.

Barnet, Richard J. *Intervention and Revolution*. New York: World, 1968.

Be'eri, Eliezer. *Army Officers in Arab Politics and Society*. New York: Praeger, 1970.

Bell, M.J.V. *Military Assistance to Independent African States*. London: Institute for Strategic Studies, December 1964.

Center for Research in Social Systems. *CRESS Work Program, Fiscal Year 1968*. Washington, D.C.: CRESS, American University, n.d.

Cleveland, Harlan, Gerard J. Mangone, and John Clark Adams. *The Overseas Americans*. New York: McGraw Hill, 1960.

Draper, Theodore. *The Dominican Revolt: A Case Study in American Policy*. New York: Commentary, 1968.

Dunn, John M. "Military Aid and Military Elites: The Political Potential of American Training and Technical Assistance Programs." Unpublished Ph.D. dissertation, Princeton University, 1961.

Finer, S.E. *The Man on Horseback: The Role of the Military in Politics*. New York: Praeger, 1962.

Fox, William, T.R. "Military Representation Abroad." In *The Representation of the United States Abroad*. The American Assembly, ed. New York: Graduate School of Business, Columbia University, 1956.

Frank, Andre Gunder. *Capitalism and Underdevelopment in Latin America*. New York: Monthly Review Press, 1966.

Fulbright, J.W. *The Pentagon Propaganda Machine*. New York: Liveright, 1970.

Furniss, Edgar S. Jr. *Some Perspectives on American Military Assistance*. Princeton, N.J.: Princeton University, June 18, 1957.

Gardner, Lloyd C. *Economic Aspects of New Deal Diplomacy*. Madison: University of Wisconsin Press, 1964.

Gimbel, John. *The American Occupation of Germany: Politics and the Military, 1945-1949*. Stanford, Calif.: Stanford University Press, 1968.

Global Defense. Washington, D.C.: Congressional Quarterly Service, 1969.

Goff, Fred, and Michael Locker. "The Violence of Domination: U.S. Power and the Dominican Republic." *Latin American Radicalism: A Documentary Report on Left and Nationalist Movements*. Irving Louis Horowitz, Josue de Castro, and John Gerassi, eds. New York: Random-Vintage, 1969.

Goss, Hilton P. *Africa Present and Potential: A Strategic Survey of the Role of the African Continent in United States Defense Planning*. Santa Barbara,

Calif.: General Electric Company Technical Military Planning Operation, RM57TMP-8, December 31, 1958.

Gutteridge, William F. *Armed Forces in New States*. London: Oxford University Press, 1962.

———. *The Military in African Politics*. London: Methuen, 1969.

———. *Military Institutions and Power in the New States*. New York: Praeger, 1965.

Harvey, T.G. *Political Socialization Research: Problems and Prospects*. Paper for presentation at the annual meeting of the Canadian Political Science Association, McGill University, Montreal, Quebec, June 1972.

Hoehn, Arthur J. *The Design of Cross-Cultural Training for Military Advisors*. Alexandria, Va.: Human Resources Research Office, George Washington University, Operating under contract with the Department of the Army, December 1966.

Holcombe, John L., and Alan D. Berg. *MAP for Security: Military Assistance Programs of the United States*. Columbia: Bureau of Business and Economic Research, University of South Carolina, April 1957.

Horowitz, David. *The Free World Colossus* New York: Hill & Wang, 1965.

Horowitz, Irving Louis. "The Military Elites." *Elites in Latin America*. Seymour M. Lipset and Aldo Solari, eds. New York: Oxford University Press, 1967.

Hovey, Harold A. *United States Military Assistance: A Study of Policies and Practices*. New York: Praeger, 1965.

Huntington, Samuel P. *Political Order in Changing Societies*. New Haven: Yale University Press, 1968.

Hurewitz, J.C. *Middle East Politics: The Military Dimension*. New York: Praeger, 1969.

Institute of International Education. *Military Assistance Training Programs of the U.S. Government*. New York: IIE, Committee on Educational Interchange Policy Statement No. 18, July 1964.

———. *Twenty Years of United States Government Programs in Cultural Relations*. New York: IIE, Committee on Educational Interchange Policy, 1959.

Intelligence and Foreign Policy. Cambridge, Mass: Africa Research Group, 1971. 38pp.

Janowitz, Morris. *The Military in the Political Development of New Nations*. Chicago: University of Chicago Press, 1964.

Johnson, John J., ed. *The Role of the Military in Underdeveloped Countries*. Princeton, N.J.: Princeton University Press, 1962.

———. "The Soldier as Citizen and Bureaucrat." *Latin American Politics: Twenty-Four Studies of the Contemporary Scene*. Robert D. Tomasek, ed. Garden City, N.Y.: Doubleday-Anchor, 1966.

Jordan, Col. Amos A. *Foreign Aid and the Defense of Southeast Asia*. New York: Praeger, 1962.

Kaplan, Stephen S. *United States Military Aid to Brazil and the Dominican Republic: Its Nature, Objectives and Impact*. Paper presented at the Symposium on Arms Control, Military Aid, and Military Rule in Latin America at the University of Chicago Arms Control and Foreign Policy Seminar, May 26-27, 1972.

Khadduri, Majid. *Republican Iraq*. London: Oxford University Press, 1969.

Kling, Merle. "Toward a Theory of Power and Political Instability in Latin America." *The Dynamics of Change in Latin American Politics*. John D. Martz, ed. Englewood Cliffs, N.J.: Prentice-Hall, 1965, pp. 130-39.

———. "Violence and Politics in Latin America." *Latin American Sociological Studies*. Paul Halmos, ed. United Kingdom: Keele University 1967, pp. 119-32.

Kolko, Gabriel. *The Politics of War: The World and United States Foreign Policy, 1943-1945*. New York: Random House, 1968.

———. *The Roots of American Foreign Policy: An Analysis of Power and Purpose*. Boston: Beacon Press, 1969.

Kurth, James R. *United States Policy and Latin American Politics: Competing Theories and Comparative Analysis*. Paper prepared for Symposium on Arms Control, Military Aid, and Military Rule in Latin America at the University of Chicago Arms Control and Foreign Policy Seminar, Center for Policy Studies, May 26-27, 1972.

Kutzger, Joseph Peter. "The Military Assistance Program: Symphysis of United States Foreign and Military Policies." Unpublished Ph.D. dissertation, University of Colorado, 1961.

LaFeber, Walter. *America, Russia, and the Cold War, 1945-1966*. New York: Wiley, 1967.

Lee, J.M. *African Armies and Civil Order*. New York: Praeger, 1969.

Lieuwen, Edwin. *Arms and Politics in Latin America*. New York: Praeger, 1961.

———. "The Changing Role of the Armed Forces: An Analysis." *Latin American Politics: Twenty-Four Studies of the Contemporary Scene*. Robert D. Tomasek, ed. Garden City, N.Y.: Doubleday-Anchor, 1966.

———. *Generals vs. Presidents*. New York: Praeger, 1964.

Lyons, Gene M. and Louis Morton. *Schools for Strategy: Education and Research in National Security Affairs*. New York: Praeger, 1965.

Magdoff, Harry. *The Age of Imperialism*. New York: Monthly Review Press, 1969.

Masland, John W. and Laurence I. Radway. *Soldiers and Scholars: Military Education and National Policy*. Princeton: Princeton University Press, 1957.

Mason, Edward S. *Foreign Aid and Foreign Policy*. New York: Harper & Row, 1964.

Nirumand, Bahman. *Iran: The New Imperialism in Action*. New York: Monthly Review Press, 1967.

Nun, Jose. "A Latin American Phenomenon: The Middle-Class Military Coup."

Latin America: Reform or Revolution? James Petras and Maurice Zeitlin, eds. New York: Fawcett Publications, 1968, pp. 148-85.

Pancake, Frank Robbins. "Military Assistance as an Element of U.S. Foreign Policy in Latin America, 1950-1968. Unpublished Ph.D. dissertation, University of Virginia, 1969.

Ra'anan, Uri. *The USSR Arms the Third World: Case Studies in Soviet Foreign Policy*. Cambridge: MIT Press, 1969.

Riddleberger, Peter B. *Military Roles in Developing Countries: An Inventory of Past Research and Analysis*. Washington, D.C.: Special Operations Research Office, AD 463188, American University, 1965.

Rockefeller, Nelson A. *The Rockefeller Report on the Americas: The Official Report of a United States Presidential Mission for the Western Hemisphere*. Chicago: Quadrangle, 1969.

Shelburne, James C., and Kenneth J. Groves. *Education in the Armed Forces* New York: Center for Applied Research in Education, 1965.

Sellers, Robert C., ed. *Armed Forces of the World*. Garden City, N.Y.: Robert C. Sellers Associates, 2d ed., 1968.

Shaw, Lawrence Joseph. "The United States Military Assistance Program: A Study in Executive-Legislative Relationships," Unpublished Ph.D. dissertation, American University, 1969.

Simons, William E. *Liberal Education in the Service Academies*. New York: Teachers College of Columbia University, 1965.

Sutton, John L., and Geoffrey Kemp. *Arms to Developing Countries*. London: Institute for Strategic Studies, 1966.

Thayer, George. *The War Business: The International Trade in Armaments.* New York: Simon and Schuster, 1969.

The University-Military Complex: A Directory and Related Documents. New York: North American Congress on Latin America, 1969.

Vagts, Alfred. *A History of Militarism: Civilian and Military*. New York: 1967.

Walterhouse, Harry F. *A Time to Build: Military Civic Action as a Medium for Economic Development and Social Reform*. Columbia: University of South Carolina Press, 1964.

Weaver, Ferry L. *Assessing the Impact of Military Rule: Alternative Approaches*. Paper presented at the Symposium on Arms Control, Military Aid and Military Rule in Latin America at the University of Chicago, May 26-27, 1972.

Williams, Benjamin H., and Harold J. Clem. *Mutual Security*. Washington, D.C.: U.S. Industrial College of the Armed Forces, 1964.

Wolf, Charles Jr. *Military Assistance Programs*. Santa Monica, Calif.: Rand Corporation, P-3240, October 1965.

Wood, David. *Armed Forces in Central and South America*. London: Institute for Strategic Studies, 1967.

Wynfred, Joshua, and Stephen P. Gibert. *Arms for the Third World: Soviet Military Aid Diplomacy*. Baltimore: Johns Hopkins Press, 1969.

Periodicals: Military and Civilian

"Army Seeks FAS Officers." *Armed Forces Journal*, 16 May 1970, p. 14.

Barnett, Frank R. "A Proposal for Political Warfare." *Military Review* 41 (March 1961), 2-10.

Bauer, Maj. Charles J., USA. "USACARIB's Biggest Little School." *Army Digest* 17 (October 1962), 24-28.

Bloomfield, L.P. and A.C. Leiss, "Arms Transfers and Arms Control." *Proceedings of the American Academy of Political Science* 29 (March 1969) 37-54.

Bogart, Maj. Gen. Theodore F., USA. "The Army's Role in Latin America." *Army* 12 (October 1961), 61-64.

——. "The United States Army Caribbean Command and Operation Friendship." *Army Digest* 17 (October 1962), 22-23.

Bondshu, Col. Lowell T. "Keeping Pace with the Future–Toward Mutual Security." *Military Review* 39 (June 1959), 58-73.

Brewster, Col. Margaret, USA. "Foreign Military Training Bolsters Free World," *Army Digest* 20 (May 1965) 16-20.

Bwy, D.P. "Political Instability in Latin America: The Cross-Cultural Test of a Causal Model." *Latin America Research Review* 3 (Spring 1968), 17-66.

Campbell, Capt. R.R., Sc, USN. "Progress and Problems in the War Colleges." *U.S. Naval Institute Proceedings* 94 (September 1968), 52-9.

Chang, David W. "Military Forces and Nationbuilding." *Military Review* 50 (September 1970), 78-88.

——. "The Military and Nationbuilding in Korea, Burma, and Pakistan." *Asian Survey* 9 (November 1969), 818-30.

C. Clapham. "The Ethiopian Coup d'Etat of December 1960." *Journal of Modern African Studies* 6 (December 1968), 495-507.

Clarke, Richard D. "U.S. Military Assistance in Latin America." *Army Digest* 21 (September 1966), 18-19.

Clinton, Richard Lee. "The Modernizing Military: The Case of Peru." *Inter-American Economic Affairs* 24 (Spring 1971), 43-66.

Clotfelter, James. "The African Military Forces." *Military Review* 48 (May 1968), 23-31.

Davison, W. Phillips. "Political Communication as an Instrument of Foreign Policy." *Public Opinion Quarterly* 27 (Spring 1963), 28-36.

Deats, Capt. John R. "Bridge of the Americas." *Army Digest* 23 (September 1968), 12-14.

"Defense College of the Americas." *Army Digest* 19 (May 1964), 36-39.

Dennon, A.R. "Political Science and Political Development." *Science & Society* 33 (Summer-Fall 1969), 285-98.

Dower, John. "The Eye of the Beholder: Background Notes on the US-Japan Military Relationship." *Bulletin of Concerned Asian Scholars* 2 (October 1969), 16-28.

Drake, St. Clair. "Nkrumah: The Real Tragedy." *Nation*. June 5, 1972, pp. 721-24.

Dunn, Maj. Matthew T. and Maj. James B. Jones, USAF. "Preventive Medicine Civic Action Training Program." *Air University Review* 18 (November-December 1966), 21-25.

Dupree, L. "Democracy and the Military Base of Power." *Middle East Journal* 22 (Winter 1968), 29-44.

Eckhardt, W., "The Factor of Militarism." *Journal of Peace Research* 2 (1969) 123-32.

——, and A.G. Newcombe. "Militarism, Personality, and other Social Attitudes," *Journal of Conflict Resolution* 13 (June 1969), 210-19.

Eckstein, Harry. "On the Etiology of Internal War." *History and Theory*. 4 (1965), 133-63.

Eisenstadt, S.N. "Breakdowns of Modernization." *Economic Development and Cultural Change* 12 (July 1964), 345-67.

Farrington, Capt. Robert F. "Military Assistance at the Crossroads." *U.S. Naval Institute Proceedings* 92 (June 1966), 68-76.

Fitzgibbon, Russell H. "What Price Latin American Armies?" *Virginia Quarterly Review* 36 (Autumn 1960), 517-32.

Gailer, Col. Frank L., Jr., USAF. "Air Force Missions in Latin America." *Air University Quarterly Review* 13 (Fall 1961), 45-58.

Gibert, Stephen P. "Soviet-American Military Aid Competition in the Third World." *Orbis* 13 (Winter 1970), 1117-31.

Gibson, Col. Baylor, USMC. "The Universities: Trial by Fire." *U.S. Naval Institute Proceedings* 96 (March 1970), 23-29.

Goshko, John M. "Latins Blame the United States for Military Coups." *Washington Post*, February 5, 1968.

Gottlieb, Dan. "Military Dictatorships: Why Rockefeller's Wrong." *The Washington Monthly*, May 1970, pp. 63-68.

Graebner, Norman A. "Global Containment: The Truman Years." *Current History* 57 (August 1969), 87-83ff.

Grand Pre, Donn R. "A Window on America: The Department of Defense Informational Program." *International Educational and Cultural Exchange* 6 (Fall 1970), 86-93.

Grant, Zalin B. "Training, Equipping the Latin American Military." *New Republic*, December 16, 1967, pp. 13-14.

Grayson, George W., Jr. "The Era of the Good Neighbor." *Current History* 56 (June 1969), 327-32ff.

Grundy, Kenneth W., and Robert L. Farlow. "Internal Sources of External Behavior: Ghana's New Foreign Policy." *Journal of Asian and African Studies* 4 (July 1969) 161-71.

Guelzo, Carl M. "The Higher Level Staff Advisor." *Military Review* 47 (February, 1967).

Hahr, James C. "Military Assistance to Latin America." *Military Review* 49 (May 1969), 13-20.

Halliday, Fred. "Class Struggle in the Arab Gulf." *New Left Review* no. 58 (November-December 1969).

Hanks, Capt. Robert J., USN. "Against All Enemies." *U.S. Naval Institute Proceedings* 96 (March 1970), 23-29.

Hanson, Simon G. "Third Year: Military." *Inter-American Economic Affairs* 18 (Spring 1965), 20-37.

Hays, S.H. "Military Training in the U.S. Today." *Current History* 55 (July 1968), 7-12, 50-51.

Heymont, Irving. "U.S. Military Assistance Programs." *Military Review* 48 (January 1968), 89-95.

Hochman, Harold M. and C. Tait Ratcliffe. "Grant Aid or Credit Sales: A Dilemma of Military Assistance Planning." *Journal of Developing Areas* 4 (July 1970).

Hughes, Lt. Col. David R., USA. "The Myth of Military Coups and Military Assistance." *Military Review* 47 (December 1967), 3-10.

I.F. Stone's By-Weekly, January 25, 1971.

"Increased Foreign Military Sales Inescapable under Nixon Doctrine." *Armed Forces Journal*, 2 November 1970, pp. 22-25, 44.

"Inter-American Defense College." *Military Review* 50 (April 1970), 20-7.

Izzett, R.R. "Authoritarianism and Attitudes toward the Vietnam War as Reflected in Behavioral and Self-Report Measures." *Journal of Personality and Social Psychology* 17 (February 1971), 145-48.

Johnson, Gen. Harold K., Chief of Staff, USA. "The Army's Role in Nation Building and Preserving Stability." *Army Digest* 20 (November 1965), 7-12.

Johnson, Kenneth F. "Causal Factors in Latin American Political Stability." *Western Political Quarterly* 17 (Summer 1964), 432-46.

Jones, Capt. Richard A., USA. "The Nationbuilder: Soldier of the Sixties." *Military Review* 45 (January 1965), 63-67.

Jordan, Col. Amos A. Jr., "Military Assistance and National Policy." *Orbis* 11 (Summer 1958), 236-44.

Karch, Brig. Gen. F.J., USMC. "Marine Corps Command and Staff College." *Marine Corps Gazette* 51 (June 1967), 27-30.

Khadduri, Majid. "The Role of the Military in the Middle East." *American Political Science Review* 47 (June 1953), 511-24.

King, Col. Edward L., USA, Ret. "Making It in the U.S. Army." *New Republic*, May 30, 1970, pp. 20-22.

Kintner, Col. William R., USA. "The Role of Military Assistance." *U.S. Naval Institute Proceedings* 87 (March 1961), 76-83.

Klare, Michael. "U.S. Arms Sales to the Third World: Arm Now Pay Later." *NACLA's Latin America & Empire Report* 6 (January 1972), 2-13.

———. "U.S. Military Operations in Latin America." *NACLA Newsletter* 2 (October 1968), p. 1-8.

Langley, Lester D. "Military Commitments in Latin America: 1960-1968." *Current History* 56 (June 1969), 346-51.

LaPalombara, J. "Macrotheories and Microapplications in Comparative Politics: A Widening Chasm." *Comparative Politics* 1 (October 1968), 52-78.

Lapham, Lewis H. "Military Theology: The Ethos of Officer Class." *Harper's Magazine*, July 1971, pp. 73-85.

Lieuwen, Edwin, "Militarism in Latin America: A Threat to the Alliance for Progress." *World Today* 19 (May 1963), 193-99.

____. "Neo-Militarism in Latin America: The Kennedy Administration's Inadequate Response." *Inter-American Economic Affairs* 16 (Spring 1963), 11-19.

Lincoln, Col. G.A. "Factors Determining Arms Aid." *Academy of Political Science Proceedings* 25 (1954), 263-72.

Lynch, John E. "Analysis for Military Assistance." *Military Review* 48 (May 1968), 41-9.

Marchetti, Victor. "CIA" The President's Loyal Tool." *Nation* (April 3, 1972), pp. 430-32.

Middleton, Drew. "Thousands of Foreign Military Men Studying in U.S." *New York Times*, November 1, 1970, p. 18.

Miksche, F.O. "Arms Race in the Third World." *Orbis* 12 (Spring 1968), 161-6.

"Military Notes: Cadet Exchange Program." *Military Review* 47 (April 1967), 100.

Millington, Thomas M. "The Latin American Military Elite." *Current History* 56 (June 1969), 352-4f.

Morton, Dr. A. Glenn. "The Inter-American Air Forces Academy." *Air University Review* 18 (November-December 1966), 13-20.

Munoz, Luis Joaquin. "Golpos de Estado en Africa: El Caso de Mali." *Revista de Politica Internacional* no. 106 (Noviembre-Diciembre 1969), 161-73.

Needler, Martin C. "The Latin American Military: Predatory Reactionaries or Modernizing Patriots?" *Journal of Inter-American Studies* 11 (April 1969), 237-44.

____. "Political Development and Military Intervention in Latin America." *American Political Science Review* 60 (September 1966), 617-5.

Nelkin, Dorothy. "The Economic and Social Setting of Military Takeovers in Africa." *Journal of Asian and African Studies* 2 (January-April 1967), 230-44.

Neufville, Richard de. "Education at the Academies." *Military Review* 47 (May 1967), 6-7.

Nielsen, Waldemar A. "Huge Hidden Impact of the Pentagon." *New York Times Magazine*, June 25, 1961, pp. 9ff.

Nordlinger, Eric A. "Soldiers in Mufti: The Impact of Military Rule upon Economic and Social Change in the Non-Western States." *American Political Science Review* 64 (December 1970), 1131-48.

O'Meara, Lt. Gen. Andrew P. "Opportunities to the South of Us." *Army* 13 (November 1962), 41-3.

Overstreet, Lewis D. "Strategic Implications of the Developing Areas." *Military Review* 46 (October 1966), 70-77.

Packenham, Robert A. "Political Development Doctrines in the American Foreign Aid Program." *World Politics* 18 (January 1966), 194-233.

Palmer, L. "The 30 September Movement in Indonesia." *Modern Asian Studies* 5 (January 1971), 1-20.

Palmer, Gen. W.B. "Military Assistance Program: A Progress Report." *Army Information Digest* 17 (April 1962), 40-6.

Peers, Maj. Gen. W.R., Asst. Deputy Chief of Staff for Special Operations. "Meeting the Challenge of Subversion." *Army* 15 (November 1964), 95-97, 158.

Perlmutter, Amos. "The Arab Military Elite." *World Politics* 22 (January 1970), 269-300.

———. "From Obscurity to Rule: The Syrian Army and the Baath Party." *Western Political Quarterly* 22 (December 1969), 827-45.

Porter, Gen. Robert W. "Look South to Latin America." *Military Review* 48 (June 1968), 82-90.

Powell, John Duncan, "Military Assistance and Militarism in Latin America." *Western Political Quarterly* 18 (June 1965), 382-92.

Price, Robert M. "A Theoretical Approach to Military Rule in New States: Reference-Group Theory and the Ghanaian Case." *World Politics* 23 (April 1971), 399-430.

Prouty, L. Fletcher. "The Secret Team and the Games They Play." *The Washington Monthly* May 1970, pp. 11-19.

Rapoport, D.C. "Political Dimensions of Military Usurpation." *Political Science Quarterly* 83 (December 1968), 551-72.

Rhodes, Robert I. "The Disguised Conservatism in Evolutionary Development Theory." *Science & Society* 32 (Fall 1968), 383-412.

Robertson, Lt. Col. Frank H. "Military Assistance Training Program." *Air University Review* 19 (September-October 1968), 27-36.

Rosser, Col. Richard F. "The Air Force Academy and the Development of Area Experts." *Air University Review* 20 (November-December 1968), 26-32.

Sampson, Lt.Col. Donald E. "Military Missionaries." *Army Digest* 22 (April 1967), 44-46.

Sanborn, Maj. Gen. Kenneth O., USAF. "United States Air Force's Southern Command." *Air Force and Space Digest* 51 (September 1968), 139-40.

"The School of the Americas Shows How Armies Can Be Builders." *Army Digest* 20 (February 1965), 16-19.

Selover, William C. "U.S. Food for Peace Converted to Armament." *Christian Science Monitor*, January 8, 1971.

Siegel, Lenny. "The Future of Military Aid," *Pacific Research & World Empire Telegram* 3 (November-December 1971), 1-8.

Singh, L.P. "Political Development or Political Decay in India." *Pacific Affairs* 44 (Spring 1971), 65-80.

Smith, Col. Stilson H. "Assignment to MAAG: How to Win Friends and Influence People." *Army* 12 (July 1962), 57-9.

Stevenson, Charles S. "The Far East MAAGs: Good Investment in Security." *Army* 11 (November 1960), 60-63.

Stewart, Edward C. "American Advisors Overseas." *Military Review* 41 (February 1965), 3-9.

Stillman, Col. Richard J., USMC. "The Inter-American Defense College." *Marine Corps Gazette* 49 (September 1965), 24-6.

Swett, Maj. Trevor. W. Jr. "Assignment: Latin America: Duty: Army Advisor; Mission: Be Ever Helpful." *Army* 14 (October 1961).

Szulc, Tad. "Influence of the U.S. Military is Expected to Wane in Latin America." *New York Times*, November 1, 1970, p. 18.

Talbot, Maj. Gen. Orwin C. "Where Nobody Ever Rests." *Army*, March 1970, pp. 35-41.

Tiffany, 1st Lt. James M., USA. "Training Foreign Students." *Ordnance* 50 (January-February 1966), 439.

Tomasek, Robert D. "Defense of the Western Hemisphere: A Need for Reexamination of United States Policy." *Midwest Journal of Political Science* 3 (November 1959), 374-78.

Tsoucalas, Constantine. "Class Struggle and Dictatorship in Greece." *New Left Review* no. 56 (July-August 1969) 7-11.

"US Army School of the Americas." *Military Review* 50 (April 1970), 88-93.

Vallance, Theodore R., and Charles D. Windle. "Cultural Engineering." *Military Review* 42 (December 1962), 60-4.

Van Der Kroef, J.M. "Interpretations of the 1965 Indonesian Coup: A Review of the Literature," *Pacific Affairs* 43 (Winter 1970-71) 557-77.

Van Rees, Cmdr. E.H., R. Neth. "Reflections of a Foreign Student." *U.S. Naval Institute Proceedings* 86 (February 1960), 36-40.

Victor, Col. A.H., Jr., USA."Military Aid and Comfort to Dictatorships." *U.S. Naval Institute Proceedings* 95 (March 1969), 43-7.

Walterhouse, Lt. Col. Harry F., USA. "Civic Action: A Counter and Cure for Insurgency." *Military Review* 42 (August 1968), 47-54.

———. "Good Neighbors in Uniform." *Military Review* 45 (February 1965), 10-21.

Welch, C.E. "Soldier and State in Africa." *Journal of Modern African Studies* 5 (November 1967), 305-22.

"Where the Soldiers Learn to Rule." *Brazilian Information Bulletin*, May 1971, pp. 5-7.

Whetten, Lawrence L. "Strategic Parity in the Middle East." *Military Review* 50 (September 1970), 24-31.

"Why U.S. Military Assistance is Given to the Argentine Dictatorship: The Explanation of the U.S. Department of Defense." *Inter-American Economic Affairs* 21 (Autumn 1967), 81-89.

Wilder, Maj. Gary, USMC. "M.A.P. Orientation." *Marine Corps Gazette* 49 (February 1965), 37-40.

Williams, Lt. Gen. Samuel T., USA, Ret. "The Practical Demands of MAAG." *Military Review* 41 (July 1961), 2-14.

Wilkins, Col. Wallace W. Jr., USA. "You are What You Say." *Military Review* 48 (June 1968), 94-95.

Windle, Charles D., and Theodore R. Vallance. "Optimizing Military Assistance Training." *World Politics* 15 (October 1962), 91-107.

Wolf, Charles. "The Political Effects of Military Programs: Some Indications from Latin America." *Orbis* 8 (Winter 1965), 871-93.

Wriggins, H. "Political Outcomes of Foreign Assistance." *Journal of International Affairs* 22 (1968), 217-30.

Congressional Hearings and Reports

U.S. Congress. House. Committee on Appropriations. *Foreign Assistance and Related Agencies Appropriations for 1966. Hearings* before a subcommittee of the Committee on Appropriations, House of Representatives, 89th Cong., 1st sess.

____. House. Committee on Appropriations. *Foreign Assistance and Related Agencies Appropriations for 1967. Hearings* before a subcommittee of the Committee on Appropriations, House of Representatives, 89th Cong., 2d sess.

____. House. Committee on Appropriations. *Foreign Assistance and Related Agencies Appropriations for 1968. Hearings* before a subcommitee of the Committee on Appropriations, House of Representatives, 90th Cong., 1st sess.

____. House. Committee on Appropriations. *Foreign Assistance and Related Agencies Appropriations for 1969. Hearings* before a subcommittee of the Committee on Appropriations, House of Representatives, 90th Cong., 2d sess.

____. House. Committee on Appropriations. *Foreign Assistance and Related Agencies Appropriations for 1970. Hearings* before the Subcommittee on Foreign Operations and Related Agencies, House of Representatives, 91st Cong., 1st sess.

____. House. Committee on Appropriations. *Foreign Operations Appropriations for 1965. Hearings* before a subcommittee of the Committee on Appropriations, House of Representatives, 88th Cong., 2d sess.

____. House. Committee on Foreign Affairs. *Castro-Communist Subversion in the Western Hemisphere*. Hearings before the Subcommittee on Inter-American Affairs of the Committee on Foreign Affairs, House of Representatives, 88th Cong., 1st sess.

____. House. Committee on Foreign Affairs. *Cuba and the Caribbean. Hearings* before the Subcommittee on Inter-American Affairs of the Committee on

Foreign Affairs, House of Representatives, 91st Cong., 2d sess., July 8-August 3, 1970.

U.S. Congress. House. Committee on Foreign Affairs. *To Amend the Foreign Military Sales Act. Hearings* before the Committee on Foreign Affairs, House of Representatives, 91st Cong., 2d sess.

____. House. Committee on Foreign Affairs. *Foreign Assistance Act of 1963. Hearings* before the Committee on Foreign Affairs, House of Representatives, 88th Cong., 1st sess.

____. House. Committee on Foreign Affairs. *Foreign Assistance Act of 1964. Hearings* before the Committee on Foreign Affairs, House of Representatives, 88th Cong., 2d sess.

____. House. Committee on Foreign Affairs. *Foreign Assistance Act of 1965. Hearings* before the Committee on Foreign Affairs, House of Representatives, 89th Cong., 1st sess.

____. House. Committee on Foreign Affairs. *Foreign Assistance Act of 1966. Hearings* before the Committee on Foreign Affairs, House of Representatives, 89th Cong., 2d sess.

____. House. Committee on Foreign Affairs. *Foreign Assistance Act of 1968. Hearings* before the Committee on Foreign Affairs, House of Representatives, 90th Cong., 2d sess.

____. House. Committee on Foreign Affairs. *Foreign Assistance Act of 1969. Hearings* before the Committee on Foreign Affairs, House of Representatives, 91st Cong., 1st sess.

____. House. Committee on Foreign Affairs. *Foreign Assistance Act of 1972. Hearings* before the Committee on Foreign Affairs, House of Representatives, 92nd Cong., 2nd sess., March 1972.

____. House. Committee on Foreign Affairs. *Winning the Cold War: The U.S. Ideological Offensive*. Hearings before the Subcommittee on International Organizations and Movements of the Committee on Foreign Affairs, House of Representatives, 88th Cong., 2d sess.

____. House. Committee on Foreign Affairs. *Foreign Assistance and Related Programs Appropriations Bill, 1970. Report*, No. 91-708, together with Supplemental and Additional Views to Accompany H.R. 15149, House of Representatives, 91st Cong., 1st sess.

____. House. Committee on Foreign Affairs. *Military Assistance Training in East and Southeast Asia. A Staff Report* prepared for the use of the Subcommittee on National Security Policy and Scientific Developments of the Committee on Foreign Affairs, House of Representatives, 91st Cong., 2d sess.

____. House. Committee on Foreign Affairs. *Military Assistance Training. Hearings* before the Subcommittee on National Security Policy and Scientific Developments of the Committee on Foreign Affairs, House of Representatives, 91st Cong., 2d sess., October-December, 1970.

____. House. Committee on Foreign Affairs. *Military Assistance Training. Report* of the Subcommittee on National Security Policy and Scientific Developments to the Committee on Foreign Affairs, House of Representatives, April 2, 1971.

____. House. Committee on Foreign Affairs. *Report of a Special Study Mission To West and Central Africa.* March 29-April 27, 1970. By Hon. Charles C. Diggs, Chairman, Subcommittee on Africa, Committee on Foreign Affairs, House of Representatives, 91st Cong., 2d sess.

____. House. Committee on Foreign Affairs. *Reports of the Special Study Mission to Latin America on Military Assistance Training and Developmental Television.* Submitted by Hon. Clement J. Zablocki, Hon. James G. Fulton and Hon. Paul Findley to the Subcommittee on National Security Policy and Scientific Developments of the Committee on Foreign Affairs, House of Representatives, 91st Cong., May 7, 1970.

____. House. *Conference Report on Foreign Assistance Act of 1969* Report No. 91-767, House of Representatives, 91st Cong., 1st sess.

____. House. *Report No. 91-708 on Foreign Assistance and Related Programs Appropriation Bill, 1970* together with Supplemental and Additional Views to Accompany H.R. 15149, 91st Cong., 1st sess., 1969.

____.Joint Economic Committee. *Economic Issues in Military Assistance. Hearings* before the Subcommittee on Economy in Government of the Joint Economic Committee, Congress of the United States, 92d Cong., 1st sess., January-February 1971.

____. Senate. Committee on the Armed Services. *Military Cold War Education and Speech Review Policies. Hearings* before the Subcommittee on Special Preparedness of the Committee on Armed Services, United States Senate, 87th Cong., 2d sess.

____. Senate. Committee on Appropriations. *Foreign Assistance and Related Agencies Appropriations for 1964. Hearings* before the Committee on Appropriations, United States Senate, 88th Cong., 1st sess.

____. Senate. Committee on Appropriations. *Foreign Assistance and Related Agencies Appropriations for 1965. Hearings* before the Committee on Appropriations, United States Senate, 2d sess.

____. Senate. Committee on Foreign Relations. *Amendments to the OAS Charter. Hearings* before the Committee on Foreign Relations, United States Senate, 90th Cong., 2d sess.

____. Senate. Committee on Foreign Relations. *Rockefeller Report on Latin America. Hearings* before the Subcommittee on Inter-American Affairs of the Committee on Foreign Relations, United States Senate, 91st Cong., November 20, 1969.

____. Senate. Committee on Foreign Relations. *United States Military Policies and Programs in Latin America. Hearings* before the Subcommittee on Western Hemisphere Affairs of the Committee on Foreign Relations, United States Senate, 91st Cong., 1st sess.

U.S. Congress. Senate. Committee on Foreign Relations. *Survey of the Alliance for Progress. Hearings* before the Subcommittee on American Republics Affairs of the Committee on Foreign Relations, United States Senate, 90th Cong., 2d sess. Februrary 27-March 6, 1968.

____. Senate. Committee on Foreign Relations. *Survey of the Alliance for Progress: The Latin American Military, A Study*. Submitted by the Subcommittee on American Republics Affairs to the Committee on Foreign Relations, United States Senate, 90th Cong., 1st sess., October 9, 1967.

____. Senate. Committee on Foreign Relations. *United States Security Agreements and Commitments Abroad: The Republic of the Philippines. Hearings* before the Subcommittee on United States Security Agreements and Commitments Abroad of the Committee on Foreign Relations, United States Senate, 91st Cong., 1st sess.

____. Senate. Committee on Government Operations. *State Department Advisors to Military Training Institutions: Analysis and Assessment*. Submitted by the Subcommittee on National Security and International Operations of the Committee on Government Operations, United States Senate, 91st Cong., 1st sess.

____. Senate. Committee on Foreign Relations. *Foreign Assistance, 1964. Hearings* before the Committee on Foreign Relations, United States Senate, 88th Cong., 2d sess.

____. Senate. Committee on Foreign Relations. *Foreign Assistance, 1965. Hearings* before the Committee on Foreign Relations, United States Senate, 89th Cong., 1st sess.

____. Senate. Committee on Foreign Relations. *Foreign Assistance, 1966. Hearings* before the Committee on Foreign Relations, United States Senate, 89th Cong., 2d. sess.

____. Senate. Committee on Foreign Relations. *Foreign Assistance Act of 1967. Hearings* before the Committee on Foreign Relations, 90th Cong., 1st sess.

____. Senate. *Foreign Assistance Act, 1969. Hearings* before the Committee on Foreign Relations, United States Senate, 91st Cong., 1st sess.

____. Senate. Committee on Foreign Relations. *Foreign Assistance Act of 1969. Report* No. 91, 603 To Accompany H.R. 14580. United States Senate, 91st Cong., 1st sess.

Executive Agency Reports and Publications

U.S. Advisory Commission on International Education and Cultural Affairs. *A Beacon of Hope: The Exchange of Persons Program*. Washington, D.C.: GPO, 1963.

U.S. Air Force University. *Allied Officer Familiarization Course Lecture Outline*. Maxwell Air Force Base, Alabama: May 1966.

U.S. Armed Forces Institute. *Informational Program on American Life and Institutions for Foreign Military Trainees: Guidance to Instructors and Training Officers*. Madison, Wisc.: 1965.

U.S. Army Command and General Staff College. *The Communist Powers: Communist China*. School Year 1969-1970. Lesson Plan Subject R2811-2/0. Fort Leavenworth, Kansas, 1969.

____. *The Communist Powers*. USSR-Lesson 1. R2811-1/0. Fort Leavenworth, Kansas.

____. *Communism in Selected Countries: Bibliography*. R2809/0. Fort Leavenworth, 1969.

____. *Nature of Communism*. Advance Sheet, R1808. Fort Leavenworth, Kansas, 1968.

____. *Nature of Communism*. Command and General Staff Officer Course. Lesson Plan R1808/9. School Year 1968-1969. Fort Leavenworth, Kansas.

____. *Strategic Subjects Handbook: RB 100-1* Fort Leavenworth, Kansas, 1 August 1969.

____. *U.S. Army English for Allied Students*. Ref. Book: RB20-8. Fort Leavenworth, Kansas, 1968.

(U.S. Army.) Escuela de las Americas. Ejercito de los EE. UU. *Como contrals el Comunismo las Ideas del Pueblo*. Folleto No. 4. Pp. 24.

____. Escuela de las Americas. Ejercito de los EE. UU. *Como controla el Comunismo la Vida Economica del Pueblo*. Folleto No. 8.

____. Escuela de las Americas. Ejercito de los EE. UU. *Como Funciona el Partido Comunista*. Folleto No. 5. Pp. 24.

____. Escuela de las Americas. Ejercito de los EE. UU. *Como logran y retienen el Poder los Comunistas*. Folleto No. 6. Pp. 22.

____. Escuela de las Americas. Ejercito de los EE. UU. *Como viven los Obreros y Campesinos bajo el Comunismo*. Folleto No. 9. Pp. 14.

____. Escuela de las Americas. Ejercito de los EE. UU. *Conquista y Colonizacion Comunista*. Folleto No. 10. Pp. 24.

____. Escuela de las Americas. Ejercito de los EE. UU. *Desarrollo de las Tacticas Revolucionarias, II Parte, Seccion I*. Fuerte Gulick, Z.C.: MI 4527, Noviembre 1965.

____. Escuela de las Americas. Ejercito de los EE. UU. *El Atractivo del Comunismo*. Fuerte Gulick, Z.C.: MI 4514, Abril 1964.

____. Escuela de las Americas. Ejercito de los EE. UU. *El Desarrollo de Tacticas Revolucionarias–Primera Parte*. Fuerte Gulick, Z.C.: HA 4526, Septiembre 1965.

____. Escuela de las Americas. Ejercito de los EE. UU. *El Desarrollo de las Tacticas Revolucionarias I*. Fuerte Gulick, Z.C.: MI 4526, Octubre 1965.

____. Escuela de las Americas. Ejercito de los EE. UU. *El Dominio del Partido Comunista en Russia*. Folleto No. 7. Pp. 14.

____. Escuela de las Americas. Ejercito de los EE. UU. *Expansion del*

Comunismo en America Latina. Fuerte Gulick, Z.C.: MI 4531A, Mayo 1968.

(U.S. Army.) Escuela de las Americas. Ejercito de los EE. UU. *Expansion Comunista en Europa Oriental*. Fuerte Gulick, Z.C.: HA 4526, Septiembre 1965.

____. Escuela de las Americas. Ejercito de los EE. UU. *Expansion en Europa: Seccion III*. Fuerte Gulick, Z.C.: HA 4525, Mayo 1964.

____. Escuela de las Americas. Ejercito de los EE. UU. *Ideologia Comunista I*. Fuerte Gulick, Z.C.: MI 4500, Octubre 1965.

____. Escuela de las Americas. Ejercito de los EE. UU. *Ideologia Comunista II*. Fuerte Gulick, Z.C.: MI 4501, Junio 1965. Pp. 40.

____. Escuela de las Americas. Ejercito de los EE. UU. J. Edgar Hoover. *Illusion Comunista y Realidad Democratica: Hoja Avanzada*. Fuerte Gulick, Z.C.: Escuela de las Americas, Marzo 1966. Pp. 17.

____. Escuela de las Americas. Ejercito de los EE. UU. J. Edgar Hoover. *La Respuesta de una Nacion al Comunismo*. Enero 1961.

____. Escuela de las Americas. Ejercito de los EE. UU. *La Democracia contra el Comunismo*. Folleto No. 1. Pp. 17.

____. Escuela de las Americas. Ejercito de los EE. UU. *La Política Exterior de la Union Sovietica: Seccion I*. Fuerte Gulick, Z.C.: MI 4535, Junio 1964. Pp. 40.

____. Escuela de las Americas. Ejercito de los EE. UU. *La Teória del Comunismo*. Fuerte Gulick, Z.C.: HA 5400-2, Abril 1966. Pp. 23.

____. Escuela de las Americas. Ejercito de los EE. UU. Seccion de Inteligencia. *Lectura Suplementaria: Princípios y Practicas del Gobierno Totalitário*. Fuerte Gulick, Z.C.: HA 4600, Febrero 1965. Pp. 25.

____. Escuela de las Americas. Ejercito de los EE. UU. *Partido Comunista de la Unión Soviética*. Fuerte Gulick, Z.C.: MI 4516, Octubre 1965.

____. Escuela de las Americas. Ejercito de los EE. UU. *Plan de Conquista del Comunismo: Seccion I*. Fuerte Gulick, Z.C.: MI 4541, Marzo 1964. Pp. 3.

____. Escuela de las Americas. Ejercito de los EE. UU. *¿Quees el Comunismo*? Folleto No. 2, Pp. 20.

____. Escuela de las Americas. Ejercito de los EE. UU. *¿Que hacen los Comunistas de la Libertad*? Folleto No. 3. Pp. 35.

____. Escuela de las Americas. Ejercito de los EE. UU. *Táctica y Técnica del Comunismo Internacional*. Fuerte Gulick, Z.C.: MI 4530, Marzo 1964.

____. Escuela de las Americas. Comite de Seguridad. Ejercito de los EE. UU. *Teoria, Practica e Instituciones de los Gobiernos Totalitarios*. Fuerte Gulick, Zona del Canal, MI 4606, Enero 1969.

____. Escuela de las Americas. Ejercito de los EE. UU. *Teorias de los Gobiernos Demócraticos, Seccion I*. Fuerte Gulick, Z.C.: MI 4600, Febrero 1965.

____. Joint Colleges: National War College, Industrial College of the Armed Forces, Armed Forces Staff College. *Information Data Sheet No. 49 to the*

Army Information Program. Reference DA Circular 11-1, 29 October 1958.

U.S. Bureau of Naval Personnel. *List of Training Manuals and Correspondence Courses*. NAVPERS 10061-AC. Washington, D.C.: GPO, 1969.

U.S. Department of the Army. *Africa Today: Army Information Support Series*. Washington, D.C.: Pamphlet 355-200-11, February 14, 1961.

——. *Alert: Heritage of Freedom, Facts for the Armed Forces*. Washington, D.C.: DA Pamphlet No. 355-140, February 1, 1963.

——. *Army Information Basic Guidance*. Washington, D.C.: Pamphlet 360-4, June 1964.

——. *Army Information Program: Supporting Materials*. Washington, D.C.: GPO, DA Pamphlet 355-201-1, July 1960.

——. *Counter-Insurgency Operations (Spanish Version): A Handbook for the Suppression of Communist Guerrilla Terrorist Operations* Fort Bragg, North Carolina: U.S. Army Special Warfare School, n.d.

——. *Foreign Countries and Nationals: Training of Foreign Personnel by the U.S. Army*. Washington, D.C.: Headquarters, Dept. of the Army, Army Regulation AR 550-50, October 1970, Effective 1 December 1970.

——. *Industrial College of the Armed Forces*: 1968-1969 Catalog. Fort McNair, Washington, D.C.

——. *Officers' Call: The Soviet Armed Forces*. Washington, D.C.: Pamphlet No. 355-22, January 1956.

——. *Report of the Thirteenth Annual U.S. Army Human Factors Research and Development Conference, October 25-27, 1967*. Fort Monmouth, N.J.: U.S. Army Signal Center & School, 1967.

——. Secretary of War. Army Information Service. *Asi es el Comunismo; Los Agentes Enemigos y Udted: Temas para Tropas*. Pp. 25.

——. *Stability Operations–U.S. Army Doctrine*. Washington, D.C.: Field Manual 31-23, December 1967.

——. *U.S. Army Handbook for Colombia*. Washington, D.C.: DA Pamphlet No. 550-26, 2d ed., June 1964.

——. *U.S. Army Area Handbook for Venezuela* Washington, D.C.: GPO, 1964.

——. *U.S. National Security and the Communist Challenge: The Spectrum of the East-West Conflict*. Washington, D.C.: Pamphlet 20-60, August 11, 1961. Pp. 93.

——. *The United States School of the Americas: 1969 Catalog*. Fort Gulick, Canal Zone.

U.S. Department of Defense. *ALERT: The Truth About Our Economic System, Facts for the Armed Forces*. Washington, D.C.: Armed Forces Information and Education, Dept. of the Army Pamphlet 355-136, May 29, 1962.

——. *Background: United States Foreign Policy in a New Age*. Washington, D.C.: Armed Forces Information and Education, Dept. of the Army Pamphlet 355-108, July 24, 1961.

U.S. Department of Defense. *Catalog of Information Materials: Publications and Motion Pictures.* Washington, D.C.: Dept. of the Army, Pamphlet 355-50, August 1961.

_____. *Directive ASD(M) No. 5410.17.* January 15, 1965.

_____. *Facts About the United States.* Washington, D.C.: Armed Forces Information and Education, Dept. of the Army Pamphlet 355-123, 1961.

_____. *Facts About the United States.* Washington, D.C.: Armed Forces Information and Education, Dept. of the Army Pamphlet 360-616, 1964.

_____. *Informational Program on American Life and Institutions for Foreign Military Trainees: Vol. II, Popular Participation in Government.* Madison, Wisconsin: United States Armed Forces Institute, 1965.

_____. *Ideas in Conflict: Liberty and Communism.* Washington, D.C.: Armed Forces Information and Education, Dept. of the Army Pamphlet 360-523, June 1967.

_____. *Know Your Communist Enemy: Communism in Red China.* Washington, D.C.: Office of Armed Forces Information and Education, No. 5, 1955.

_____. *Military Assistance: Data for Debate* Washington, D.C.: Office of the Assistant Secretary of Defense for International Security Affairs, August 1966.

_____. *Military Assistance Facts: 1 May 1966.* Washington, D.C.: Office of the Assistant Secretary of Defense for International Security Affairs, 1966.

_____. *Military Assistance Facts: March 1968.* Washington, D.C.: Office of the Assistant Secretary of Defense for International Security Affairs, 1968.

_____. *Military Assistance Facts: May 1969* Washington, D.C.: Office of the Assistant Secretary of Defense for International Security Affairs, 1969.

_____. *Military Assistance and Foreign Military Sales Facts: March 1970* Washington, D.C.: Office of the Assistant Secretary of Defense for International Security Affairs, 1970.

_____. *Military Assistance and Foreign Military Sales Facts: March 1971* Washington, D.C.: Office of the Assistant Secretary of Defense for International Security Affairs, 1971.

_____. *Republic of South Africa: Capsule Facts for the Armed Forces.* Washington, D.C.: Armed Forces Information Service, Dept. of the Army Pamphlet 360-764, January 1969.

_____. Unclassified printout data supplied by Col. Arthur Aguilar, USA, Chief, Organization and Training Division, Directorate for Operations, Military Assistance and Sales, Office of the Assistant Secretary of Defense for International Security Affairs, 3 December 1969. Bindery Control-Job Number P500-88-5500117, 20 October 1969.

U.S. Department of the Navy. "Informational Program for FMT (Foreign Military Trainees) and Visitors in the United States." OPNAVINIST 4950 ID Ch-1, 14 Dec. 1966.

U.S. Department of State. *A Guide to U.S. Government Agencies Involved in*

International Educational and Cultural Activites. Bureau of Educational and Cultural Affairs, International Information and Cultural Series 97, Dept. of State Publication 8405. Washington, D.C.: GPO, September, 1968.

U.S. Industrial College of the Armed Forces. *National Security Seminar, 1968-1969: Presentation Outlines and Reading List.* Washington, D.C.: Fort Lesley J. McNair, December 1968.

U.S. President. *The Foreign Assistance Program: Annual Report to the Congress for Fiscal Year 1967.* Washington, D.C.: GPO, 1968.

____. *The Foreign Assistance Program: Annual Report to the Congress for Fiscal Year 1968.* Washington, D.C.: GPO, 1969.

____. *The Foreign Assistance Program: Annual Report to the Congress for Fiscal Year 1970.* Washington, D.C.: GPO, 1971.

U.S. President's Committee to Study the United States Military Assistance Program. *Composite Report.* Vol. I. Washington, D.C.: GPO, 1959.

____. *Supplement to the Composite Report.* Vol. II. Washington, D.C.: GPO, 1959.

U.S. Task Force on International Development. *U.S. Foreign Assistance in the 1970's: A New Approach.* Report to the President from the Task Force on International Development, March 4, 1970. Washington, D.C.: GPO, 1970.

Index

About the Author

Miles D. Wolpin was born in Mount Vernon, New York in 1937. After attending the London School of Economics, he received his B.S. from the University of Pennsylvania. Following graduate study, Wolpin was awarded a J.D., and a Ph.D. in international politics by Columbia University. He conducted research in Chile on a Ford Foundation grant administered by the Latin American Teaching Fellowship Program at the Fletcher School of Law and Diplomacy. He formerly taught at Marlboro College and St. Francis Xavier University, while at the present time, he is a Visiting Assistant Professor of International Relations at the University of New Mexico.